Milestones in Drug Therapy

More information about this series at http://www.springer.com/series/4991

Carlos A. Zarate Jr. • Husseini K. Manji
Editors

Bipolar Depression: Molecular Neurobiology, Clinical Diagnosis, and Pharmacotherapy

Second Edition

 Springer

Editors
Carlos A. Zarate Jr.
Experimental Therapeutics and
 Pathophysiology Branch, National
 Institute of Mental Health, National
 Institutes of Health
Bethesda, MD, USA

Husseini K. Manji
Janssen Pharmaceuticals of Johnson and
 Johnson, Janssen Research and Development
Titusville, NJ, USA

ISSN 2296-6056   ISSN 2296-6064   (electronic)
Milestones in Drug Therapy
ISBN 978-3-319-31687-1   ISBN 978-3-319-31689-5   (eBook)
DOI 10.1007/978-3-319-31689-5

Library of Congress Control Number: 2016958584

Printed on acid-free paper

This Springer imprint is published by Springer Nature
The registered company is Springer International Publishing AG Switzerland

Preface

Depression and mania are mankind's oldest known brain disorders and were the first brain disorders conceptualized by Hippocrates as a part of medicine; mania and depression as two parts of the same disease were first described in the first century AD. Yet even today, a diagnosis of bipolar disorder (BD) is often missed in patients who report depressive rather than manic symptoms, and the disorder is still frequently unrecognized, misdiagnosed, and mismanaged.

Worldwide, BD is a major public health issue. It is a leading cause of disability, much of which occurs during the depressive phase of the disorder. Clinically, BD continues to be one of the most debilitating medical illnesses. Afflicted patients generally still experience high rates of relapse, a chronic recurrent course, lingering residual symptoms, functional impairment, psychosocial disability, and diminished well-being. Strikingly, almost every drug developed for BD initially entered the market as a treatment for acute mania. Thus, in contrast to the manic phase of the illness where a fairly large variety of effective treatments are available—antiepileptic agents in particular—effective therapeutics are scarce in bipolar depression. A large, 26-week study funded by the National Institute of Mental Health (NIMH) found no benefit to antidepressant use for patients with BD-I or BD-II depression (Sachs et al. 2007). And when currently available therapeutics for depression—both for major depressive disorder (MDD) and BD—do work, they are often poorly tolerated or associated with a delayed onset of action of several weeks; this latency period significantly increases risk of suicide and self-harm and is a key public health issue (Machado-Vieira et al. 2009).

This book focuses on this underdeveloped area: bipolar depression. We have brought together an international array of major leaders from a broad swath of interrelated disciplines—from clinical phenomenology to basic molecular and cellular neurobiology, genetics, neuroimaging, and circadian physiology—to offer the most up-to-date information about the diagnosis, treatment, and research surrounding bipolar depression.

Since the first edition of this book was published in 2008, we are heartened to note some significant advances in this field, many of which are highlighted in the chapters that follow. These include altered neurobiology in patients suffering from BD (Chap. 6), new information about the pathophysiology of BD (Chap. 8), genetic findings from diverse studies (Chap. 5), and solid advances in the treatment of bipolar depression with already available (Chaps. 9–11) and novel (Chap. 12) agents. These chapters underscore how research over the past few decades—and even over the last five years—has made great advances toward improving our understanding of this illness so that better treatments can be developed. A better understanding of the neurobiological underpinnings of this condition, informed by preclinical and clinical research, will be essential for the future development of targeted therapies that are more effective, act more rapidly, and are better tolerated than currently available treatments. Importantly, advancements in our understanding of the neurobiology of BD continue to be made and do hold promise for future clinical applications.

Despite the devastating impact of bipolar depression on the lives of millions worldwide, there has been a dearth of knowledge concerning its underlying etiology and pathophysiology. This dearth, in turn, has undoubtedly contributed to the lack of development of improved therapeutic strategies for bipolar depression. Collectively, many questions remain unanswered regarding the diagnosis and treatment of patients with bipolar depression. However, as the information contained in this book shows, new strides are being made daily in this field. It is our hope that readers will be able to extract integrated themes and useful insights from the material contained in these diverse chapters.

Bethesda, MD   Carlos A. Zarate Jr.
Titusville, NJ   Husseini K. Manji

References

Machado-Vieira R, Salvadore G, Ibrahim LA, Diaz-Granados N, Zarate CA, Jr (2009) Targeting glutamatergic signaling for the development of novel therapeutics for mood disorders. Curr Pharm Des 15(14):1595–1611

Sachs GS, Nierenberg AA, Calabrese JR, Marangell LB, Wisniewski SR, Gyulai L, Friedman ES, Bowden CL, Fossey MD, Ostacher MJ, Ketter TA, Patel J, Hauser P, Rapport D (2007) Effectiveness of adjunctive antidepressant treatment for bipolar depression. N Engl J Med 356:1711–1722

Acknowledgments

The editors gratefully acknowledge the contributions of our book project coordinator, Ioline Henter. Her editing abilities, organizational assistance, and attention to detail were integral to the preparation of this volume. We also wish to acknowledge the support of Jutta Lindenborn and Detlef Klueber at Springer, whose professionalism made them a pleasure to work with. Our thanks also go out to the research staff of the 7 Southeast Unit and the Outpatient Clinic at the National Institute of Mental Health (NIMH-NIH).

Contents

Part I
Bipolar Depression: Clinical Phenotype and Course of Illness

Chapter 1
The History of Bipolar Disorder

Philip B. Mitchell

Abstract Our contemporary construct of bipolar disorder (BD) is only relatively recent in terms of the history of psychiatry. In this chapter, we begin by tracing the origins of this concept from the early writings of the classic Greek and Roman medical writers, through to the observations of the first European academic 'alienists' such as Pinel, Esquirol, and Griesinger in the eighteenth and nineteenth centuries. We then consider the first claims for delineation of a distinct condition akin to BD by the French psychiatrists Falret and Baillarger, who described, respectively, *la folie circulaire* and *la folie à double forme*. The journey from those accounts to our modern concept of BD is then traced through the nineteenth and twentieth century works of the German academic clinicians Kraepelin, Wernicke, Kleist, and Leonhard, with later critical validations being undertaken by Angst, Perris, and Winokur in the 1960s. We then explore the incorporation of BD into the modern operationalised diagnostic criteria such as the Research Diagnostic Criteria and DSM-III in the late twentieth century. In considering this history of the concept of BD, several broad themes become apparent. First, the consistency of the clinical descriptions stands out, from the earliest classical observations to those of the eighteenth and nineteenth century alienists to contemporary operationalised criteria. Second, one cannot help but be struck by the oscillation over time between narrow and broad concepts of this condition: from the very circumscribed descriptions of Falret and Baillarger to the all-encompassing 'manic-depressive insanity' of Kraepelin. In many ways, the contemporary controversy between the narrow definition of BD as reflected in DSM-IV and -5 and the breadth of the 'soft bipolar spectrum disorder' recapitulates this historical tension. Third, however, there is at the same time a clear linearity in the development of ideas, with each 'generation' of academic psychiatrists being influenced by, and refining the concepts of, their predecessors.

Keywords Bipolar disorder • Bipolar depression • History • Diagnosis • Nosology

---

P.B. Mitchell (✉)
School of Psychiatry, University of New South Wales, Sydney, NSW 2052, Australia

Black Dog Institute, Sydney, Australia
e-mail: phil.mitchell@unsw.edu.au

© Springer International Publishing Switzerland 2016
C.A. Zarate Jr., H.K. Manji (eds.), *Bipolar Depression: Molecular Neurobiology,*
Clinical Diagnosis, and Pharmacotherapy, Milestones in Drug Therapy,
DOI 10.1007/978-3-319-31689-5_1

3

1.1   Overview

Our contemporary construct of bipolar disorder (BD) is only relatively recent in terms of the history of psychiatry. In this chapter, we begin by tracing the origins of this concept from the early writings of the classic Greek and Roman medical writers, through to the observations of the first European academic 'alienists' such as Pinel, Esquirol, and Griesinger in the eighteenth and nineteenth centuries. We then consider the first claims for delineation of a distinct condition akin to BD by the French psychiatrists Falret and Baillarger, who described, respectively, *la folie circulaire* and *la folie à double forme*. The journey from those accounts to our modern concept of BD is then traced through the nineteenth and twentieth century works of the German academic clinicians Kraepelin, Wernicke, Kleist, and Leonhard, with later critical validations being undertaken by Angst, Perris, and Winokur in the 1960s. We then explore the incorporation of BD into the modern operationalised diagnostic criteria such as the Research Diagnostic Criteria (Spitzer et al. 1978) and DSM-III (American Psychiatric Association 1980) in the late twentieth century and discuss the more recent and controversial broadening of this concept in the proposal by Akiskal and Mallya for a 'bipolar spectrum disorder' (1987).

In considering these historical developments, we need to be cognizant that the meaning of the term 'mania' has changed dramatically over the course of the centuries (Berrios 1996). Its use prior to the late nineteenth and early twentieth century encompassed an overactivity or frenzy that may be as likely due to what we would now term schizophrenia or delirium as resulting from a 'true' manic episode of BD.

Further, particularly whilst contemplating progress since the seminal French descriptions of the mid-nineteenth century, one cannot help but be struck by the oscillation over time between narrow and broad concepts of this condition: from the very circumscribed descriptions of Falret and Baillarger to the all-encompassing 'manic-depressive insanity' of Kraepelin. In many ways, the contemporary controversy (Healy 2008; Mitchell 2012) between the narrow definition of BD as reflected in DSM-IV (American Psychiatric Association 1993) and -5 (American Psychiatric Association 2013) and the breadth of the 'soft bipolar spectrum disorder' recapitulates this historical tension.

1.2   'Mania' in Classical Greece and Rome

While the writings on 'mania' by the authors of the classical Greek and Roman period provide their own fascination and intrigue, as discussed above, they need to be considered with circumspect, as the term 'mania' undoubtedly also encompassed over-activity resulting from a multiplicity of medical and psychiatric causes. Furthermore, some of the descriptions appear to have reflected behaviours related to enthusiasm and even fanaticism, not necessarily arising from 'illness', so it is not appropriate to enforce a modern medical model upon all such accounts. Nevertheless, some of those accounts do appear consistent with current descriptions of BD.

We would note in passing that, for access to the medical writings of this classical era, we are largely indebted to the seminal work of the Italian psychiatrist Giuseppe Roccatagliata (1986).

Hippocrates (460–370 BC) was the central figure in the transition of Greek thought from the influence of the 'physician-priests' to a secularised understanding of mental disturbances, with his conceptualisation that these illnesses were due to humoural effects on the brain. He posited 'furor' as the opposite of melancholy, as characterised by excitement, euphoria, aggressiveness, and megalomanic ideas and explained this behaviour as due to the effect of yellow bile upon the brain (Roccatagliata 1986, p. 171).

Aristotle (384–322 BC), the disciple of Socrates and Plato, also interpreted such behaviour in terms of disturbances of the humours, with cooling of the black bile leading to '*slowing down, depression, and anxiety*', whereas its heating caused '*euphory, excitement, and exhilaration*' (Roccatagliata 1986, p. 109).

Aretaeus of Cappadocia, who lived in Rome in the second century AD, is often described as the 'father (or clinician) of mania', whose writings were later acknowledged by the great French academic psychiatrist Pinel (1806). Arateus was an adherent of the 'pneumatic' school, which claimed that behavioural disturbances were due to vicissitudes of the 'pneuma'—the physical soul. Believing that the heart was the seat of the soul, he thought that there was a secondary effect on the mind, as the heart was '*connected by sympathy with the brain*'. Mania was expressed, he said, in '*furor, excitement, and cheerfulness*', in severe forms of which the person '*sometimes kills and slaughters the servants*'. He also observed more expansive or exalted forms, where the patient: '*. . .. Without being cultivated says he is a philosopher . . . and the incompetent say they are good artisans . . .*'. He described transitions between phases of melancholy and mania: '*Some patients after being melancholic have fits of mania. . .*', '*. . . they show off in public with crowned heads as if they were returning victorious from the games; sometimes they laugh and dance all day and all night*'. Conversely, after mania, a patient '*has a tendency to melancholy; he becomes, at the end of the attack, languid, sad, taciturn, he complains that he is worried about the future, he feels ashamed*' (Roccatagliata 1986, p. 229, 230).

Posidonius, who lived in the fourth century AD, was an adherent of the 'eclectic' school, which eschewed imposition of preconceived explanatory paradigms. While ascribing 'mania' to yellow bile acting on the frontal area of the brain, his 'clinical' descriptions are surprisingly contemporary: '*The patient laughs, sings, dances . . . he bites himself . . . sometimes he is wicked and kills . . . sometimes he is seized by terror or hate . . . sometimes he is abulic [avolitional, indecisive]*'. Posidonius goes on to describe (in terms later echoed by Falret) the course of this presentation as '*. . . an intermittent disease, which proceeds through a periodical circuit . . . it repeats itself once a year or more often . . . melancholy occurs in Autumn whereas mania in Summer . . .*' (Roccatagliata 1986, p. 143).

1.3   Descriptions from the Eighteenth and Nineteenth Centuries

While the assertion that a condition akin to our concept of BD represented a specific distinct mental illness was only first made in the mid-nineteenth century by Falret and Baillarger, there were undoubted descriptions of similar patients reported prior to that time. However, such shifts in patients' activity and mood had been deemed to be either variants of 'madness' or transitional states on the path to more severe and incurable illness. Furthermore, hitherto, 'madness' had been largely considered to reflect disturbances of intellectual capacities rather than being primarily defects in other functions such as emotions (Berrios 1996).

It is clear that the famous French psychiatrist Pinel (1806) was very familiar with this presentation:

> Nothing appears more inexplicable, at the same time that nothing can be more certain, than that melancholia with delirium, presents itself in two very opposite forms. Sometimes it is distinguished by an exalted sentiment of self-importance, associated with chimerical pretensions to unbounded power or inexhaustible riches. At other times, it is characterized by great depression of spirits, pusillanimous apprehensions, and even absolute despair. Lunatic asylums afford numerous instances of those opposite extremes.

Esquirol, a similarly eminent protégé of Pinel in Paris, and mentor to Falret and Baillarger, also described such patients in his monograph '*Mental Maladies. A Treatise on Insanity*' (1845):

> There are persons who are hypochondriacal, and buried in a profound melancholy for some hours, days, and even months, before the explosion of mania; while others sink into a deep stupor, appearing to be deprived of every thought and idea. They move not, but remain where they are placed. . . . Suddenly, mania bursts forth, in all the strength of its delirium and agitation. (p. 384)

In the USA, Rufus Wyman (1830)—physician superintendent of the McLean Hospital in Massachusetts—described in a discourse:

> Exaltation, and depression of passion, are sometimes manifested alternately in the same individual.

In the UK, James Prichard (1835) described:

> A state of gloom and melancholy depression occasionally gives way after an uncertain period to an opposite condition of preternatural excitement . . . .

In Germany, Griesinger (1867) noted in his chapter on mania (p. 276):

> We have more than once had occasion to remark that, in the majority of cases, melancholic states precede the maniacal, and that the latter is engendered by the former. In the more chronic cases we often have the opportunity of following the whole course of the disease, and of seeing in melancholics the mental suffering and anxiety increase from day to day, at first manifesting itself merely by extreme restlessness, but gradually passing into complete mania.

1.4 A Distinct Psychiatric Disorder? *'La folie circulaire'* and *'la folie à double forme'*

Many authorities ascribe the origins of the modern concept of BD to the Parisian psychiatrists Falret and Baillarger who each claimed in 1854 that they were the first to recognise and describe this condition as a separate disease entity. It is clear from the foregoing section that others had previously described and recognised such patients. However, those other writers had conceived of this as just one manifestation of a broad concept of 'madness'. What distinguished the contributions of Falret and Baillarger was the claim that this was a distinct illness entity.

The claims and counterclaims of Falret and Baillarger to originator status were acrimonious and not resolved within their lifetimes (Pichot 1995, 2006). They were both protégés of Esquirol, gained positions at the prestigious Hospice of La Salpêtrière in Paris, and presented their papers within a few weeks of each other in 1854 to the eminent French Imperial Academy of Medicine. We will not focus here on that aspect of the history, rather examine the evidence presented by each for the delineation of a new illness entity.

Falret termed this syndrome *'la folie circulaire'* (the circular madness). Importantly, he defined this in terms of the course and pattern of the illness. This decision was critical to our understanding of the importance of his observations, as prior to this time the various psychiatric disorders had been largely defined in terms of *'single cardinal symptoms'*, such as *'the degree of delirium, sadness, or agitation'* (Falret 1854):

> We even gave it a name because, in our opinion, it is not a mere variant but a genuine form of mental illness. We called it circular insanity because the unfortunate patients afflicted with this illness live out their lives in a perpetual circle of depression and manic excitement interrupted by a period of lucidity, which is typically brief but occasionally long lasting.

> ... whether the course is rapid or slow, the tempo of the disorder does not alter its general nature: the disease remains the same in its general features as well as in its principal details. That is why we believe it constitutes a genuine form of mental illness, because it consists of a group of physical and mental symptoms which stay the same for any of the respective phases which succeed one another in a determinate order so that, once the symptoms are identified, the subsequent evolution of the illness can be predicted.

Falret was also impressed by the familial nature of this disorder:

> The evidence we have just presented from the parents and from our own observations leads us to believe that circular insanity is very hereditary.

Unfortunately, Baillarger's paper and later publications have not been translated into English, so we are dependent upon the accounts of his writings by authorities such as Pichot (1995) and Haustgen and Akiskal (2006). Baillarger proposed the name *'folie à double forme'* for *'a type of madness different from monomania, melancholia, and mania'*. He described a typical phase consisting of two periods, one of depression, the other of excitement, in immediate succession, without any intermediate free interval, with the switch being either sudden or more gradual.

Baillarger considered the manic period to be a 'reaction' to the preceding melancholy and proportional to that. Haustgen and Akiskal (2006) further recount that Baillarger described the two periods of mania and melancholia as being bound together ('*soudées*') and thereby constituting one 'attack' ('*accès*'). These 'attacks', said Baillarger, could be isolated, intermittent, or follow each other without interruption. This was contrasted with Falret's *folie circulaire,* for which the cycle was comprised of three periods—mania, depression, and the free interval (*intervalle lucide*), with this cyclic course, often over a lifetime, being the main distinguishing criterion.

The influence of Baillarger dominated in France, possibly due to his surviving Falret by 20 years, and his later prestigious appointment to the presidency of the Imperial Academy of Medicine—one of very few psychiatrists to have been honoured in that role (Pichot 1995). However, ultimately it was Falret who had the most impact on international thought. His *folie circulaire* was reported by Kahlbaum to have influenced his critical conceptualisation of catatonia (see below), and Kraepelin's earlier textbooks refer to this.

Perhaps, the greatest contribution of Falret was not in terms of formative clinical descriptions of mental phenomena (as was the legacy of Baillarger), but rather his focus on the search for the '*discovery of disease entities (espèces naturelles), characterised by a set of physical and psychological symptoms, and by a special evolution*' (Ritti 1864, as translated by Pichot 1995). This focus on 'disease evolution', i.e. the specifics of the longitudinal course of an illness, was to strongly influence Kahlbaum and, later, Kraepelin.

1.5   'Manic-Depressive Insanity' and 'Bipolar Disorder'

Before we consider the seminal contributions of Kraepelin and Leonhard, we need to be aware of the critical influence on their thinking of their German predecessor Kahlbaum, who worked in Dresden in the mid- to late nineteenth century. Kahlbaum, similarly to Falret and Baillarger, focused on the centrality of the longitudinal course of the illness. He urged abandonment of the more unitary views of psychotic disorders favored by the leading German academic theorists of his time such as Griesinger (Baethge et al. 2003). Kahlbaum's approach to nosology strongly influenced Kraepelin's views, and he detailed a number of conditions that were to be later incorporated in the diagnostic systems of Kraepelin and Leonhard.

First, in 1874, Kahlbaum described 'catatonia'. He detailed overactive and underactive forms of this condition, which he saw as a 'motility psychosis' and usually episodic (Healy 2008). This concept of catatonia as an episodic psychotic disorder strongly influenced the later thinking of Wernicke, Kleist, and Leonhard, though later evaluations of the cases described in his writings suggest that their actual clinical presentations were not consistent with his definition of catatonia as cyclic presentations with consecutive symptoms of melancholia, mania, stupor, confusion, and ultimate dementia (Berrios 1996, p. 383).

Second, Kahlbaum described (and coined the term) *cyclothymia*, contrasting this condition with *vesania typica circularis* or *circular insanity* (Kahlbaum 1882) which he inferred to be a widely held concept. According to Kahlbaum, *vesania typica circularis* was usually psychotic, leading to hospitalisation for manic and depressive episodes. Cyclothymia showed similarly cyclical, though less severe, variations in mood (Berrios 1996, p. 311), reminiscent of many current patients diagnosed with BD-II (see below).

Details of Kahlbaum's concept of cyclothymia were later elaborated upon by his younger colleague Hecker (1898):

> In his lecture on cyclic insanity, published in "Der Irrenfreund" in 1882, Kahlbaum distinguishes two forms of circular psychosis. In his opinion, these forms are considerably dissimilar. He describes a form that he termed Vesania typica circularis as having a tendency to end up in a demented state after several episodes of melancholic and manic episodes have occurred. A contrasting form was termed by Kahlbaum as cyclothymia which, he proposed, never proceeds to confusion and dementia, even after lifelong duration. In the first form, typica circularis, we observe that all the major mental functions are affected, whereas in cyclothymia we see only a fluctuation of two opposed mood states, namely dysthymia and hyperthymia. In cyclothymia reasoning is more or less completely unaffected, indicating that it is a pure mood disorder. Kahlbaum rightly draws our attention to the fact that it is not rare for cyclothymia to occur with such a mild intensity that most cases are not admitted to mental asylums.

These developments now bring us to Emil Kraepelin, one of the most influential figures on modern thinking concerning BD, with his description of (and coining of the term) 'manic-depressive insanity'. Kraepelin, whose main work was undertaken in Heidelberg and Munich, was a leading German academic psychiatrist of the late nineteenth and early twentieth century. His major contribution was to distill—from the hitherto morass of ill-classified and poorly defined severe mental illnesses—two major forms of psychotic disorder, i.e. 'dementia praecox' (schizophrenia) and 'manic-depressive insanity'. His manic-depressive insanity was defined mainly on the basis of the recurrent or phasic nature of the condition (reflecting the influence of Falret's and Kahlbaum's emphases on a longitudinal approach), in conjunction with the symptoms being primarily disturbances of mood. He counterposed this condition against 'dementia praecox' which was characterised by a chronic and usually deteriorating course and reflective of disturbed thinking.

He first employed the term 'manic-depressive insanity' in the sixth edition of his widely influential textbook on psychiatry (Kraepelin 1899). His 'manic-depressive insanity' was a much broader concept than our modern 'bipolar disorder'. This diagnostic grouping became influential in the English speaking world after publication of the 1921 English translation of '*Manic-depressive Insanity and Paranoia*'—a selection from the multivolume eighth edition of his textbook that was published between 1909 and 1915 (Kraepelin 1921). In his introduction to this volume, Kraepelin stated:

> Manic-depressive insanity ... includes on the one hand the whole domain of periodic and circular insanity, on the other hand simple mania, the greater part of the morbid states termed melancholia, and also a not inconsiderable number of cases of amentia (confusional or delirious insanity). Lastly, we include here certain slight and slightest colourings of

mood, some of them periodic, some of them continuously morbid, which on the one hand
are to be regarded as the rudiment of more severe disorders, on the other hand pass over
without sharp boundary into the domain of personal predisposition. In the course of the
years I have become more and more convinced that all the above-mentioned states only
represent manifestations of a single morbid process. (p. 1).

In that text, he goes on to emphasise '*certain common fundamental features*'
which recur. Further, these morbid forms '*not only pass over the one into the other
without recognisable boundaries, but that they may even replace each other in one
and the same case. . . . it is fundamentally and practically quite impossible to keep
apart in any consistent way simple, periodic, and circular cases; everywhere there
are gradual transitions.*' (p. 2).

He then articulates a further critical delineation from dementia praecox:

A further common bond ... is their uniform prognosis. ... the universal experience is
striking, that the attacks of manic-depressive insanity within the delimitation attempted
here never lead to profound dementia, not even when they continue throughout life almost
without interruption. (p. 3).

The centrality of the course of the illness to his concept is articulated later in the
1921 text: '*The starting point of the conception of the disease is formed by the
concept of the periodic or ... intermittent psychic disorders. This doctrine was
elaborated principally by the French alienists.*' (p. 185).

He argued strongly against delineating further subgroups within his 'omnibus'
disorder:

The different varieties of course taken by manic-depressive insanity ... have been analysed
into a series of clinical sub-varieties, specially by Falret and Baillarger, who first made us
intimately acquainted with this disease; these sub-varieties are intermittent mania and
melancholia, regular and irregular type, folie alterne, folie à double forme, folie circulaire
continue. I think that I am convinced that that kind of effort at classification must of
necessity wreck on the irregularity of the disease. (p. 139).

Kraepelin's other major argument for the integrity of his concept was the
apparent shared genetic origins: '*As a last support of the view here represented of
the unity of manic-depressive insanity the circumstance may be adduced, that the
various forms which it comprehends may also apparently mutually replace one
another in heredity. In members of the same family we frequently enough find side
by side pronounced periodic or circular cases, occasionally isolated cases of ill
temper or confusion, lastly very slight, regular fluctuations of mood or permanent
conspicuous colouration of disposition.*" (p. 3). Later in that text he stated: '*I could
demonstrate (hereditary taint) in about 80 per cent of cases observed in Heidelberg.*' (p. 165).

It should be noted here, however, that Kraepelin held varying opinions at
different points of his career about the breadth of his 'manic-depressive insanity'.
His main uncertainty related to the status of 'involutional melancholia' which was
not incorporated in the broader illness through the seventh edition of his textbook,
though finally included in the eighth (and most influential) edition. This has been
expanded upon in detail by Trede and colleagues (2005) who commented:

However, our findings indicate a complex evolution of Kraepelin's MDI concept in the 1880s and 1890s, his use of more creative and less empirical clinical methods than traditionally believed, and his considerable personal uncertainty about making clear distinctions among MDI, dementia præcox, intermediate conditions, and paranoid disorders—an uncertainty that persisted to the end of his career in the 1920s. ... Modern international "neo-Kraepelinian" enthusiasm for descriptive, criterion-based diagnosis should be tempered by Kraepelin's own appreciation of the tentative and uncertain nature of psychiatric nosology, particularly in classifying illnesses with both affective and psychotic features.

Despite these uncertainties and vacillations, Kraepelin's 'manic-depressive insanity' was widely taken up internationally, largely through the influence of the English translations of his textbooks and monographs. In his home country of Germany, however, there was scepticism over the 'simplicity' of his proposed dichotomy of the major psychotic disorders.

Our modern delineated concept of BD arises largely from a strand of German academic psychiatry distinct from that of Kraepelin, though clearly cognizant of his work. While the origin of the modern term 'bipolar disorder' is largely ascribed to Karl Leonhard (1957), he himself was strongly influenced by the writings of Wernicke (the neuropsychiatrist who described Wernike's area in the brain—the specific receptive speech area—and Wernicke–Korsakoff syndrome) and his acolyte, Kleist.

Wernicke, who worked in Halle, was a contemporary of Kraepelin, born eight years earlier, and prominent due to his major neurological discoveries. Wernicke was also influenced by Kahlbaum and endorsed his call for the distinction of syndromes based on clinical observation (Lanczik et al. 1995). Wernicke's contention, unlike Kraepelin whose approach was largely clinical, was that classification should take into account neurological localisation. Wernicke considered that it was not possible at that time to identify separate psychiatric disorders and remained an ardent adversary of Kraepelin, whose dichotomising of the major psychiatric disorders he viewed as overly simplistic. Essentially, Wernicke attempted to understand clinical observations from an abstract notion of neuropathological functioning. Like Kahlbaum, he categorised psychiatric disorders on the neurological concepts of function, i.e. hyper-, hypo-, and para-function. He also described a 'motility symptom complex', regarding motility disturbances as a 'symptom complex' that could occur in different psychotic disorders (Perris 1995).

Wernicke's classification of the psychoses was complex, based as it was on his concept of the three major contributants to consciousness—the environment, self, and body (Lanczik et al. 1995). He, therefore, created an overarching classification system for a variety of endogenous psychoses: autopsychoses (resulting from having false impressions of oneself), allopsychoses (having false impressions of the environment), and somatopsychoses (having a distorted experience of one's own body). He conceived of melancholia and mania as examples of autopsychoses, associated with symptoms of *intrapsychic afunction* and *hyper-function*, respectively.

Kleist was one of the most prominent students of Wernicke, similarly attempting to understand clinical observations from a neuropathological perspective. He maintained the independence of a group of 'motility psychoses' distinct from other psychotic disorders (Perris 1995). He went on to coin the term 'cycloid psychoses' (zykoloiden Psychosen) in 1926, which he distinguished from, on the one hand, the 'mood psychoses' comprising mania, melancholia, and manic-depressive insanity, and schizophrenia on the other. Kleist conceived of a broad group of 'bipolar phasophrenias' which included the cycloid psychoses.

It is from Leonhard (1957)—one of Kleist's followers—however, that modern use of the term 'bipolar disorder' is derived. It should be recognised, though, that this concept was one component of a highly complex nosological system of 'endogenous' psychoses, which comprised 38 distinct clinical illnesses. First, Leonhard subdivided the affective psychoses into bipolar (manic-depressive) and unipolar (manic and depressive) forms. He also categorised the cycloid psychoses into three subtypes (anxiety–happiness psychosis, confusion psychosis, and motility psychosis). He also considered these cycloid psychoses to be 'bipolar' disorders, with the bipolar manifestations occurring more frequently *within episodes* rather than as distinct episodes of differing polarity (Perris 1995).

As with Wernicke, Leonhard was scathing of the Kraepelinian dichotomy of the psychoses: '*Kraepelin's classification into only two forms has been damaging. He himself attempted many finer distinctions with great enthusiasm and continued open-mindedness, but his followers ignored this; they only saw the coarse division into dementia praecox and the manic-depressive disease.*' (Leonhard, in his Introduction to the 1979 English translation of his 1957 monograph; p. xv).

Leonhard placed a major emphasis on bipolar presentations being 'polymorphic' as opposed to the 'pure' unipolar depressive and manic presentations:

> ... in bipolar cases no clear syndromes can be described since there are many transitions between various formations and the picture may even be distorted during one phase. ... The opposite phase will merely hint at itself not infrequently though it cannot fail to be recognised. ... With the essentially unipolar forms, there are no signs of lability toward the other poles ...

Leonhard also presented some (very basic) evidence of genetic validation of this distinction between unipolar and bipolar disorders, reporting that those with mania were more likely to have relatives with manic episodes and a lack of such history of mania in those with unipolar disorders. His overall system was, however, highly complex and overly detailed. For example, he subdivided the unipolar disorders further into 'pure melancholia' and 'pure mania' and 'pure depressions' and 'pure euphorias'. There was a further level of categorisation, with pure depressions being grouped into 'harried', 'hypochondriacal', 'self-torturing', 'suspicious', and 'non-participatory' depressions. Pure euphorias were grouped into 'unproductive', 'hypochondriacal', enthusiastic', 'confabulatory', and 'non-participatory' forms.

1.6   Validation of the Concept of Bipolar Disorder

Leonhard's proposal for distinct bipolar and unipolar disorders was validated in the mid-1960s by independent studies from Scandinavia (Perris 1966), Switzerland (Angst 1966), and the USA (Winokur and Clayton 1966).

Perris was an Italian psychiatrist working in Sweden. His study specifically tested Leonhard's proposition that the various depressive disorders should be distinguished on the basis of 'polarity'. Perris undertook a series of genetic, social, clinical, neurophysiological, and therapeutic studies, though the most critical of these were the genetic investigations, based as they were on Leonhard's contention that '*heredity plays a more dominant part in the bipolar than in the unipolar forms*'. (Perris 1966, p. 11):

> As a result of our division into bipolar and unipolar cases, we have been able to confirm Leonhard's observation that the overall heredity for psychoses is greater in bipolar probands... Moreover, the results of our investigation seem to indicate a specificity in the heredity of depressive psychoses. This is supported by the high morbidity risk for the same form of illness and the low for the other one within each group. Manic probands are closer to the bipolar than the unipolar probands. (Perris 1966, p. 41).

The latter observation led to what Leonhard termed 'unipolar mania' being incorporated into the broader concept of BD, though Angst has continued to contend that it should be considered as a separate entity (Angst and Grobler 2015).

The findings of the eminent Swiss psychiatrist Jules Angst were published in a monograph in German several months prior to Perris (Angst 1966) and were in fact referred to in the aforementioned publication by Perris and seen as broadly similar in its findings. Angst later summarised some of the major findings of his own study (Angst and Marneros 2001): (1) unipolar depression differs significantly from BD in many characteristics such as genetics, gender, course, and premorbid personality and (2) late-onset depression ('involutional melancholia') seems to belong to unipolar depression. George Winokur, in the USA, published analogous findings to those of Perris and Angst (Winokur and Clayton 1966).

David Healy, a thoughtful contemporary commentator on the concept of BD and its history (Healy 2008, 2010) has further contended that the therapeutic studies on lithium in the 1940s, 1950s, and 1960s by Cade (Cade 1949; Mitchell and Hadzi-Pavlovic 1999) and Schou (Baastrup and Schou 1967)—which confirmed its efficacy in mania (both acutely and prophylactically)—accelerated interest in this condition and the rapid acceptance of the new diagnostic proposals.

1.7 Incorporation of 'Bipolar Disorder' into Operationalised Diagnostic Criteria

In 1965, the broad 'Kraepelinian' concept of manic-depressive psychosis was accepted into the WHO classificatory system in the eighth edition of the International Classification of Diseases (ICD-8) as a single disease with three symptomatic forms: manic, depressive, and circular (Pichot 2006). The term 'bipolar disorder' was not used in formal diagnostic systems until the late 1970s when it was included in the US Research Diagnostic Criteria (RDC; Spitzer et al. 1978), the operational definitions of which were largely incorporated into the DSM-III (American Psychiatric Association 1980). The specific link between Leonhard, Perris, and Angst on the one hand and the RDC and DSM-III nosologies on the other was George Winokur. Winokur was one of the leaders of the prominent research group at Washington University in St. Louis, Missouri that originally developed the Feighner Criteria (Feighner et al. 1972) and subsequently the Research Diagnostic Criteria—the first operationalised diagnostic systems in psychiatry. BD has continued to be incorporated into succeeding iterations of the DSM, though it was not until 1992 that this term was first used in the International Classification of Diseases (in the ICD-10).

1.8 Bipolar I and II Disorders and the Bipolar Spectrum Disorders

BD was further categorised into BD-I and BD-II by Dunner and colleagues (1976) with the latter condition being defined in terms of severe depression (leading to hospitalisation) in conjunction with hypomanic episodes. This distinction between BD-I and BD-II was included in the Research Diagnostic Criteria, and subsequently the various editions of the DSM, though the criteria for BD-II have changed considerably over time. A more recent development is the proposal for a 'soft bipolar spectrum disorder' (Akiskal and Mallya 1987), a concept largely premised on a broad model of affective instability. Since the original descriptions of the soft bipolar spectrum disorder in the late 1980s and the derived research focus on patients with 'subthreshold BD' (particularly arising from the work of Angst and colleagues), there has been a major shift in focus in some sectors of both the clinical and research communities towards patients who would not have been historically regarded as fulfilling criteria for this condition. The concept of the soft bipolar spectrum has become increasingly influential, despite a paucity of empirical validation.

1.9 Conclusions

In concluding this overview of the history of the evolution of the concept of BD, several broad themes are apparent. First, the consistency of the clinical descriptions clearly stands out, from the earliest classical observations to those of the eighteenth and nineteenth century alienists to contemporary operationalised criteria. Second (as detailed above), one cannot help but be struck by the oscillation over time between narrow and broad concepts of this condition: from the very circumscribed descriptions of Falret and Baillarger to the all-encompassing 'manic-depressive insanity' of Kraepelin. In many ways, the contemporary controversy between the narrow definition of BD as reflected in DSM-IV and -5 and the breadth of the 'soft bipolar spectrum disorder' recapitulates this historical tension. Third, however, there is at the same time a clear linearity in the development of ideas, with each 'generation' of academic psychiatrists being influenced by, and refining the concepts of, their predecessors.

References

Akiskal HS, Mallya G (1987) Criteria for the "soft" bipolar spectrum: treatment implications. Psychopharmacol Bull 23(1):68–73

American Psychiatric Association (1980) Diagnostic and statistical manual of mental disorders, 3rd edn. American Psychiatric Association, Washington, DC

American Psychiatric Association (1993) Diagnostic and statistical manual of mental disorders, 4th edn. American Psychiatric Association, Arlington, VA

American Psychiatric Association (2013) Diagnostic and statistical manual of mental disorders, 5th edn. American Psychiatric Association, Arlington, VA

Angst J (1966) Zur Ätiologie und Nosologie endogenor depressiver Psychosen. Eine genetische, soziologische und klinische Studie. Springer, Berlin

Angst J, Grobler C (2015) Unipolar mania: a necessary diagnostic concept. Eur Arch Psychiatry Clin Neurosci 265(4):273–280

Angst J, Marneros A (2001) Bipolarity from ancient to modern times: conception, birth and rebirth. J Affect Disord 67(1–3):3–19

Baastrup PC, Schou M (1967) Lithium as a prophylactic agents. Its effect against recurrent depressions and manic-depressive psychosis. Arch Gen Psychiatry 16(2):162–172

Baethge C, Salvatore P, Baldessarini RJ (2003) Cyclothymia, a circular mood disorder. Harv Rev Psychiatry 11(2):78–90

Berrios GW (1996) The history of mental symptoms. Cambridge University Press, Cambridge

Cade JF (1949) Lithium salts in the treatment of psychotic excitement. Med J Aust 2(10):349–352

Dunner DL, Stallone F, Fieve RR (1976) Lithium carbonate and affective disorders. V: A double-blind study of prophylaxis of depression in bipolar illness. Arch Gen Psychiatry 33(1):117–120

Esquirol E (1845) Des Maladies Mentales. Lea and Blanchard, Philadelphia, PA, English translation Mental Maladies. A Treatise on Insanity by Hunt EK

Falret JP [(1854) 1983] Memoir on circular insanity. Am J Psychiatry 140:1127–1133. English translation by Sedler MJ and Dessain EC

Feighner JP, Robins E, Guze SB, Woodruff RA, Winokur G, Munoz R (1972) Diagnostic criteria for use in psychiatric research. Arch Gen Psychiatry 26:57–63

Griesinger W (1867) In Die Pathologie und Therapie der Psychischen Krankheiten. The New Sydenham Society, London, English translation Mental Pathology and Therapeutics by Robertson CL and Rutherford J

Haustgen T, Akiskal H (2006) French antecedents of "contemporary" concepts in the American Psychiatric Association's classification of bipolar (mood) disorders. J Affect Disord 96 (3):149–163

Healy D (2008) Mania: a short history of bipolar disorder. The Johns Hopkins University Press, Baltimore, MD

Healy D (2010) From mania to bipolar disorder. In: Yatham LN, Maj M (eds) Bipolar disorder: clinical and neurobiological foundations. Wiley, New York.

Hecker E [(1898) 2003] Cyclothymia, a circular mood disorder. Hist Psychiatry 14(55 Pt 3):377–399. English translation by Baethge C, Baldessarini RJ

Kahlbaum KL [(1882) 2003] On cyclic insanity. Harv Rev Psychiatry 11(2):78–90. English translation and commentary. Baethge C, Salvatore P, Baldessarini RJ

Kraepelin E (1899) Ein Lehrbuch für Studirende und Aertze. Barth, Leipzig

Kraepelin E (1921) Manic-depressive insanity and paranoia. E & S Livingstone, Edinburgh, English translation by Barclay RM, Edited by Robertson GM

Lanczik MH, Beckmann H, Keil G (1995) Wernicke. In: Berrios GE, Porter R (eds) A history of clinical psychiatry. Athlone Press, London

Leonhard K (1957) Aufteilung der Endogenen Psychosen. Akademie-Verlag, Berlin, English translation as The Classification of Endogenous Psychoses, 5th Edition by Berman R, Edited by Robins E. Irvington Publishers, New York

Mitchell PB (2012) Bipolar disorder: the shift to overdiagnosis. Can J Psychiatry 57(11):659–665

Mitchell PB, Hadzi-Pavlovic D (1999) John Cade and the discovery of lithium treatment for manic depressive illness. Med J Aust 171(5):262–264

Perris C (1966) A study of bipolar (manic-depressive) and unipolar recurrent depressive psychoses. Acta Psychiatr Scand 42(Suppl 194):1–189

Perris C (1995) Leonhard and the cycloid psychoses. In: Berrios GE, Porter R (eds) A history of clinical psychiatry. Athlone Press, London

Pichot P (1995) The birth of the bipolar disorder. Eur Psychiatry 10(1):1–10

Pichot P (2006) Tracing the origins of bipolar disorder: from Falret to DSM-IV and ICD-10. J Affect Disord 96(3):145–148

Pinel P (1806) Traite Medico-Philosophique sur l'Alienation Mentale. W. Todd, Sheffield, English translation A Treatise on Insanity by Davis DD

Prichard J [(1835) 1965] Reproduced in "Three hundred years of psychiatry 1535–1860" by Hunter R and Macalpine I. Oxford University Press, London, p 836

Ritti A [(1864) 1879] Folie a double forme. In: Dechambre A (ed) Dictionaire encyclopedique des sciences medicales – 4th Series. T. 3. Masson/Asselin, Paris, pp 321–339

Roccatagliata G (1986) A history of ancient psychiatry. Greenwood Press, Westport, CT

Spitzer RL, Endicott J, Robins E (1978) Research diagnostic criteria: rationale and reliability. Arch Gen Psychiatr 35:773–782

Trede K, Salvatore P, Baethge C, Gerhard A, Maggini C, Baldessarini RJ (2005) Manic-depressive illness: evolution in Kraepelin's textbook, 1883–1926. Harv Rev Psychiatry 13(3):155–178

Winokur G, Clayton P (1966) Family history studies. I. Two types of affective disorders separated according to genetic and clinical factors. Recent Adv Biol Psychiatry 9:35–50

Wyman R [(1830) 1965] Reproduced in "Three hundred years of psychiatry 1535–1860" by Hunter R and Macalpine I. Oxford University Press, London, p 810

Chapter 2
The Clinical Diagnosis of Bipolar Depression

Gordon Parker

Abstract This chapter overviews considerations as to the nature of bipolar depression, an issue of some importance because of potential treatment implications. Representative studies indicate that those with a bipolar I disorder (BD-I) are somewhat likely to experience psychotic depression during depressive episodes, while for the remainder, a melancholic depressive state is most likely to be experienced. In contrast, for those with a bipolar II disorder (BD-II), episodes of psychotic depression are extremely rare, and most are more likely to experience melancholic depressive episodes. For both BD subtypes, 'bipolar depression' is rarely non-melancholic in nature, although as non-melancholic depressive episodes can be experienced by any individual as a consequence of life stressors, those with BD are also likely to acknowledge such episodes as well. Identification of the bipolar depressive subtype is therefore best addressed in relation to the individual's prototypic episodes. The high rates of nonpsychotic and psychotic melancholic depression in those with BD invite consideration as to whether such episodes differ from similar states experienced by those with equivalent unipolar states. Several studies indicate that certain symptoms, such as the 'atypical features' of hypersomnia and hyperphagia, may be more frequent in bipolar than unipolar melancholia, but the general conclusion is more one of similarity than of differences in symptom patterns. As bipolar depression is principally of the 'melancholic' type, clinical features weighting a diagnosis of melancholia are considered in some detail. Finally, several management nuances in managing bipolar depression are briefly noted.

Keywords Bipolar disorder • Bipolar depression • Melancholia

G. Parker, MD, PhD (✉)
School of Psychiatry, University of New South Wales, Sydney, NSW, Australia

Black Dog Institute, Prince of Wales Hospital, Sydney, NSW, Australia
e-mail: g.parker@unsw.edu.au

© Springer International Publishing Switzerland 2016   17
C.A. Zarate Jr., H.K. Manji (eds.), *Bipolar Depression: Molecular Neurobiology, Clinical Diagnosis, and Pharmacotherapy*, Milestones in Drug Therapy,
DOI 10.1007/978-3-319-31689-5_2

2.1   Introduction

The bipolar disorders (both bipolar I (BD-I) and bipolar II (BD-II)) are mood disorders principally marked by oscillations in mood and energy across the hypo/manic and depressive phases. How features of the depressed phase (i.e. bipolar depression) differ from unipolar depression is not simply of definitional interest but should assist diagnosis and, as a consequence, management too.

A number of studies (reviewed shortly) have considered how the depressed phases experienced by those with BD differ from 'unipolar depression'. On theoretical grounds, such a question is unlikely to able to be answered with any precision when unipolar depression is a dimensional construct encompassing quite heterogeneous depressive states. If a dimensional model is applied, unipolar depression includes major and minor clinical depressive conditions (also very heterogeneous constructs). If a subtyping approach is adopted, candidate unipolar depressive conditions include psychotic depression, melancholic depression, and a heterogeneous mix of residual non-melancholic conditions. Thus, rather than question how bipolar depression differs from unipolar depression, a more productive question is how bipolar depression may correspond to or differ from distinctive unipolar depression subtypes. In essence, does bipolar depression correspond most closely to psychotic depression, melancholic depression, or the residual category of unipolar non-melancholic depression?

In this chapter, I will first overview previous studies that have explored the nature of bipolar depression, and that suggest that 'bipolar depression' corresponds most closely to unipolar melancholic and psychotic depression. I will then detail features that distinguish those depressive subtypes and make some brief comments about management models that respect such findings.

2.2   Previous Studies

Goodwin and Jamison (2007) tabulated relevant studies published over the preceding 50 years. They noted that most studies compared unipolar and BD-I patients and observed that bipolar-unipolar differences appear clearer when the BD-II group was excluded from such examinations. They concluded that the most widely replicated findings were that BD-I (compared to unipolar) depressed patients were more likely to show mood lability, psychotic features, and psychomotor retardation and to report more comorbid substance abuse; in contrast, 'typical unipolar patients in these studies had more anxiety, agitation, insomnia, physical complaints, anorexia and weight loss' (p. 17). They also noted that so-called atypical features (i.e. hypersomnia, increased weight, and appetite) were more likely to be reported in BD-II than in unipolar depressed patients. They too emphasised that heterogeneity in samples of unipolar patients provided a source of diverse findings, while differences between BD-I depressive and BD-II depressive states needed to be identified in any study examining for clinical differences. In their synthesis, they judged that those with BD-II (1) were more likely to be female, (2) were more likely to experience comorbid anxiety and alcohol abuse, (3) had less severe but more frequent and chronic depressions, and (4) had shorter inter-episode intervals. In contrast, the BD-I patients were more likely to have severe and prolonged episodes and higher rates of psychosis and hospitalizations.

A number of studies undertaken by our group will now be reviewed as they consider the nature of the diagnostic subtype(s) in bipolar depression rather than differences in clinical features.

Parker and colleagues (2000) reported a study of patients attending a tertiary referral mood disorder unit, with the sample comprising 83 bipolar and 904 unipolar depressive patients. DSM diagnoses established a higher rate of psychotic depression (19.3 % vs. 10.4 %) as well as of melancholic depression (68.7 % vs. 36.9 %) in the BD participants and therefore a distinctly lower rate of non-melancholic depression (12.0 % vs. 52.7 %) when compared to the unipolar participants. Examining symptoms reported during the depressed phase in the two groups (which, importantly, did not differ significantly in terms of age or gender) established that those in the BD group were more likely when depressed to report appetite loss, slowed thinking, indecision, being slowed physically, a loss of interest, anticipatory anhedonia, non-reactivity to both friends and events, a non-variable mood, pathological guilt, and psychotic features. These features are historically weighted to melancholic and—in relation to delusions—to psychotic depression. Another study nuance was that decisions in regard to melancholic or non-melancholic depression status were made by three differing diagnostic strategies. Across all three definitions of melancholia, it was established that bipolar depression was distinctly more likely to be 'melancholic' in terms of its clinical features, while those with bipolar depression were also somewhat more likely to have psychotic depression.

Another study (Parker et al. 2006) that explored differences between bipolar and unipolar depression examined BD-I and BD-II subjects separately (and not as an overall 'bipolar' group) in comparison with a residual set of unipolar depressed patients. Study findings allowed the generation of an 'isomer model' for differentiating BD-I and BD-II from each other. Such differentiation is important because recent DSM manuals (including DSM-5) define mania and hypomania (and thus BD-I and BD-II) with very similar criteria sets. In essence, DSM symptoms are identical for both mania and hypomania—as is the cut-off score for their presence—so that the two conditions are essentially only differentiated across duration, severity, and hospitalisation parameters. The study involved comparing those who were assigned a BD-I diagnosis, respecting DSM-IV decision rules for mania (other than imposing any duration criterion), but required manic episodes to be associated with either (1) distinct impairment, (2) psychotic features at any time, or (3) hospitalisation during a high. A BD-II diagnosis was assigned to those who met DSM-IV decision rules for hypomania (again ignoring any duration criterion) but did not meet criteria on any of the three BD-I defining features listed in the previous sentence. Consecutive recruitment of 157 patients attending the clinic assigned 49 as having BD-I, 52 as having a BD-II, and 56 as having a unipolar depressive disorder. The groups did not differ significantly by age or gender, nor did they differ in terms of social class, family history of BD, or age of onset at either their initial hypo/manic episode or initial depressive episode. By diagnostic assignment rules, none of the BD-II subjects had experienced psychotic features or been hospitalised when high, while two-thirds of the BD-I subjects reported psychotic features when high, and one-third had required hospitalisation. Importantly, 41 % of the BD-I subjects had experienced psychotic features when depressed compared to 0 % of the BD-II subjects, a finding that generated the 'isomer model' (described shortly) as a consequence of the specificity of psychotic features to BD-I in both hypo/manic and depressed moods.

In this study, several sets of analyses failed to establish any significant differences between the severity of manic and hypomanic symptoms, suggesting that the nature and severity of the 'core' mood/energy increase in hypomanic and manic states differs only marginally. Thus, levels of mood and energy are not clearly helpful in differentiating BD-I and BD-II states. We concluded that both disorders are unlikely to be successfully modelled or measured by any strategy that merely assesses the core construct of increased mood/energy and therefore proposed an 'isomer model'. This model assumes that the elevated mood/energy state is the 'core' construct to BD, being shared across BD-I and BD-II states, but is somewhat more severe in BD-I conditions. Mood and energy decreases are viewed as the core construct for melancholic and psychotic depressive states, with that core component being somewhat more severe in psychotic depression than in melancholic depression. In the same way, the model assumes that the presence of psychotic features provides a psychotic 'mantle' distinguishing BD-I from BD-II.

The principal advantage of the model is that it argues that 'mania' is reserved simply for those BD individuals who experience psychotic features during an elevated mood state (at some period in their lifetime). In contrast, those who have never been psychotic at such times are assigned to a BD-II category. The model is underpinned by the empirical data embedded in this paper and that goes to the thrust of this chapter—that those who have BD and have had psychotic features when high (BD-I according to the model) are moderately likely to have psychotic features when depressed (41 % in this study). In contrast, those who have never experienced psychotic features when high (and had been assigned a BD-II diagnosis) are unlikely to have experienced psychotic features when depressed (quantified as 0 % in this study). In essence, a mood/energy construct forms the core construct for the oscillations (and is not in and of itself differentiating), while the presence or absence of a 'mantle' of psychotic features provides categorical differentiation of BD-I and BD-II, respectively. In relation to the focus of this chapter, the message is that BD-I depression is likely to be psychotic or melancholic in its nature, while BD-II depression is virtually never psychotic in nature but likely to be melancholic in its type. This study further underlines the earlier point about the need to study BD-I and BD-II separately (rather than simply examining amalgamated samples of BD-I and BD-II subjects) when studying the nature of bipolar depression.

Parker and Fletcher (2009) focused on patterns of depression in BD-II patients only and when compared to unipolar depressed patients. They studied a consecutive group of those attending a tertiary referral centre and reported data returned by 119 BD-II and 275 unipolar depressed patients. No significant differences were found between the groups in terms of gender or state depression severity scores, but the BD-II subjects were significantly younger. Comparison of depressive symptoms experienced during depressed phases identified only two differences: the BD-II subjects were more likely to report their thinking as slowed and felt less 'need to be close to people'. In a refined analysis after the groups had been matched for age, the groups showed minimal differences in depressive patterns. In further analyses, depressive symptoms in the BD-II group were compared with subsets of those with a unipolar melancholic and those with a unipolar non-melancholic depression to determine whether the clinical depressive pattern for those with BD-II approximated more to a melancholic or to a non-melancholic pattern of depressive features. The unipolar melancholic subjects were more likely to return higher scores on classical melancholic symptoms (e.g. anticipatory and consummatory anhedonia, mood non-reactivity, psychomotor slowing, and weight loss), supporting their diagnostic subtype assignment and thus allowing the secondary analyses to proceed. Those analyses showed that the depressive pattern in the BD subjects approximated more to the unipolar melancholic than to the unipolar non-melancholic subset, specifically in terms of psychomotor and cognitive slowing. The study also examined the impact of age on symptom ratings and established a number of trends, one of which was significant. This nuance is important in suggesting that studies seeking to determine differences between bipolar and unipolar depression should control for the age of participants when making comparisons.

Parker and colleagues (2013) employed the SMPI (Sydney Melancholia Prototype Index) in another comparative study. The SMPI provides 12 prototypic features of melancholic depression and 12 features of non-melancholic depression. These 'features' include symptoms and illness correlates—including premorbid functioning, personality factors, distal and proximal stressors, and ongoing emotional dysregulation—all of which capture historical differences between melancholic and non-melancholic depression. This study used both the self-report and the clinician-rated SMPI measure to investigate whether bipolar depression is prototypically closer to melancholic than non-melancholic depression. The sample comprised 901 subjects, with 468 having a BD condition (46 BD-I and 422 BD-II) and 433 having a unipolar longitudinal pattern. The composite group of BD patients was more likely to be female, have a younger onset of their depression, be older at initial assessment, more likely to have an illicit drug problem, and more likely to have a lifetime anxiety disorder in comparison to both unipolar melancholic and non-melancholic patients. Comparison of the BD and composite unipolar groups on the 12 SMPI prototypic melancholic clinical features (as rated by clinicians) indicated higher prevalences in the BD group for most items; five were significant for the self-report measure, and nine of the 12 were significant for the clinician-rated measure. A converse pattern was suggested for the 12 SMPI non-melancholic prototypic features, with the BD group reporting significantly lower prevalences for eight of the 12 items.

In essence, the BD patients were more likely than the comparator unipolar patients to report their depression as being disproportionately severe compared to their circumstances, for depression to be less consistent with circumstances, for there to be less likelihood of a clear cause for their depression, and for their depressed mood to be less reactive to positive events and to support. They were also more likely to report anergia, anhedonia, physical slowing, impaired concentration, and for their depression to be more likely to come 'out of the blue'. In essence, BD participants were distinctly more likely to report prototypic melancholic clinical features and a more endogenous onset, again supporting a conclusion that bipolar depression is more likely to be melancholic in nature. Additional analyses compared item prevalences in the BD group with those in the unipolar melancholic and non-melancholic depressive subtypes to determine whether the prototypic pattern for bipolar depression corresponded to either pattern. The bipolar depressed group differed distinctly from the non-melancholic unipolar subset on 22 of the 24 comparisons but less distinctly from the melancholic subjects (with only 11 of 24 being significant), indicating again that bipolar depression was phenotypically closer to melancholic than to non-melancholic depression. Differentiation using this prototypic measure appeared more distinctive than findings from previous studies examining differences on the basis of symptoms only.

A self-report depression severity measure included in this study also allowed consideration as to whether atypical features of hypersomnia and hyperphagia were over-represented in those with bipolar depression as had been suggested from clinical observations. The BD patients were more likely than the unipolar to affirm hyperphagia (46.9 % vs. 39.2 %) and hypersomnia (56.8 % vs. 38.2 %). However, while such atypical features are relatively common in BD patients, they appear to also be common in those with a unipolar depressive condition, with rates in this study suggesting a relatively similar prevalence of such symptoms in the melancholic and non-melancholic subsets. In an earlier paper considering the nature of 'atypical depression' (Parker et al. 2002), we proposed that such atypical features may be better conceptualised as reflecting a homeostatic mechanism seeking to reset the depressed individual's level of emotional dysregulation. Such data suggest that—whether homeostatic or not—they are not specific to any depressive type per se but that they are likely to be more prevalent in bipolar than unipolar depressive states.

Frankland and colleagues (2015) compared 202 patients with a DSM-IV diagnosis of BD-I, 44 patients with BD-II, and 120 patients with a unipolar major depressive disorder diagnosis. In comparison with the unipolar depressive group, the BD-I patients were significantly more likely to report terminal insomnia, hypersomnia, psychomotor retardation, difficulty thinking, morning worsening, and psychotic features (indicating a depressive pattern suggestive of psychotic and melancholic depression). Compared to the unipolar group, the BD-II patients were more likely to report initial insomnia, excessive guilt, difficulty thinking, and morning worsening (suggestive of a melancholic pattern) and also to report more 'mixed' features. This study had the advantage of comparing both BD-I and BD-II subsets against those with a unipolar depression and effectively reported that while melancholic and psychotic depressive patterns were more likely in the BD-I participants, only melancholic features were more common in the BD-II participants.

2.3   Differences Between Unipolar and Bipolar Melancholic Patients

Such studies—indicating that bipolar depression is likely to be more melancholic than non-melancholic in type—allow that conclusion to be pursued by comparing unipolar and BD patients who are judged to experience melancholic depressive episodes. Mitchell and colleagues (1992) recruited depressed patients who met three differing criterion measures (DSM-III, RDC, and CORE) for a diagnosis of melancholia. Of the 138 patients, 27 were rated as having BD (17 manic, 10 hypomanic); after matching for age and gender, this group was compared with 27 unipolar 'melancholic' subjects (i.e. given a melancholia diagnosis by all three measures). Age of onset was similar. The BD patients reported significantly briefer depressive episodes, but the two groups did not differ on either self-reported or clinician-rated depression severity measures, while similar percentages (15 % BD vs. 22 % unipolar) were judged to have a clinical diagnosis of psychotic melancholia. In terms of mental state signs, there was a general trend for BD patients to be rated as showing more psychomotor agitation and less psychomotor retardation, to show a greater loss of appetite, and to be more likely to report subjective agitation and more 'vegetative' abnormalities such as early morning wakening, and diurnal variation with mood and energy being worse in the morning.

In a replication study, Mitchell and Sengoz (1996) adopted a similar methodology and compared 25 DSM-III-R-defined bipolar melancholic patients with a similar number of unipolar melancholic patients, with the groups matched by age and gender. The groups did not differ by age of onset of the first episode of depression, but there were trends for the duration of the current episode of depression to be briefer in the BD subjects and for them to have had more previous episodes. The two groups did not differ by depression severity nor by prevalence of psychotic features, but there was a trend for BD patients to be more likely to be psychotic during a previous episode. In terms of mental state signs, the only formal difference was for the bipolar melancholic patients to demonstrate shortened verbal responses (suggesting psychomotor retardation). When compared across 18 symptoms, the BD patients were less likely to report initial insomnia and suicidal thoughts but more likely to report hypersomnia. Mitchell and Sengoz concluded (p. 178) that analyses across the two studies indicated that there were more similarities than differences between the BD and unipolar patients in terms of cross-sectional clinical features and that the close phenotypic resemblance argued for 'a commonality of biological dysfunction at some level'.

The studies reviewed in the first two sections suggest that those with BD-I are likely to have a psychotic or melancholic depressive pattern during depressed phases while those with BD-II are unlikely, when depressed, to have a psychotic depressive pattern. They are, however, likely to have melancholic depression; consequently 'bipolar depression' overall rarely evidences a non-melancholic depressive pattern. The lack of absolute specificity is likely to reflect two key factors (apart from measurement error). First, even if BD patients are most likely to experience (psychotic or nonpsychotic) melancholic depression as their proto-typic depressive pattern, this does not mean they are not vulnerable to non-melancholic depressive episodes as might be experienced by any individual in response to some major stressors. Second, it may well be that a percentage of BD patients do experience episodes of non-melancholic depression as their standard depressive phenotype, but how common or rare this is remains unclear. Thus, if bipolar depression is principally melancholic or psychotic depression, what might be the best 'signals' for identifying such depressive subtypes?

2.4 Clinical Features Indicative of a Melancholic or Psychotic Depression

If BD patients are most likely to experience episodes of depression marked by melancholic features, it is important to be able to identify melancholia on the basis of clinical symptoms. Historically, a number of so-called endogeneity symptoms (e.g. anhedonia, mood non-reactivity) and vegetative symptoms (e.g. appetite loss, terminal insomnia) have been weighted. In our own studies, we have found that, while a number of such symptoms are common in melancholic depression, they are also common in those with non-melancholic depression. I therefore now detail features that we find clinically—as well as in our research studies—to be distinctly more common in melancholic than in non-melancholic depression; a number are included in the SMPI measure detailed earlier.

2.4.1 Psychomotor Disturbance

Historically (see Jackson 1986; Parker and Hadzi-Pavlovic 1996; Taylor and Fink 2006), melancholia was viewed more as a disorder of movement than of mood, with motor components including retardation and/or agitation. Retardation may be evidenced by a slowing of walking and talking, as well as facial immobility, postural slumping, a monotonous voice, scarcity of speech, and a loss of light in the eyes. Agitation can often be seen physically or mentally, with the patient often speaking in sharp and abrupt sentences, appearing preoccupied and unable to settle, experiencing multiple worrying thoughts, or having physical symptoms such as churning in the stomach and a fairly characteristic importuning refrain (i.e. 'What's going to become of me'). In both the retarded and agitated expressions, concentration is likely to be significantly impaired (with 'foggy thinking' and fewer thoughts). An impaired capacity to absorb information occurs in both those with retardation and agitation often due to the scarcity and slowness of thought in retardation and multiple racing thoughts in those with agitation. Those with the 'retarded' form of melancholia tend to show such features consistently (although there may be a diurnal variation with retardation improving in late mornings), while those with the 'agitated' form of melancholia tend to have a base of retardation and superimposed epochs of agitation, with agitation also generally worse in the mornings. Psychomotor disturbance is generally more distinctive in older subjects with melancholia and may be one of the most distinctive phenomenological features in psychiatry in terms of its specificity to melancholia, but it is only distinctive in a small percentage of younger patients (i.e. those under the age of 40). In essence, when distinctive, it strongly supports a diagnosis of melancholia; however, as melancholia can present without substantive psychomotor disturbance, its absence does not reject a diagnosis of melancholia. Such an age impact on this phenotypic disturbance may reflect those with melancholia progressively recruiting more monoaminergic circuits (especially dopaminergic ones). As those with melancholia grow older, this age-related change in phenotype may perhaps also explain why those with melancholia tend to report a progressive lack of response to narrow-action antidepressants over the years. The 'psycho' component of psychomotor disturbance is considered next.

2.4.1.1   Impaired Concentration

Many measures of melancholia weight the presence of distractible thoughts and poor concentration. However, these constructs need to be assessed carefully as impaired concentration is actually common in melancholia and non-melancholic depression. However, those with a non-melancholic depression tend to report lots of racing and/or worrying thoughts that impair their concentration and make them distractible. In melancholia, the individual is much more likely to report their thinking as 'foggy' with fewer and foggier thoughts as well as difficulty absorbing information, so that reading a book or preparing for an examination can become distinctly compromised.

2.4.1.2   Anergia

I view it as important to distinguish between anergia, fatigue, and a lack of motivation. While fatigue and amotivation are common in those with melancholia, they are also common in non-melancholic depression; in contrast, anergia (or lack of physical energy) is far more distinctive in melancholic patients. As a consequence, rather than asking about fatigue or amotivation, I ask 'Do you find it difficult to get out of bed in the morning and to get going? Possibly even failing to have a bath or shower'? Those with distinct anergia will generally affirm this probe question, detailing that they may stay in bed for many hours or, if they get out of bed, they may only move to the lounge, without bathing or washing for days or even weeks.

2.4.1.3   Anhedonic and Non-reactive Mood

Earlier studies suggested that those with melancholia were more likely to differ from those with a non-melancholic depression in reporting anticipatory rather than consummatory anhedonia, but our studies have not found support for any such differential; as a consequence, we tend to judge anhedonia as an overall construct. Thus, we ask the extent to which the patient finds a lack of pleasure in daily activities or in those activities which might generally give them pleasure, while a non-reactive mood is defined by an inability to be cheered up in social circumstances or when experiencing a pleasant life event. While both are over-represented in melancholic patients, they are difficult to quantify for several reasons. First, it is extremely rare to find a melancholic individual who describes such features as absolutely categorically present. For example, a melancholic patient may acknowledge that they are not getting any pleasure out of anything but, when pressed, note that when they see their grandchild, they may be cheered up briefly or superficially. Thus, such features are extremely helpful when they are absolute or clear-cut but, at minor levels, their specificity to melancholia tends to be low.

2.4.1.4   Diurnal Variation of Mood and Energy

Most individuals with a melancholic depression will report a diurnal variation, with their mood and energy levels being worst in the morning, and that they improve later in the morning or early afternoon. There is also a small percentage of those with seeming true melancholia who report mood and energy dropping late in the day (and usually when the sun is setting). The former is worth weighting, but the latter is only modestly differentiating.

2.4.2   Appetite and/or Weight Loss

While both are relatively common in melancholic depression, they are also commonly reported by patients with non-melancholic depression. Conversely, a percentage of both will experience food cravings and therefore report appetite increase (usually for specific foods such as carbohydrates and chocolates) and weight gain. As many depressed patients are taking weight-gaining medications when assessed, this can compromise the assessment of these constructs.

2.4.3 Insomnia

Those with melancholic depression commonly report early morning wakening with the classic time being around 3 AM. However, early morning wakening is a common feature in any anxiety or depressive condition marked with physiological arousal, so that early morning wakening may be quite common in those with mixed anxiety/depression and with grief states as well as in those with other expressions of non-melancholic depression.

2.4.4 Psychotic Features

The nature of psychotic depression remains somewhat unclear, with the two commonest models viewing it either as a more severe form of melancholia or as melancholia with superimposed (and categorical) features of delusions and/or hallucinations. Such psychotic features may be 'mood congruent' (e.g. feeling that the world is so bleak that the individual would be better off dead, viewing themselves as facing penury) or 'mood incongruent' (i.e. without any seeming depressive overtone or theme) with both expressions almost equally likely (Parker et al. 1996). If the patient does not volunteer such symptoms, they can generally be best elicited by pursuing any sense of guilt or shame directly or via the individual feeling that they 'deserve to be punished'.

2.4.5 Insular and Asocial Behaviours

While this is a relatively vague construct and territory, it is part of the melancholic terrain, although not specific to melancholia. It can be particularly useful in judging whether melancholia exists or not in an adolescent or young adult. Those with melancholia tend to become quite asocial, going or staying in their room, not phoning or returning telephone calls, and retreating from those around them.

2.4.6 Impairment

While those with a non-melancholic depression can clearly be impaired, impairment tends to be more severe in those with a melancholic depression—where patients are much more likely to report that their depression makes it a struggle to get to work, physically get out of bed, engage in normal exercise, relate to a partner, keep up hygiene, and maintain work performance.

2.4.7 Relationship to Stress

While melancholic depression was long described as 'endogenous depression' (i.e. it appeared to come from within rather than being caused by external factors or stressors), empirical research (e.g. Brown et al. 1994) indicated that both melancholic and non-melancholic depressive disorders were commonly preceded by a stressor, a finding that challenged the very construct of 'endogeneity'. Nevertheless, the role of stressors has a number of important diagnostic nuances. The first episode of a melancholic depression is commonly preceded by a stressor. Over time, the illness tends to become more autonomous, with the individual experiencing episodes of melancholic depression without stressors or in response to only minor stressors. More importantly, those with melancholia generally judge the depressive condition as more 'severe' than warranted by any antecedent stressor and, additionally, that it tends to persist far longer than might be expected for the stressor or persist when the stressor is no longer present or operative.

2.4.8 Disease-Like

In one of our studies (Parker et al. 2015), we quantified that some 70 % of those with a melancholic depression and less than 30 % of those with a non-melancholic depression were likely to judge their depression as akin to a 'disease'. Further, when asked to assign the predominant cause (as either biological, psychological, and/or environmental), they were likely to weight a biological contribution, whereas the non-melancholic subjects were distinctly more likely to view their depression as almost entirely environmental.

2.4.9 Family History

Those with a melancholic depression (unipolar or bipolar) are likely to report a family history of depression. In those with BD, they are also more likely to report a family history of depression, BD, and/or suicide in a first-degree or second-degree relative.

The clinical diagnosis of bipolar depression simply requires that the patient has BD and that they are currently distinctly depressed. We judge, however, that it is important to determine the depressive subtype during such states, as management is likely to be influenced by whether the patient is experiencing a psychotic, melancholic, or non-melancholic depressive episode.

2.5   Some Management Nuances

As management options are considered extensively in treatment guidelines for BD and in other chapters of this book, only some brief observations are provided here in relation to managing bipolar depression, but which build on our 'core and mantle' (or 'isomer') model for differentiating BD-I and BD-II and on the depressive subtype experienced by the patient.

My personal view (Parker 2012) is that the management of BD-II (including its depressive phase) should differ from the management of BD-I, in part reflecting the respective nonpsychotic and psychotic status of those two conditions. In managing a patient who presents with significant clinical depression and who has BD, as recommended in most treatment guidelines, I will initiate a mood stabiliser. However, the choice of the mood stabiliser varies considerably between BD-I and BD-II diagnoses. For BD-II, I will generally trial lamotrigine, and only if this failed would I consider other mood stabilisers such as lithium or valproate. For BD-I, I favour lithium, and if this is unsuccessful, I will try valproate next and tend not to find lamotrigine as likely to be effective. As most patients with BD present during an acute episode—and almost invariably the depressed phase—I will seek to address that state (as covered below) and not simply rely on a mood stabiliser.

The use of antidepressant medication in a patient who presents with a distinct episode of bipolar depression is controversial. Most treatment guidelines argue against the use of antidepressant medication on the basis that antidepressants can cause switching (into highs) and mixed states (where the patient is sometimes experiencing a mix of hypo/manic and depressive features or, equally commonly, experiencing a state of agitation not unlike a serotoninergic reaction) or worsen the course of the illness. Each of these propositions has been considered by a number of commentators (see Parker 2012), while the International Society for Bipolar Disorders (ISBD) has reviewed the benefits and concerns about the use of antidepressants in managing an individual with BD (Pacchiarotti et al. 2013); that review was more sanguine about the use of antidepressants. My personal practice is to introduce an antidepressant together with a mood stabiliser in a patient who presents with a significant episode of (bipolar) melancholic depression, but I continue to find the choice a difficult issue. Selective serotonin reuptake inhibitors (SSRIs) tend not to be particularly effective, dual-action antidepressants are more effective but are most likely to cause switching, tricyclics are commonly effective but have also been incriminated as causing switching, and monoamine oxidase inhibitors (MAOIs, long used in the management of bipolar depression) require judicious use in BD patients in light of the need to avoid certain foods. Surprisingly, psychostimulants such as methylphenidate can be useful in managing those with bipolar depression as we reported (Parker and Brotchie 2010) in a clinical case study, either as a single antidepressant agent or augmenting a more orthodox agent. The risk of such a stimulant switching the patient into a high has to be conceded but, as quantified in our report, was a relatively rare event. If the patient with a bipolar melancholic depression has not improved in the next week, I tend to introduce a low-dose atypical antipsychotic medication as an augmenting strategy—and then try to cease it when the depression has been brought under control. In managing an individual with a (bipolar) psychotic episode, I would favour the immediate use of an antipsychotic in conjunction with the introduction of a mood stabiliser and possibly an antidepressant. If the individual is experiencing a (bipolar) non-melancholic depressive episode, I would introduce a mood stabiliser for the bipolar condition and weight a nondrug strategy initially to determine if this assisted their depressive episode. Management of the bipolar depressive episode is therefore somewhat contingent on identifying the bipolar subtype (I or II) and the depressive subtype (i.e. psychotic, melancholic, or non-melancholic). In essence, the subtyping model—and particularly the nature of the bipolar depressive condition—shapes management priorities.

Acknowledgements My thanks to Amelia Paterson for manuscript assistance and with this report supported by an NHMRC Program Grant (1037196).

References

Brown G, Harris T, Hepworth C (1994) Life events and endogenous depression: a puzzle reexamined. Arch Gen Psychiatry 51(7):525–534

Frankland A, Cerrillo E, Hadzi-Pavlovic D et al (2015) Comparing the phenomenology of depressive episodes in bipolar I and II disorder and major depressive disorder within bipolar disorder pedigrees. J Clin Psychiatry 76(1):32–39

Goodwin FK, Jamison KR (2007) Manic-depressive illness: bipolar disorders and recurrent depression. Oxford University Press, Oxford

Jackson SW (1986) Melancholia and depression: from hippocratic times to modern times. Yale University Press, New Haven, CT

Mitchell P, Sengoz A (1996) Phenotypic expression of melancholia contrasted for those with bipolar and unipolar illness courses. In: Parker G (ed) Melancholia: a disorder of movement and mood, 1st edn. Cambridge University Press, Cambridge, pp 172–178

Mitchell P, Parker G, Jamieson K et al (1992) Are there any differences between bipolar and unipolar melancholia? J Affect Disord 25(2):97–105

Pacchiarotti I, Bond DJ, Baldessarini RJ et al (2013) The International Society for Bipolar Disorders (ISBD) task force report on antidepressant use in bipolar disorders. Perspectives 170(11):1249–1262

Parker G (ed) (2012) Bipolar II disorder: modelling, measuring and managing. Cambridge University Press, Cambridge

Parker G, Brotchie H (2010) Do the old psychostimulant drugs have a role in managing treatment-resistant depression? Acta Psychiatr Scand 121(4):308–314

Parker GB, Fletcher K (2009) Is bipolar II depression phenotypically distinctive? Acta Psychiatr Scand 120(6):446–455

Parker G, Hadzi-Pavlovic D (eds) (1996) Melancholia: a disorder of movement and mood. Cambridge University Press, Cambridge

Parker G, Hickie I, Hadzi-Pavlovic D (1996) Psychotic depression: clinical definition, status and the relevance of psychomotor disturbance to its definition. In: Parker G (ed) Melancholia: a disorder of movement and mood, 1st edn. Cambridge University Press, Cambridge, pp 172–178

Parker G, Roy K, Wilhelm K et al (2000) The nature of bipolar depression: implications for the definition of melancholia. J Affect Disord 59:217–224

Parker G, Roy K, Mitchell P et al (2002) Atypical depression: a reappraisal. Am J Psychiatry 159 (9):1470–1479

Parker G, Hadzi-Pavlovic D, Tully L (2006) Distinguishing bipolar and unipolar disorders: an isomer model. J Affect Disord 96(1):67–73

Parker G, McCraw S, Hadzi-Pavlovic D, Hong M, Barrett M (2013) Bipolar depression: proto-typically melancholic in its clinical features. J Affect Disord 147(1):331–337

Parker G, Paterson A, Hadzi-Pavlovic D (2015) Cleaving depressive diseases from depressive disorders and non-clinical states. Acta Psychiatr Scand 131(6):426–433

Taylor MA, Fink M (2006) Melancholia: the diagnosis, pathophysiology and treatment of depressive illness. Cambridge University Press, Cambridge

Chapter 3
Course and Outcome of Bipolar Disorder: Focus on Depressive Aspects

Rodrigo Escalona and Mauricio Tohen

Abstract The presence of depressive symptoms dominates the longitudinal course of bipolar disorder (BD) and predicts functional impairment. Despite great progress in understanding the biological basis of BD, the course and outcome of the illness can only be predicted using clinical variables. This chapter summarizes the main factors that predict course and outcome in BD with a focus on depressive symptoms. The natural course of the illness, the impact of the first episode, the impact of the depressive phase, cycle length, onset, age, gender, type of illness, personality traits, temperament, comorbidity, family history, life events, and outcome features will be reviewed. Conceptual models such as disease staging and their prognostic value will also be discussed.

Keywords Bipolar disorder • Course • Outcome • Predictors • Mixed depression • Clinical diagnosis

3.1 Introduction

Despite great progress in understanding the biological basis of bipolar disorder (BD), the course and outcome of the illness can still only be predicted using clinical variables. However, clinical features are not always reliable or available, and their impact cannot be applied directly to predict the outcome of an individual patient. The assessment is further complicated by the fact that BD represents a dimensional condition within a full spectrum of mood disorders and is often accompanied by psychiatric comorbidity (Kessler et al. 2006).

R. Escalona
Department of Psychiatry & Behavioral Sciences, Health Sciences Center, University of New Mexico, Albuquerque, NM 87131, USA

New Mexico Veteran Affairs Health Care System, Albuquerque, NM, USA

M. Tohen (✉)
Department of Psychiatry & Behavioral Sciences, Health Sciences Center, University of New Mexico, Albuquerque, NM 87131, USA
e-mail: mtohen@salud.unm.edu

© Springer International Publishing Switzerland 2016   33
C.A. Zarate Jr., H.K. Manji (eds.), *Bipolar Depression: Molecular Neurobiology, Clinical Diagnosis, and Pharmacotherapy*, Milestones in Drug Therapy,
DOI 10.1007/978-3-319-31689-5_3

Several methodological issues affect the study of the natural course of BD (Wittchen et al. 2003). Despite a considerable amount of research, the course and outcome of BD still remain highly unpredictable. Likewise, it is difficult to determine the effect of treatment on the natural course of an illness that, despite best evidence-based treatment, still involves multiple relapses and impaired psychosocial functioning (Goldberg et al. 1995). There is, however, agreement among researchers that BD is a severe, chronic, and disabling lifelong condition and that breakthrough depression usually presents higher risk for long-term functional impairment than mania. While identifying and treating the illness early in its time course may improve the chance of a better prognosis, there are several barriers to early intervention. These include the well-known delay of approximately 10 years from the first episode of illness to a diagnosis of BD (Hirschfeld et al. 2003). This is particularly true for the many patients who present first with depressive episodes and much later with mania or hypomania, making the BD diagnosis impossible until later in the illness. Looking at the other clinical variables in patients who present with depression—such as melancholic or psychotic features, family history, and age of onset—may help identify BD earlier so that appropriate treatment can be begun. This delay in diagnosis poses a threat for the effectiveness of early treatment interventions, especially because data suggest that beginning lithium therapy within the first 10 years of illness may provide better outcomes than beginning prophylaxis later in life for patients with BD (Franchini et al. 1999). Furthermore, a history of multiple previous episodes may be associated with poor response to lithium (Tohen et al. 1990a; Swann et al. 1999), although these findings are limited by the lack of a comparator and the inclusion of subjects who had previously failed to respond to lithium. Similarly, long-term divalproex (Calabrese et al. 2005), non-pharmacological therapies (Scott et al. 2007), and maintenance therapy with olanzapine (Ketter et al. 2006) have been found to be less effective in preventing relapses in patients with a high number of previous episodes.

The very high degree of comorbidity and treatment resistance in outpatients with BD highlights the need to develop new treatment approaches, much earlier illness recognition, diagnosis, and intervention in an attempt to reverse or prevent this illness burden (Post et al. 2003). Although full symptomatic remission does not guarantee functional recovery (Tohen et al. 1990b, 2000, 2003), it may have a favorable impact on long-term prognosis.

This chapter will summarize the main factors that predict course and outcome in BD with a focus on depressive symptoms. The natural course of the illness, the impact of the first episode, the impact of the depressive phase, cycle length, age of onset, age, gender, type of illness, personality traits, temperament, comorbidity, family history, life events, and outcome features will be reviewed. Conceptual models such as staging and outcome dimensions and their prognostic value will also be discussed.

3.2 Natural Course

Researchers agree that bipolar spectrum disorders are severe chronic conditions that should be considered lifelong disabilities (Post et al. 2003). The impact of modern treatment on the natural course of the illness is uncertain. In addition, high diagnostic instability is considered a feature of BD. A recent naturalistic study found a high prevalence of misdiagnosis and diagnostic shift from other psychiatric disorders to BD (Salvatore et al. 2007). More than half of severe mood disorders become BD, and the risk of depression developing into BD is lifelong (Salvatore et al. 2007; Angst et al. 2005). Naturalistic and long-term studies showed that patients with BD develop persistent functional impairment; that is, patients experienced some degree of disability during most of their long-term follow-up including 19–23 % of the time with moderate impairment and 7–9 % of the time with severe overall impairment (Baca-Garcia et al. 2007; Judd et al. 2008). One study found that bipolar I (BD-I) patients were completely unable to carry out work role functions during 30 % of the assessed months, which was significantly higher than rates in individuals with major depressive disorder (MDD) or bipolar II (BD-II) (21 % and 20 %, respectively). Neuropsychological impairment persists during euthymic states, but it is confounded partly by mild affective symptoms in remitted patients. The clinical representations of these persistent alterations are related to the degree of disability (Marneros et al. 1991).

The recurrence risk of BD is about twice that of MDD. Furthermore, recovery is more frequent among MDD than BD patients, although five-year remission rates were found to be independent of the number of episodes (Angst et al. 2005). There appears to be a constant risk of recurrence over the life-span up to the age of 70 or more, even 30–40 years after onset (Angst et al. 2003). This long-term course usually causes significant handicaps and problems in the lives of patients and, in many cases, leads to disability (Nolen et al. 2004).

3.3 Illness Recurrence and the Course of Syndromal and Functional Recovery

Most of the evidence from both the pre-lithium and modern eras suggests that the index episode tends to predict the polarity of the subsequent major mood episode: a manic index episode tends to predict a manic relapse, whereas a depressive index episode predicts a depressive relapse (Calabrese et al. 2004); indexed mixed episodes have been found to predict relapse into a depressive episode (Tohen et al. 2003). The presence of at least two manic/hypomanic symptoms in the index episode is associated with increased family history of BD-I, a higher score for suicidal thoughts during the episode, a longer duration of the episode, and a higher affective morbidity during the observation period (Maj et al. 2006).

The McLean-Harvard First-Episode Project has systematically followed large numbers of patients with BD and other psychotic disorders from their first hospitalization. The project's findings indicate that the course of BD-I is much less favorable than had formerly been believed, despite modern clinical treatment with mood-stabilizing and other pharmacological agents. Full functional recovery from initial episodes was uncommon, and full symptomatic recovery was much slower than early syndromal recovery; most early morbidity was depressive-dysphoric, as reported in midcourse, and initial depression or mixed states predicted an increased number of depressive episodes and overall morbidity, whereas initial mania or psychosis predicted later mania and a better prognosis (Baca-Garcia et al. 2007).

Within four years of first lifetime hospitalization for mania, prospective data show that most subjects achieved syndromal recovery by two years, but 28 % remained symptomatic, only 43 % achieved functional recovery, and 57 % switched phases or had new illness episodes after achieving recovery (Tohen et al. 2003). In this study, factors associated with a shorter time to syndromal recovery for 50 % of the subjects were female sex, shorter index hospitalization, and lower initial depression ratings. The 43 % who achieved functional recovery were more often older and had shorter index hospitalizations. Within two years of syndromal recovery, 40 % experienced a new episode of mania (20 %) or depression (20 %), and 19 % switched phases without recovery. Predictors of manic recurrence were initial mood-incongruent psychotic features, lower premorbid occupational status, and initial manic presentation. Predictors of depression onset were higher occupational status, initial mixed presentation, and any comorbidity (Tohen et al. 2003).

Targeting residual symptoms in maintenance treatment may represent an opportunity to reduce the risk of recurrence of BD. Another two-year follow-up study of the clinical features associated with risk of recurrence in patients with BD receiving treatment found that 58 % of patients subsequently achieved recovery (Perlis et al. 2006). For up to two years of follow-up, half of these individuals experienced recurrences, with more than twice as many developing depressive episodes versus manic, hypomanic, or mixed episodes. Residual depressive or manic symptoms at recovery and proportion of days spent depressed or anxious in the preceding year were significantly associated with shorter time to depressive recurrence. Residual manic symptoms at recovery and the proportion of days of elevated mood in the preceding year were significantly associated with shorter time to manic, hypomanic, or mixed episode recurrence (Perlis et al. 2006). Another recent report (Perlis et al. 2006) suggested that it is not just chronic subsyndromal symptoms that predict shorter time to a new episode, but rather their emergence, particularly the emergence of depressive symptoms.

Although most adolescents with BD experience syndromic recovery following their first hospitalization, the rates of symptomatic and functional recovery are much lower (Tohen et al. 1990b, 2000, 2006). Few studies have examined the clinical, neuropsychological, and pharmacological factors involved in the functional outcome of BD. The variable that appears to best predict psychosocial functioning in BD patients is verbal memory; low-functioning patients are cognitively more impaired than high-functioning patients on verbal recall and executive functions (Martinez-Aran et al. 2007). Few studies have examined whether comorbid personality disorders and other clinical factors can predict functional morbidity in BD. However, residual depression predicts poorer residential and social/leisure outcomes independent of personality disorders or maladaptive traits (Loftus and Jaeger 2006).

The mortality of patients with BD is considerably higher than that of the general population. At least 25–50 % of patients with BD attempt suicide at least once in their lives (Jamison 2000). The polarity of a patient's first reported mood episode suggests that depression-prone subtypes have a greater probability of suicidal acts (Chaudhury et al. 2007). Patients with mood disorders in general have a higher risk of death by suicide (15–30 %) than healthy people; however BD-II patients may be more likely to attempt suicide than BD-I patients. Comorbid anxiety disorders may also elevate the risk for suicidal ideation and attempts (Simon et al. 2007a). The rates of mixed depression among BD and non-BD depressive suicide attempters are much higher than previously reported among non-suicidal BD-II and MDD outpatients, suggesting that suicide attempters come mainly from mixed depressives who predominantly have BD-II (Balazs et al. 2006). Recent findings show that while modest changes in the severity of depression are associated with statistically and clinically significant changes in functional impairment and disability in patients with BD, changes in the severity of mania or hypomania are not consistently associated with differences in functioning (Simon et al. 2007b).

3.4 Prognostic Staging Models

During the last 10 years, prognostic staging models for BD have attracted growing attention by raising the possibility of defining stage-specific strategies for treatment (Kapczinski et al. 2014).

The proposed models use findings from clinical studies of treatment and functioning to stage the illness and integrate the potential role of neurocognitive, neuroimaging, and peripheral biomarkers. Most studies to date indicate that the progression to late stages of the illness predicts worse overall prognosis and poorer response to standard treatment. Berk and colleagues suggested a staging model to predict outcome (stages 0–5) (Berk et al. 2007), suggesting that BD begins with an at-risk, asymptomatic period. Patients then begin to exhibit mild or nonspecific symptoms and usually progress to manifest the range of prodromal patterns that have been described in the literature. The first threshold episode may then be followed by a first relapse, subsequently followed by a pattern of periods of euthymia and recurrences. Some patients may have syndromal or symptomatic recovery, while others may have an unremitting or treatment refractory course. It is possible that all these stages require specific therapeutic interventions, and the impact of comorbidity, specific treatment, personality, adherence, and response to therapy could differ in each stage. In general, a greater number of episodes imply a progression to a later stage with poorer treatment response and prognosis. The persistence of neurocognitive impairment is associated with poorer psychosocial functioning at any stage. Peripheral biomarkers of inflammation are more likely present at the end stage of BD, consistent with the hypothesis of a neuroprogression of the illness (Kauer-Sant'Anna et al. 2009). However, significant medical and substance abuse comorbidities must be considered as potential confounders. Some neuroimaging findings also suggest some kind of disease progression, but this needs confirmation.

Additional research is needed to clarify the usefulness of the staging model to complement existing classifications of BD, with an emphasis on a longitudinal dimension instead of a merely cross-sectional view. Also, in the near future, improved staging models may include biological markers in addition to clinical variables.

3.5   The Impact of Treatment on the Course of Illness

The 10-year delay from the first episode of illness to a diagnosis of BD is an important impediment to early treatment intervention and possibly a better prognosis. Studies have shown that BD outcome worsens as the number of manic episodes increases (Tohen et al. 1990b), suggesting that prevention of recurrent episodes early during the disorder could improve long-term prognosis.

The initial prodrome of BD has received very little attention to date, and there are no prodromal features that clearly distinguish between patients who go on to develop BD and those who develop schizophrenia (Thompson et al. 2003). Several authors point out that pharmacological treatment of the early phase of BD lacks specific guidelines (Conus et al. 2006). Knowledge is limited as to how to distinguish prodromal BD from MDD, but even mania is frequently misdiagnosed. This is key because the outcome of mania is not as good as was formerly believed (Conus and McGorry 2002).

Although the impact of different treatment options for BD is discussed elsewhere in this book, it is worth noting that few effective treatments exist for acute bipolar depression and prevention of recurrent episodes. Furthermore, the effectiveness and safety of specific treatments such as standard antidepressant agents for depressive episodes associated with BD have not been well studied. Because episodes of depression are the most frequent cause of disability among patients with BD, it is important to determine whether adjunctive antidepressant therapy reduces symptoms of bipolar depression without increasing the risk of mania and therefore changing the course and outcome of the disorder.

A recent, double-blind, controlled trial showed that the use of adjunctive, standard antidepressant medication, as compared to the use of mood stabilizers, was not associated with increased efficacy or with increased risk of treatment-emergent affective switch (Sachs et al. 2007). It is also important to determine the benefits of the continued use of typical antipsychotic agents following remission from an acute manic episode. Studies show that there are no short-term benefits associated with continued use of a typical antipsychotic after achieving remission from an episode of acute mania. In fact, its continued use may be associated with detrimental effects, including relapse into depression in some cases (Zarate and Tohen 2004). On the other hand, despite the recent FDA approval of two atypical antipsychotics for the treatment of bipolar depression (quetiapine and lurasidone) as monotherapy and adjunctive treatment to mood stabilizers, the long-term effect on illness prognosis is unknown (Loebel et al. 2014). Finally, a recent report suggested that early-stage (but not intermediate- or later-stage) patients had a significantly lower rates of relapse/recurrence of manic/mixed episodes with some treatments but not with others (Ketter et al. 2006). Subsyndromal symptoms are common during maintenance treatment and appear to be associated with relapse into an episode of the same polarity (Tohen et al. 2006; Frye et al. 2006). Comorbid anxiety symptoms in patients with bipolar depression have a negative impact on treatment outcome, so treatment interventions should focus on reducing both depressive and anxiety symptoms in these patients (Tohen et al. 2007).

3.6 Predictive Factors Affecting Prognosis

3.6.1 Age at Onset and Gender

The average age of onset of a first manic episode is 21 years, but onset may occur at any age from childhood to old age. Childhood-onset BD usually has a poorer prognosis and is associated with long delays to first treatment, averaging more than 16 years. Patients with childhood or adolescent onset retrospectively report more episodes, more comorbidities, and rapid cycling; prospectively, they demonstrate more severe mania, depression, and fewer days well (Leverich et al. 2007).

Data have consistently shown that 70–100 % of children and adolescents with BD will eventually recover from their index episode; however, despite ongoing treatment, up to 80 % will experience recurrences after recovery (Birmaher and Axelson 2006). BD has a considerable effect on the normal psychosocial development of the child and increases the risk for academic, social, and interpersonal (family, peers, work) problems, as well as for healthcare utilization. Some studies suggest that approximately 30 % of preadolescents with MDD experience a manic episode and manifest BD within five years (Geller et al. 1994).

Mania in the elderly appears to be a heterogeneous disorder. In elderly patients with first-episode mania who were followed for three to 10 years, men had a higher risk of mortality. Compared to elderly patients with early onset and multiple episodes of mania, elderly patients with first-episode mania were twice as likely to have a comorbid neurological disorder (Tohen et al. 1994).

Previous findings suggest that men have a significantly earlier onset of first-episode mania and BD associated with childhood antisocial behavior; women have more depressive episodes than manic episodes and higher incidence rates of BD-II throughout adult life, except for early life, and a greater likelihood of rapid cycling (Marneros 2006). More men than women report mania at the onset of BD-I, and men also have higher rates of comorbid alcohol abuse/dependence, cannabis abuse/dependence, pathological gambling, and conduct disorder (Kawa et al. 2005). Women report higher rates of comorbid eating disorders, weight change, appetite change, and middle insomnia during depressive episodes (Kawa et al. 2005). However, no gender differences appear to exist between male and female subjects in time to remission from the index episode, number of recurrences, and time spent with any clinical or subclinical mood symptom over a 48-week period, at least when similar treatment strategies are adopted (Benedetti et al. 2007).

3.6.2   Type of Onset and Type of Disorder

The length of untreated individual illness episodes in BD varies from several weeks to several months and depends on the type of episode. There are significant differences in time to recovery in patients with BD by episode subtype (Sachs et al. 2007; Keller et al. 1986). Based on a median follow-up of 18 months, the life-table estimate of the probability of remaining ill for at least one year was 7 % for pure manic patients compared with 32 % for patients who entered the study with episodes that were mixed or cycling. Purely depressed patients had a 22 % probability of remaining ill, approximating rates found in patients without BD who have episodes of depression. However, the duration of individual episodes also depends on response to treatment, and 15–30 % of patients with mood disorders suffer from persisting alterations of personality or social interaction or from persisting symptoms. Rapid cycling and mixed states are associated with a poorer prognosis and nonresponse to antimanic agents. Risk factors for rapid cycling include biological rhythm dysregulation, antidepressant or stimulant use, hypothyroidism, and premenstrual and postpartum states (American Psychiatric Association 2002). Patients with BD have an average of four episodes during the first 10 years of their illness (Tsai et al. 2001; Meeks 1999). After that, the average length of time between episodes is between one and two years. In both BD-I and BD-II, 60–70 % of manic episodes occur immediately before or after a major depressive episode, and the interval between episodes tends to decrease as the individual ages. Differentiation of mood congruence of psychotic features in mania evidently has prognostic validity. Mood-incongruent psychotic features during the index manic episode predicted shorter time in remission at four years (Tohen et al. 1992). Higher occupational status, initial mixed presentation, and any comorbidity predicted depressive rather than manic onset (Tohen et al. 2003). Increased number of hospitalizations and less rapid cycling were associated with BD-I as compared to BD-II (Coryell et al. 1989, 1992).

3.6.3 Personality Traits and Temperament

Personality and temperament are thought to impact prognosis and the clinical manifestation of BD. Studies have suggested that mixed episodes may result from a mixture of inverse temperamental factors to a manic syndrome (Rottig et al. 2007). Some studies question the current categorical split of mood disorders into bipolar and depressive disorders, suggesting that two highly unstable personality features, i.e., the cyclothymic temperament and borderline personality disorder, have more in common with BD-II than MDD (Benazzi 2006). Several research findings that are in line with current familial-genetic models of this disorder suggest that the characterization of BD-II must include a greater emphasis on temperamentally based mood and anxious reactivity (Akiskal et al. 2006a). Such phenotypic characterization may assist in genotyping; however its predictive value on outcome still requires more research (Akiskal et al. 2006b).

3.6.4 Family History and Genetics

The application of genomics to clinical practice is limited at present, but is expected to grow rapidly. Despite some recent successes, identifying genes for BD through classic human genetic studies is not consistent; the main issue is the lack of replication of the findings in this field (Kato 2007). There are many possible reasons for this relatively slow discovery. BD is a complex polygenic disorder, with variable penetrance and phenotypic heterogeneity, and it overlaps and is interdependent with other neuropsychiatric disorders.

   In addition, the effects of environmental factors (epigenetic modifications, effects of stress, infections, drugs, medications) on the expression of the phenotype are not fully understood nor factored into human genetic linkage studies (Le-Niculescu et al. 2007). There is increasing evidence that genome-wide association studies represent a powerful approach to the identification of genes involved in common human diseases (Wellcome Trust Case Control Consortium 2007). The first genome-wide association study of BD showed that several genes, each of modest effect, reproducibly influence disease risk (Baum et al. 2008).

3.7 Bipolar II Depression, Subsyndromal Depression, and Mixed Depression

Mixed depression is probably a key component of the continuum concept of mood disorders, and it might have a predictive role in the course of BD. Recent findings suggest that the prevalence of mixed depression is high in patients with BD. Mixed depression is defined by the combination of depression (a major depressive episode) and non-euphoric, usually subsyndromal, manic, or hypomanic symptoms (Benazzi 2007). The reemerging concept of mixed depression also influences how we see the boundaries between bipolar and depressive disorders. A major addition in the DSM-5 is the introduction of mixed features in patients with MDD, defined as the presence of at least two criteria of the opposite pole (American Psychiatric Association 2013).

BD-II and mixed depression are relatively understudied, despite a prevalence of about 5 % in the community and about 50 % in depressed outpatients (Benazzi 2007). Prospective studies have shown that the longitudinal weekly symptomatic course of BD-I is chronic, that the symptomatic structure is primarily depressive rather than manic, and that subsyndromal and minor affective symptoms predominate, although symptom severity levels fluctuate (Judd et al. 2002).

Depressive episodes and symptoms, which dominate the course of BD-I and BD-II, appear to be more disabling than corresponding levels of manic or hypomanic symptoms. Table 3.1 summarizes the predictive value of depressive symptoms on the course of BD. Subsyndromal depressive symptoms, but not subsyndromal manic or hypomanic symptoms, are associated with significant impairment, and subsyndromal hypomanic symptoms appear to enhance functioning in BD-II (Benazzi 2007). Subsyndromal symptoms in BD impair functioning and diminish quality of life. Findings suggest that the presence of subsyndromal depressive symptoms during the first two months significantly increases the likelihood of depressive relapse (Tohen et al. 2006). Patients with psychotic features and those with a greater number of previous depressive episodes were more likely to experience subsyndromal depressive symptoms.

As noted previously, because a substantial number of patients with BD present with an index depressive episode, it is likely that many are misdiagnosed with MDD. Whether or not antidepressants worsen the course of BD is still being debated, because misdiagnosed patients are often treated with antidepressants, which, if used improperly, are known to induce mania and provoke rapid cycling (Goldberg 2003). Furthermore, it appears that a first depressive rather than manic episode in BD might lead to a subsequent course with a greater burden of depressive symptoms (Perlis et al. 2005). Depressive-onset BD is significantly associated with more lifetime depressive episodes and a greater proportion of time with depression and anxiety in the year prior to assessment. However, the quantity and severity of weeks in symptomatic affective states are possibly greater predictors of affective burden in BD-I patients than the quantity and direction of affective switches (Mysels et al. 2007). Analysis of assessments in clinical trials revealed that over 80 % of the treatment effect was attributable to the indirect effects of improvements in the depressive factors of the Montgomery-Asberg Depression Rating Scale (MADRS) like sadness, negative thoughts, detachment, and neurovegetative symptoms, and changes in factor scores were highly correlated with changes in clinical improvement (Williamson et al. 2006).

Table 3.1 Predictive value of depressive symptoms in the course of bipolar disorder[a]

Onset and index episode
• Index mixed episodes have been found to predict relapse into a depressive episode
• It is likely that many patients presenting with an index depressive episode are misdiagnosed with major depressive disorder
• One third of preadolescents with major depressive disorder experience a manic episode and manifest bipolar disorder within five years
• Lower initial depression ratings are associated with shorter time to syndromal recovery
• A depressive onset is predicted by higher occupational status, initial mixed presentation, and any comorbidity
• Polarity of patients' first reported mood episode suggests a depression-prone subtype with a greater probability of past suicide attempt
• Depressive-onset bipolar disorder is significantly associated with more lifetime depressive episodes and a greater proportion of time with depression and anxiety

Course, number, and length of episodes
• The symptomatic structure of bipolar II disorder is primarily depressive rather than manic
• Twice as many patients develop depressive episodes as manic, hypomanic, or mixed episodes
• Residual depressive or manic symptoms at recovery and proportion of days depressed are significantly associated with shorter time to depressive recurrence
• The longest duration of episodes was found for mixed episodes, while depressive episodes have an intermediate duration and manic episodes are the shortest
• 60–70 % of manic episodes occur immediately before or after a major depressive episode, and manic episodes often precede or follow major depressive episodes
• Shorter time to a depressive recurrence can be predicted if residual depressive or manic symptoms are still present at recovery
• Rapid cycling can be related to a higher number of prior depressive episodes
• Patients with psychotic features and those with a greater number of previous depressive episodes were more likely to experience subsyndromal depressive symptoms
• 80 % of the treatment effect is attributable to the indirect effects of improvements in depressive symptoms

Risk for long-term prognosis
• Breakthrough depression represents higher risks for long-term treatment than mania
• Every new episode of depression brings a new risk for mania
• The risk of depression developing into bipolar disorder remains lifelong
• Subsyndromal depressive symptoms during the first two months after recovery significantly increase the likelihood of depressive relapse

Functional recovery—outcome
• Depressed patients are more impaired than euthymic or hypomanic patients on tests of verbal recall and fine motor skills
• Residual depression predicts poorer residential and social/leisure outcomes independent of personality disorders or maladaptive traits
• Suicide attempters come mainly from mixed depressives with predominantly bipolar II base
• Subsyndromal depressive symptoms, but not subsyndromal manic or hypomanic symptoms, are associated with significant impairment

[a]All the statements in this table are referenced in the text

3.8 Comorbidity

BD has frequent comorbidities that worsen prognosis, especially in association with substance-use disorders (Tohen 1999; Tohen et al. 1998). The relative age at onset of alcohol use and BD is associated with differences in the course of both conditions. A first hospitalization for mania is associated with a period of recovery from comorbid alcohol abuse (Strakowski et al. 1992, 2005). Those patients with alcohol-use problems prior to BD are usually older, more likely to recover, and more likely to recover quickly than those whose alcohol problems occur after their diagnosis of BD. In contrast, those who have BD first spend more time with affective episodes and symptoms of an alcohol-use disorder during follow-up. Comorbid alcoholism is also usually related to poorer psychosocial adjustment (Coryell et al. 1989). Slower recovery has been associated with comorbid drug abuse (Strakowski et al. 1998; Baethge et al. 2005). Attention deficit/hyperactivity disorder and anxiety disorders, including those present during relative euthymia, also predict a poorer course of BD (Otto et al. 2006). Comorbid panic disorder is associated with a higher likelihood of rapid cycling (Coryell et al. 1992). One study showed that anxiety comorbidity impacts health-related quality of life in patients with BD-I but not BD-II (Albert et al. 2008).

Little is known about the treatment of psychiatric comorbidities in BD because their treatment is largely empirically based rather than based on controlled data (Singh and Zarate 2006). Many studies have examined the prevalence and predictive validity of personality disorders among MDD patients, but few have examined these issues among BD patients (George et al. 2003). Findings suggest that clinicians should be more vigilant for comorbid personality disorders and BD and less reluctant to diagnose them (George et al. 2003; Barbato and Hafner 1998). However, when structured assessments of personality disorders are performed during clinical remission of BD, fewer than one in three BD patients meet full syndromal criteria for a personality disorder (Paris et al. 2007). For instance, borderline personality disorder and BD can often co-occur, but their relationship is not consistent or specific. Existing data fail to support the conclusion that borderline personality disorder and BD exist on a spectrum, but allow for the possibility of partially overlapping etiologies and syndromatic presentation (Stromberg et al. 1998).

BD patients with lifetime smoking are more likely to have an earlier age of onset of mood disorder, greater severity of symptoms, poorer functioning, history of a suicide attempt, and a lifetime history of comorbid anxiety and substance-use disorders. Smoking may also be independently associated with suicidal behavior in BD (Ostacher et al. 2006). The effects of the sequence of onset of BD and cannabis-use disorders are less pronounced than observed in co-occurring alcohol-use disorders and BD (Strakowski et al. 2007). Cannabis use is associated with more time spent in affective episodes and with rapid cycling. Most cannabis-use disorders remit immediately after hospitalization, followed by rapid rates of recurrence. Individuals with BD are disproportionately affected by several stress-sensitive medical disorders such as circulatory disorders, obesity, and diabetes mellitus. Individuals with respiratory disorders, infectious diseases, epilepsy, multiple sclerosis, migraine, and circulatory disorders may also have a higher prevalence of BD (McIntyre et al. 2007). The increased medical burden in BD is not simply a result of psychiatric symptoms and their corresponding dysfunction (Kupfer 2005).

Medical comorbidity is often associated with earlier onset of BD symptoms, more severe course, poorer treatment compliance, and worse outcomes related to suicide and other complications. It is still uncertain whether the medical comorbidities are a consequence of BD, another manifestation of the condition, or adverse effects of its pharmacological treatment (Krishnan 2005). To ensure prompt, appropriate intervention while avoiding iatrogenic complications, the clinician must evaluate and monitor patients with BD for the presence and the development of comorbid psychiatric and medical conditions.

3.9 Life Events

Stressful life events can unfavorably alter the course of the illness and negatively influence adherence to maintenance treatment. They have been associated with slower recovery and higher relapse rates. Stress is linked to changes in mood symptoms among BD adolescents, although correlations between life events and symptoms vary with age (Kim et al. 2007). There is no significant interaction between stress and episode number when predicting BD recurrence, and the interaction of early adversity severity and stressful life events significantly predicts recurrence in a manner consistent with the sensitization hypothesis (Dienes et al. 2006).

Few studies have examined the prognostic value of family factors on the course of BD. Patients who were more distressed by their relatives' criticisms had more severe depressive and manic symptoms and proportionately fewer days well (Miklowitz et al. 2005). Besides associations between high emotionality and MDD, studies that examined the relationship between temperament, recent and remote life events, and psychopathology among the offspring of parents with BD found an association between psychopathology and the number of recent negative life events, but no association between psychopathology and the number of early losses (Duffy et al. 2007). In this population, any effect of undesirable life events would appear to be mediated through the association with emotionality. Childhood adversity may be a risk factor for vulnerability to early onset illness, and an array of stressors may be relevant not only to the onset, recurrence, and progression of affective episodes, but the highly prevalent substance abuse comorbidities as well (Post and Leverich 2006).

3.10   Neurocognition

Recent analyses have revealed modest impairment in executive functioning, memory, and attention in both hypomanic and depressed BD patients, with additional fine motor skills impairment in the latter (Malhi et al. 2007). BD depressed and hypomanic patients differ with respect to the nature of their memory impairment. Depressed patients are more impaired compared to euthymic patients on tests of verbal recall and fine motor skills. Psychosocial functioning is impaired across all three patient groups, but only in depressed and hypomanic patients does this correlate significantly with neuropsychological performance. These cognitive difficulties, especially related to verbal memory, may help explain the impairment regarding daily functioning, even during remission (Martinez-Aran et al. 2004), and these are in line with findings that full symptomatic recovery (remission) does not guarantee functional recovery (Tohen et al. 1990b, 2003). While considerable evidence suggests that neurocognition declines steadily over the early course of schizophrenia but is more stable in BD, very little is known about the longitudinal trait stability of neurocognitive performance in BD. One study found that patients with BD showed stability over time in attentional measures but greater variability in other domains over a five-year period (Burdick et al. 2006). Impaired insight and other neurocognitive dysfunctions correlate among symptomatic as well as remitted BD patients (Varga et al. 2006). Cognitive impairment seems to be related to a worse clinical course and poor functional outcome; however, further studies are needed to clarify whether a severe course of illness is associated with more pronounced cognitive disorders and whether psychotic symptoms during the acute phase of the illness can predict cognitive deficits in patients with BD later in the illness. Recent findings suggest that patients with BD lose hippocampal, fusiform, and cerebellar gray matter at an accelerated rate compared with healthy control subjects. This tissue loss can be associated with deterioration in cognitive function and illness course (Moorhead et al. 2007).

3.11   Future Trends and Needs

Despite considerable research efforts in this area, the psychiatric interview and an examination focusing on the longitudinal course specifiers remain the main source of prognostic information to guide physicians in their assessment of BD. Although tailored therapies are the preferred future goal of an individual treatment plan, more research is needed to establish better and more reliable course predictors for individual patients. Besides pharmacogenomic evaluation of subject data from long-term naturalistic studies, more dimensional descriptions of the disorder are warranted to maximize subtype homogeneity. The predictive value and use of mood-congruent versus mood-incongruent psychotic symptoms, mixed episodes, cognitive symptoms, and predominant polarities are limited by current specifiers of BD (Vieta 2006).

Future diagnostic classification systems need to reconsider relying solely on categorical descriptors and include dimensional measures of the different phases of BD, thus further stimulating and refining research in the field (Kupfer et al. 2007). New studies using the RDoC (research domain criteria) in addition to categorical diagnostic constructs may prove useful in providing biological markers to aid disease staging and better treatment matching to improve outcomes.

Depressive symptoms dominate the longitudinal course of BD and predict functional impairment. Therefore they deserve more attention in the clinical assessment of patients, on treatment decisions, and in future studies. Finally, we hope that in the not too distant future, biomarkers such as brain imaging will become a tool in the selection of treatment and the prediction of outcome in patients suffering from this devastating condition.

Disclosures Dr. Tohen was a full-time employee at Lilly (1997–2008). He has received honoraria from, or consulted for, Abbott, AstraZeneca, Bristol-Myers Squibb, GlaxoSmithKline, Lilly, Johnson & Johnson, Otsuka, Merck, Sunovion, Forest, Gedeon Richter Plc, Roche, Elan, Alkermes, Lundbeck, Teva, Pamlab, Wyeth, and Wiley Publishing. His spouse was a full-time employee at Lilly (1998–2013). Dr. Escalona has no conflict of interest to disclose, financial or otherwise.

References

Akiskal HS, Akiskal KK, Perugi G et al (2006a) Bipolar II and anxious reactive "comorbidity": toward better phenotypic characterization suitable for genotyping. J Affect Disord 96 (3):239–247

Akiskal HS, Kilzieh N, Maser JD et al (2006b) The distinct temperament profiles of bipolar I, bipolar II and unipolar patients. J Affect Disord 92(1):19–33

Albert U, Rosso G, Maina G et al (2008) Impact of anxiety disorder comorbidity on quality of life in euthymic bipolar disorder patients: differences between bipolar I and II subtypes. J Affect Disord 105(1–3):297–303

American Psychiatric Association (2002) Practice guideline for the treatment of patients with bipolar disorder (revision). Am J Psychiatry 159(4):1–50

American Psychiatric Association (2013) Diagnostic and statistical manual of mental disorders, 5th edn. American Psychiatric Association, Arlington, VA

Angst J, Gamma A, Sellaro R et al (2003) Recurrence of bipolar disorders and major depression. A life-long perspective. Eur Arch Psychiatry Clin Neurosci 253(5):236–240

Angst J, Sellaro R, Stassen HH et al (2005) Diagnostic conversion from depression to bipolar disorders: results of a long-term prospective study of hospital admissions. J Affect Disord 84 (2–3):149–157

Baca-Garcia E, Perez-Rodriguez MM, Basurte-Villamor I et al (2007) Diagnostic stability and evolution of bipolar disorder in clinical practice: a prospective cohort study. Acta Psychiatr Scand 115(6):473–480

Baethge C, Baldessarini RJ, Khalso HM et al (2005) Substance abuse in first-episode bipolar I disorder: indications for early intervention. Am J Psychiatry 162(5):1008–1010

Balazs J, Benazzi F, Rihmer Z et al (2006) The close link between suicide attempts and mixed (bipolar) depression: implications for suicide prevention. J Affect Disord 91(2–3):133–138

Barbato N, Hafner RJ (1998) Comorbidity of bipolar and personality disorder. Aust N Z J Psychiatry 32(2):276–280

Baum AE, Akula N, Cabanero M et al (2008) A genome-wide association study implicates diacylglycerol kinase eta (DGKH) and several other genes in the etiology of bipolar disorder. Mol Psychiatry 13(2):197–207

Benazzi F (2006) Does temperamental instability support a continuity between bipolar II disorder and major depressive disorder? Eur Psychiatry 21(4):274–279

Benazzi F (2007) Bipolar disorder–focus on bipolar II disorder and mixed depression. Lancet 369 (9565):935–945

Benedetti A, Fagiolini A, Casamassima F et al (2007) Gender differences in bipolar disorder type 1: a 48-week prospective follow-up of 72 patients treated in an Italian tertiary care center. J Nerv Ment Dis 195(1):93–96

Berk M, Hallam KT, McGorry PD (2007) The potential utility of a staging model as a course specifier: a bipolar disorder perspective. J Affect Disord 100(1–3):279–281

Birmaher B, Axelson D (2006) Course and outcome of bipolar spectrum disorder in children and adolescents: a review of the existing literature. Dev Psychopathol 18(4):1023–1035

Burdick KE, Goldberg JF, Harrow M et al (2006) Neurocognition as a stable endophenotype in bipolar disorder and schizophrenia. J Nerv Ment Dis 194(4):255–260

Calabrese JR, Vieta E, El-Mallakh R et al (2004) Mood state at study entry as predictor of the polarity of relapse in bipolar disorder. Biol Psychiatry 56(12):957–963

Calabrese JR, Shelton MD, Rapport DJ et al (2005) A 20-month, double-blind, maintenance trial of lithium versus divalproex in rapid-cycling bipolar disorder. Am J Psychiatry 162 (11):2152–2161

Chaudhury SR, Grunebaum MF, Galfalvy HC et al (2007) Does first episode polarity predict risk for suicide attempt in bipolar disorder? J Affect Disord 104(1–3):245–250

Conus P, McGorry PD (2002) First-episode mania: a neglected priority for early intervention. Aust N Z J Psychiatry 36(2):158–172

Conus P, Berk M, McGorry PD (2006) Pharmacological treatment in the early phase of bipolar disorders: what stage are we at? Aust N Z J Psychiatry 40(3):199–207

Coryell W, Keller M, Endicott J et al (1989) Bipolar II illness: course and outcome over a five-year period. Psychol Med 19(1):129–141

Coryell W, Endicott J, Keller M (1992) Rapidly cycling affective disorder: demographics, diagnosis, family history, and course. Arch Gen Psychiatry 49(2):126–131

Dienes KA, Hammen C, Henry RM et al (2006) The stress sensitization hypothesis: understanding the course of bipolar disorder. J Affect Disord 95(1–3):43–49

Duffy A, Alda M, Trinneer A et al (2007) Temperament, life events, and psychopathology among the offspring of bipolar parents. Eur Child Adolesc Psychiatry 16(4):222–228

Franchini L, Zanardi R, Smeraldi E et al (1999) Early onset of lithium prophylaxis as a predictor of good long-term outcome. Eur Arch Psychiatry Clin Neurosci 249(5):227–230

Frye MA, Yatham LN, Calabrese JR et al (2006) Incidence and time course of subsyndromal symptoms in patients with bipolar I disorder: an evaluation of 2 placebo-controlled maintenance trials. J Clin Psychiatry 67(11):1721–1728

Geller B, Fox LW, Clark KA (1994) Rate and predictors of prepubertal bipolarity during follow-up of 6- to 12-year-old depressed children. J Am Acad Child Adolesc Psychiatry 33(4):461–468

George EL, Miklowitz DJ, Richards JA et al (2003) The comorbidity of bipolar disorder and axis II personality disorders: prevalence and clinical correlates. Bipolar Disord 5(2):115–122

Goldberg JF (2003) When do antidepressants worsen the course of bipolar disorder? J Psychiatr Pract 9(3):181–194

Goldberg JF, Harrow M, Grossman LS (1995) Course and outcome in bipolar affective disorder: a longitudinal follow-up study. Am J Psychiatry 152(3):379–384

Hirschfeld RM, Lewis L, Vornik LA (2003) Perceptions and impact of bipolar disorder: how far have we really come? Results of the national depressive and manic-depressive association 2000 survey of individuals with bipolar disorder. J Clin Psychiatry 64(2):161–174

Jamison KR (2000) Suicide and bipolar disorder. J Clin Psychiatry 61(9):47–51

Judd LL, Akiskal HS, Schettler PJ et al (2002) The long-term natural history of the weekly symptomatic status of bipolar I disorder. Arch Gen Psychiatry 59(6):530–537

Judd LL, Schettler PJ, Solomon DA et al (2008) Psychosocial disability and work role function compared across the long-term course of bipolar I, bipolar II and unipolar major depressive disorders. J Affect Disord 108(1–2):49–58

Kapczinski F, Mahgalhaes PVS, Balanza-Martinez V et al (2014) Staging systems in bipolar disorder: an International Society for Bipolar Disorders Task Force Report. Acta Psychiatr Scand 130(5):354–363

Kato T (2007) Molecular genetics of bipolar disorder and depression. Psychiatry Clin Neurosci 61 (1):3–19

Kauer-Sant'Anna M, Kapczinski F, Andreazza AC et al (2009) Brain-derived neurotrophic factor and inflammatory markers in patients with early- vs. late-stage bipolar disorder. Int J Neuropsychopharmacol 12(4):447–458

Kawa I, Carter JD, Joyce PR et al (2005) Gender differences in bipolar disorder: age of onset, course, comorbidity, and symptom presentation. Bipolar Disord 7(2):119–125

Keller MB, Lavori PW, Coryell W et al (1986) Differential outcome of pure manic, mixed/cycling, and pure depressive episodes in patients with bipolar illness. JAMA 255(22):3138–3142

Kessler RC, Akiskal HS, Angst J et al (2006) Validity of the assessment of bipolar spectrum disorders in the WHO CIDI 3.0. J Affect Disord 96(3):259–269

Ketter TA, Houston JP, Adams DH et al (2006) Differential efficacy of olanzapine and lithium in preventing manic or mixed recurrence in patients with bipolar I disorder based on number of previous manic or mixed episodes. J Clin Psychiatry 67(1):95–101

Kim EY, Miklowitz DJ, Biuckians A et al (2007) Life stress and the course of early-onset bipolar disorder. J Affect Disord 99(1–3):37–44

Krishnan KR (2005) Psychiatric and medical comorbidities of bipolar disorder. Psychosom Med 67(1):1–8

Kupfer DJ (2005) The increasing medical burden in bipolar disorder. JAMA 293(20):2528–2530

Kupfer DJ, First MB, Regier DA (2007) A research agenda for DSM-V. American Psychiatric Publishing, Washington, DC

Le-Niculescu H, McFarland MJ, Mamidipalli S et al (2007) Convergent functional genomics of bipolar disorder: from animal model pharmacogenomics to human genetics and biomarkers. Neurosci Biobehav Rev 31(6):897–903

Leverich GS, Post RM, Keck PE Jr et al (2007) The poor prognosis of childhood-onset bipolar disorder. J Pediatr 150(5):485–490

Loebel A, Cucchiaro J, Silva R et al (2014) Lurasidone monotherapy in the treatment of bipolar I depression: a randomized, double-blind, placebo-controlled study. Am J Psychiatry 171 (2):160–168

Loftus ST, Jaeger J (2006) Psychosocial outcome in bipolar I patients with a personality disorder. J Nerv Ment Dis 194(12):967–970

Maj M, Pirozzi R, Magliano L et al (2006) Agitated "unipolar" major depression: prevalence, phenomenology, and outcome. J Clin Psychiatry 67(5):712–719

Malhi GS, Ivanovski B, Hadzi-Pavlovic D et al (2007) Neuropsychological deficits and functional impairment in bipolar depression, hypomania and euthymia. Bipolar Disord 9(1–2):114–125

Marneros A (2006) Mood disorders: epidemiology and natural history. Psychiatry 4:119–122

Marneros A, Deister A, Rohde A (1991) Phenomenologic constellations of persistent alterations in idiopathic psychoses. An empirical comparative study. Nervenarzt 62(11):676–681

Martinez-Aran A, Vieta E, Reinares M et al (2004) Cognitive function across manic or hypomanic, depressed, and euthymic states in bipolar disorder. Am J Psychiatry 161(2):262–270

Martinez-Aran A, Vieta E, Torrent C et al (2007) Functional outcome in bipolar disorder: the role of clinical and cognitive factors. Bipolar Disord 9(1–2):103–113

McIntyre RS, Soczynska JK, Beyer JL et al (2007) Medical comorbidity in bipolar disorder: reprioritizing unmet needs. Curr Opin Psychiatry 20(4):406–416

Meeks S (1999) Bipolar disorder in the latter half of life: symptom presentation, global functioning and age of onset. J Affect Disord 52(1–3):161–167

Miklowitz DJ, Wisniewski SR, Miyahara S et al (2005) Perceived criticism from family members as a predictor of the one-year course of bipolar disorder. Psychiatry Res 136(2–3):101–111

Moorhead TW, McKirdy J, Sussmann JE et al (2007) Progressive gray matter loss in patients with bipolar disorder. Biol Psychiatry 62(8):894–900

Mysels DJ, Endicott J, Nee J et al (2007) The association between course of illness and subsequent morbidity in bipolar I disorder. J Psychiatr Res 41(1–2):80–89

Nolen WA, Luckenbaugh DA, Altshuler LL et al (2004) Correlates of 1-year prospective outcome in bipolar disorder: results from the Stanley Foundation Bipolar Network. Am J Psychiatry 161 (18):1447–1454

Ostacher MJ, Nierenberg AA, Perlis RH et al (2006) The relationship between smoking and suicidal behavior, comorbidity, and course of illness in bipolar disorder. J Clin Psychiatry 67 (12):1907–1911

Otto MW, Simon NM, Wisniewski SR et al (2006) Prospective 12-month course of bipolar disorder in out-patients with and without comorbid anxiety disorders. Br J Psychiatry 189:20–25

Paris J, Gunderson J, Weinberg I (2007) The interface between borderline personality disorder and bipolar spectrum disorders. Compr Psychiatry 48(2):145–154

Perlis RH, Delbello MP, Miyahara S et al (2005) STEP-BD investigators. Revisiting depressive-prone bipolar disorder: polarity of initial mood episode and disease course among bipolar I systematic treatment enhancement program for bipolar disorder participants. Biol Psychiatry 58(7):549–553

Perlis RH, Ostacher MJ, Patel JK et al (2006) Predictors of recurrence in bipolar disorder: primary outcomes from the Systematic Treatment Enhancement Program for Bipolar Disorder (STEP-BD). Am J Psychiatry 163(2):217–224

Post RM, Leverich GS (2006) The role of psychosocial stress in the onset and progression of bipolar disorder and its comorbidities: the need for earlier and alternative modes of therapeutic intervention. Dev Psychopathol 18(4):1181–1211

Post RM, Leverich GS, Altshuler LL et al (2003) An overview of recent findings of the Stanley Foundation Bipolar Network (Part I). Bipolar Disord 5(5):310–319

Rottig D, Rottig S, Brieger P et al (2007) Temperament and personality in bipolar I patients with and without mixed episodes. J Affect Disord 104(1–3):97–102

Sachs GS, Nierenberg AA, Calabrese JR et al (2007) Effectiveness of adjunctive antidepressant treatment for bipolar depression. N Engl J Med 356(17):1711–1722

Salvatore P, Tohen M, Khalsa HM et al (2007) Longitudinal research on bipolar disorders. Epidemiol Psichiatr Soc 16(2):109–117

Scott J, Colom F, Vieta E (2007) A meta-analysis of relapse rates with adjunctive psychological therapies compared to usual psychiatric treatment for bipolar disorders. Int J Neuropsychopharmacol 10(1):123–129

Simon GE, Bauer MS, Ludman EJ et al (2007a) Mood symptoms, functional impairment, and disability in people with bipolar disorder: specific effects of mania and depression. J Clin Psychiatry 68(8):1237–1245

Simon NM, Zalta AK, Otto MW et al (2007b) The association of comorbid anxiety disorders with suicide attempts and suicidal ideation in outpatients with bipolar disorder. J Psychiatr Res 41 (3–4):255–264

Singh JB, Zarate CA Jr (2006) Pharmacological treatment of psychiatric comorbidity in bipolar disorder: a review of controlled trials. Bipolar Disord 8(6):696–709

Strakowski SM, Tohen M, Stoll AL et al (1992) Comorbidity in mania at first hospitalization. Am J Psychiatry 149(4):554–556

Strakowski SM, Keck PE Jr, McElroy SL et al (1998) Twelve months outcome after a first hospitalization for affective psychosis. Arch Gen Psychiatry 55(1):49–55

Strakowski SM, DelBello MP, Fleck DE et al (2005) Effects of co-occurring alcohol abuse on the course of bipolar disorder following a first hospitalization for mania. Arch Gen Psychiatry 62 (8):851–858

Strakowski SM, DelBello MP, Fleck DE et al (2007) Effects of co-occurring cannabis use disorders on the course of bipolar disorder after a first hospitalization for mania. Arch Gen Psychiatry 64(1):57–64

Stromberg D, Ronningstam E, Gunderson J et al (1998) Brief communication: pathological narcissism in bipolar disorder patients. J Personal Disord 12(2):179–185

Swann AC, Bowden CL, Calabrese JR et al (1999) Differential effect of number of previous episodes of affective disorder on response to lithium or divalproex in acute mania. Am J Psychiatry 156(8):1264–1266

Thompson KN, Conus PO, Ward JL et al (2003) The initial prodrome to bipolar affective disorder: prospective case studies. J Affect Disord 77(1):79–85

Tohen M (ed) (1999) Comorbidity in affective disorders. Marcel Decker, New York, NY

Tohen M, Waternaux CM, Tsuang MT (1990a) Outcome in mania. A 4-year prospective follow-up of 75 patients utilizing survival analysis. Arch Gen Psychiatry 47(12):1106–1111

Tohen M, Waternaux CM, Tsuang MT et al (1990b) Four year follow-up of twenty-four first-episode manic patients. J Affect Disord 19(2):79–86

Tohen M, Tsuang MT, Goodwin DC (1992) Prediction of outcome in mania by mood-congruent or mood-incongruent psychotic features. Am J Psychiatry 149(11):1580–1584

Tohen M, Shulman KI, Satlin A (1994) First-episode mania in late life. Am J Psychiatry 151 (1):130–132

Tohen M, Greenfield SF, Weiss RD et al (1998) The effect of comorbid substance use disorders on the course of bipolar disorder. Harv Rev Psych 6(3):133–141

Tohen M, Hennen J, Zarate C et al (2000) Two-year syndromal and functional recovery in 219 cases of major affective disorders with psychotic features. Am J Psychiatry 157 (2):220–228

Tohen M, Zarate CA, Hennen J et al (2003) The McLean-Harvard First-Episode Mania Study: prediction of recovery and first recurrence. Am J Psychiatry 160(12):2099–2107

Tohen M, Bowden CL, Calabrese JR et al (2006) Influence of sub-syndromal symptoms after remission from manic or mixed episodes. Br J Psychiatry 189:515–519

Tohen M, Calabrese J, Vieta E et al (2007) Effect of comorbid anxiety on treatment response in bipolar depression. J Affect Disord 104(1–3):137–146

Tsai SM, Chen C, Kuo C et al (2001) 15-Year outcome of treated bipolar disorder. J Affect Disord 63(1–3):215–220

Varga M, Magnusson A, Flekkoy K et al (2006) Insight, symptoms and neurocognition in bipolar I patients. J Affect Disord 91(1):1–9

Vieta E (2006) On bipolar disorder. In: Deconstructing psychosis. Future of psychiatric diagnosis: refining the research agenda. Conference on DSM-V. American Psychiatric Association, Arlington, VA, pp 16–17

Wellcome Trust Case Control Consortium (2007) Genome-wide association study of 14,000 cases of seven common diseases and 3,000 shared controls. Nature 447(7145):661–678

Williamson D, Brown E, Perlis RH et al (2006) Clinical relevance of depressive symptom improvement in bipolar I depressed patients. J Affect Disord 92(2–3):261–266

Wittchen HU, Mhlig S, Pezawas L (2003) Natural course and burden of bipolar disorders. Int J Neuropsychopharmacol 6(2):145–154

Zarate CA Jr, Tohen M (2004) Double-blind comparison of the continued use of antipsychotic treatment versus its discontinuation in remitted manic patients. Am J Psychiatry 161 (1):169–171

Chapter 4
Suicide and Bipolar Disorder

Zoltán Rihmer and Péter Döme

Abstract Bipolar disorders are common but frequently under-referred, underdiagnosed, and undertreated illnesses with markedly elevated premature mortality. Up to 15 % of patients with bipolar disorder (BD) die by suicide, and about half of them make at least one suicide attempt in their lifetime. The suicide rate of (untreated) BD patients is 20–25 times higher than the same rate in the general population. Suicidal behavior (completed suicide and suicide attempt) in BD patients occurs almost exclusively during the severe major (often mixed) depressive episode and less frequently in mania with mixed features but very rarely during euphoric mania, hypomania, or euthymia; this suggests that suicidal behavior in BD patients is a state- and severity-dependent phenomenon. However, since the majority of BD patients never commit suicide (and up to 50 % of them never attempt it), risk factors other than BD itself also play a significant contributory role. This chapter summarizes the clinically most relevant suicide risk and protective factors in BD and also briefly explores the most effective suicide prevention strategies.

Keywords Bipolar disorder • Unipolar depression • Major depressive episode • Suicide • Suicide attempt • Suicide ideation • Suicide risk factors • Suicide protective factors • Suicide prevention

4.1 Introduction

Bipolar disorder (BD) is associated with a substantial burden of illness-related health and economic problems. Given the 1.3–5.0 % lifetime prevalence of BD-I and BD-II (Rihmer and Angst 2009), they are among the most frequent psychiatric

Z. Rihmer, MD, PhD, DSc (✉) • P. Döme, MD, PhD
Faculty of Medicine, Department of Clinical and Theoretical Mental Health, Semmelweis University, Kutvolgyi ut 4, 1125 Budapest, Hungary

Laboratory for Suicide Research and Prevention, National Institute of Psychiatry and Addictions, Lehel u. 59, 1135 Budapest, Hungary
e-mail: rihmer.zoltan@med.semmelweis-univ.hu

© Springer International Publishing Switzerland 2016   53
C.A. Zarate Jr., H.K. Manji (eds.), *Bipolar Depression: Molecular Neurobiology, Clinical Diagnosis, and Pharmacotherapy*, Milestones in Drug Therapy,
DOI 10.1007/978-3-319-31689-5_4

illnesses (Hawton et al. 2005; Tondo et al. 2003; Rihmer 2005; Goodwin and Jamison 2007; Dunner 2003; Angst et al. 2005; Schaffer et al. 2015). BD is also among the most potentially life-threatening psychiatric disorders, since life expectancy for patients with BD is about nine to 14 years less than for individuals of the general population. The standardized mortality ratio (SMR) for all-cause mortality for BD is around two (Fiedorowicz et al. 2014; Crump et al. 2013; Chang et al. 2011; Hayes et al. 2015; Kessing et al. 2015). Increased mortality among patients is attributed to both unnatural (suicide, unintentional injuries) and natural (cardiovascular disorders, diabetes mellitus, COPD, influenza, or pneumonia) causes of death (Hawton et al. 2005; Rihmer 2005; Goodwin and Jamison 2007; Dunner 2003; Angst et al. 2005; Fiedorowicz et al. 2014; Crump et al. 2013; Chang et al. 2011; Hayes et al. 2015; Kessing et al. 2015). Surprisingly, leading causes for lost life years in BD are natural rather than unnatural ones (Kessing et al. 2015; Weiner et al. 2011).

In spite of the great clinical and public health significance of BD, it is still under-referred, underdiagnosed, and undertreated (Rihmer and Angst 2009; Dunner 2003). As successful acute and long-term treatment with mood stabilizers and other psychotropics markedly reduces the risk of attempted and completed suicides in BD-I and BD-II (Tondo et al. 2003; Rihmer 2005, 2007a; Angst et al. 2005; Akiskal 2007; Rihmer and Gonda 2013), the early recognition and appropriate acute and long-term treatment of BD patients are key elements in suicide prevention for this population. In spite of the fact that suicidal behavior is very rare in the absence of current major mental disorders (Hawton et al. 2005; Tondo et al. 2003; Goodwin and Jamison 2007), suicide is not the linear consequence of them; it is a very complex and multicausal human behavior involving also several psychosocial, demographic, and cultural components. This chapter summarizes the clinically most relevant suicide risk and protective factors in BD and briefly reviews the most effective prevention strategies.

4.2   Suicidal Behavior in Major Mood Disorders

In their meta-analysis of studies on suicide risk in all psychiatric disorders, Harris and Barraclough (1997) separately analyzed the risk of completed suicide in patients with an index diagnosis of major depressive disorder (MDD) (23 reports, more than 8000 patients) or BD (14 reports, more than 3700 patients). The patients in some of these studies had been followed for many decades. They found that the risk of completed suicide (i.e., the SMR) was about 20-fold for patients with an index diagnosis of MDD. For BD, the SMR was 15. In their recently published meta-review, similar figures were reported by Chesney and colleagues (2014): SMRs for suicide were 19.7 and 17.1 for MDD and BD, respectively. However, this type of analysis cannot provide a precise estimation of separate suicide risk in MDD and BD; specifically, it overestimates the risk for MDD and underestimates the same risk for BD. The main source of this is that the index diagnosis frequently changes during the long-term course of illness from MDD to BD-I or BD-II (Angst et al. 2005; Goldberg et al. 2001) and, in the studies reviewed by Harris and Barraclough (1997), the diagnostic category of BD-II depression (major depression with a history of hypomania but not mania), which is the most common form of BD (Rihmer and Angst 2009; Dunner 2003), was not considered separately. Therefore, it is very likely that most BD-II patients in these studies were included in the MDD subgroup. Moreover, recent findings showed that up to 50 % of unipolar depressions were found to be bipolar depressions when lifetime and/or current intra-depressive subthreshold hypomanic symptoms as well as bipolar spectrum disorders (i.e., "unipolar" major depression with bipolar family history, treatment-associated hypomania/mania in major depression, and "unipolar" major depression with mixed features/agitated depression) were also considered (Rihmer and Angst 2009; Dunner 2003; Akiskal et al. 2005; Benazzi 2006; Angst et al. 2010). Indeed, the most recent meta-analysis of 28 reports, published between 1945 and 2001 (including only patients with an index diagnosis of BD without long-term lithium treatment), by Tondo and colleagues (Tondo et al. 2003) found that during an average 10 years of follow-up, the SMR for completed suicide in BD patients was as high as 22. This is higher than the same figures for MDD and BD reported by Harris and Barraclough (1997) and by Chesney and colleagues (2014). These authors also calculated that suicide rates in BD patients average 0.4 % per year, which is more than 25 times higher than the same rate in the general population (Tondo et al. 2003). Another very recent meta-analysis of 15 studies (more than 46,000 patients) reported an SMR value of 14.44 for suicide in BD (13 for males and 16 for females) (Hayes et al. 2015).

However, in a 40–44-year prospective follow-up study of 406 formerly hospitalized patients with major mood disorders (186 MDD and 220 BD) in which the unipolar–bipolar conversion was carefully considered during the follow-up, Angst and colleagues (Angst et al. 2005) found that 14.5 % of MDD and 8.2 % of BD (I+II) patients committed suicide; the SMRs for suicide in unipolar and BD patients were 26 and 12, respectively. On the other hand, in their very recent long-term prospective follow-up study (average 11 years) of 1983 MDD and 843 BD (I+II) patients, Tondo and colleagues (Tondo et al. 2007) found much higher rates of completed suicide in BD-I and BD-II patients than in MDD patients (0.25 % of patients/year vs. 0.05 % of patients/year). In a 35-year long follow-up study of 4441 formerly hospitalized psychiatric patients, Sani and colleagues (2011) also found that BD-II patients had the highest risk for suicide; the authors reported that 2.8 % of 1163 BD-I and 4.2 % of 602 BD-II patients completed suicide while the rate was 1.9 % in 1142 MDD patients. Similarly, in the STEP-BD study (4360, mostly pharmacologically treated BD patients, mean follow-up 18 months), the rate of completed suicide was more than twofold in BD-II (0.34 %) than BD-I patients (0.14 %) (Dennehy et al. 2011). Investigating the absolute risk of completed suicide in a total Danish national cohort of patients with first psychiatric discharge ($n = 176,347$), Nordentoft and colleagues (2011) found that the lifetime risk for suicide in mood disorders was higher for BD (7.8 % for males and 4.8 % for females) than MDD (6.7 % for males and 3.8 % for females).

Previous suicide attempt is the most powerful single predictor of future completed suicide, particularly in patients with major mood disorders (Hawton et al. 2005; Rihmer 2005, 2007b; Goodwin and Jamison 2007; Harris and Barraclough 1997). Considering only the 10 clinical studies (including more than 3100 patients) published between 1993 and 2005 in which MDD and BD (I+II) patients were analyzed separately, it was found that the lifetime rate of prior suicide attempt(s) was much higher in BD (I+II) patients (mean: 28 %, range: 10–61 %) than in unipolar depressives (mean: 13 %, range: 9–30 %) (Rihmer 2005). A long-term prospective study also found that the rate of suicide attempts during the follow-up was more than double in BD (I+II) than MDD patients (Tondo et al. 2007). A recently published 18-month follow-up study also found that suicide attempts were two times more frequent among patients with BD than among MDD patients (20 % vs. 9.5 %); suicide attempts prior to study baseline were also more frequent in the BD than in the MDD group (Holma et al. 2014). Community-based epidemiological studies from the USA (Chen and Dilsaver 1996; Kessler et al. 1999) and from Hungary (Szadoczky et al. 2000) also showed that the lifetime rate of prior suicide attempts was 1.5- to 2.5-fold higher in BD (I+II) patients than in MDD patients. In agreement with the above, it has also been reported that current suicidal ideation, the major precursor of suicidal behavior (Hawton et al. 2005; Rihmer 2007b), was more frequent in BD-I and BD-II (36–64 %) than in MDD (32–46 %) inpatients and outpatients (Hantouche et al. 1998; Benazzi 2005; Sato et al. 2005).

It is important to emphasize that in BD (similarly to some other psychiatric disorders, e.g., MDD, schizophrenia), the ratio of attempted and completed suicides is about 10 times lower than the corresponding proportion of the general population; the fact that BD patients choose suicide methods with high lethality may explain this finding (Pompili et al. 2009; Costa Lda et al. 2015; Radomsky et al. 1999).

4.3   Suicide Risk Factors in Bipolar Disorders

Suicidal behavior (completed suicide, suicide attempt) and suicidal ideation in BD patients occur mostly during the severe pure or mixed major depressive episodes, less frequently in mania with mixed features, and very rarely during euphoric mania, hypomania, and euthymia (Hawton et al. 2005; Goodwin and Jamison 2007; Schaffer et al. 2015; Rihmer 2007b; Holma et al. 2014; Valtonen et al. 2005; Sokero et al. 2006; Judd et al. 2012; Isometsa et al. 2014), indicating that suicidal behavior in BD patients is a state- and severity-dependent phenomenon. In accordance with the above, some authors speculate that—in addition to the fact that BD patients use more frequently violent (highly lethal) suicide methods— elevated suicide risk for BD (when compared to MDD) may also be related to the fact that BD patients spend more time in depressive and mixed episodes than patients with MDD (Holma et al. 2014). Depressive polarity of the most recent episode also carries an additional risk for suicidal behavior in patients with BD

(Schaffer et al. 2015). However, since the majority of BD patients never commit suicide (and up to 50 % of them never attempt it) (Hawton et al. 2005; Tondo et al. 2003; Rihmer 2005; Chen and Dilsaver 1996; Kessler et al. 1999; Szadoczky et al. 2000; Hantouche et al. 1998; Benazzi 2005; Sato et al. 2005), risk factors other than BD itself also play a significant contributory role; these may include special clinical characteristics as well as some personality, familial, and psychosocial risk and protective factors (Rihmer 2005; Tondo et al. 2003; Hawton et al. 2005; Akiskal 2007; Akiskal et al. 2005). The majority of suicide risk factors in BD are related to the acute phases of the illness (mostly major depressive and mixed affective episodes), but there are several historical and personality-related factors that can help clinicians identify highly suicidal BD patients.

The clinical condition which is the most alarming for suicidal behavior in BD is the recent suicide attempt and the severe (mostly melancholic) major depressive episode, frequently accompanied by hopelessness, guilt, few reasons for living, and suicidal ideation (Hawton et al. 2005; Tondo et al. 2003; Rihmer 2005, 2007b; Akiskal 2007; Holma et al. 2014; Valtonen et al. 2005) as well as agitation, insomnia (Tondo et al. 2003; Akiskal et al. 2005; Rihmer 2007b; Rihmer and Akiskal 2006), and psychotic features (Angst et al. 2005; Rihmer 2007b). According to the results of Sanchez-Gistau and colleagues (2009), atypical features of the last depressive episode are significantly associated with history of suicide attempts among patients with BD. Recent results strongly suggest that mixed depressive episodes labeled as major depressive episodes with mixed features in DSM-5 (MDD plus three or more co-occurring intra-depressive hypomanic symptoms, which highly corresponds to the category of "agitated depression") that is present in up to 60 % of BD-I and BD-II depressives (Akiskal et al. 2005; Benazzi 2005, 2006; Sato et al. 2005; Judd et al. 2012) substantially increases the risk of both attempted and committed suicides. Some results further suggest that mixed affective episodes carry a higher suicide risk than "pure" depressive ones (Schaffer et al. 2015; Akiskal 2007; Akiskal et al. 2005; Angst et al. 2010; Rihmer 2007b; Holma et al. 2014; Benazzi 2005; Sato et al. 2005; Judd et al. 2012; Isometsa et al. 2014; Rihmer and Akiskal 2006; Balazs et al. 2006; Musil et al. 2013; Takeshima and Oka 2013). In addition, some studies raised the possibility that the presence of more mixed and/or depressive affective episodes in the medical *history* is also associated with elevated suicidality in patients with BD (Undurraga et al. 2012; Tidemalm et al. 2014). These results offer an explanation for the rarely occurring "antidepressant-induced" suicidal behavior: antidepressant monotherapy, unprotected by mood stabilizers or atypical antipsychotics, particularly in BD and bipolar spectrum disorder (including "unipolar" depressives with mixed features) can favor not only hypomanic/manic switches and rapid cycling but also worsen the preexisting mixed state or generate de novo mixed conditions, making the clinical picture more serious and ultimately leading to self-destructive behavior (Akiskal et al. 2005; Benazzi 2005; Rihmer and Akiskal 2006; Musil et al. 2013; Rihmer et al. 2007). The role of mood instability in suicidal behavior was also supported by a study showing that a history of rapid mood switching and panic attacks was associated with increased likelihood of history of self-reported suicidal thoughts or action in patients with BD (MacKinnon et al. 2005).

However, suicidal behavior in BD patients is not exclusively restricted to depressive and mixed episodes (see above). In contrast to classic (euphoric) mania, where suicidal tendencies are extremely rare, suicidal thoughts and attempts are relatively common in patients with manic episodes with mixed features (Hawton et al. 2005; Tondo et al. 2003, 2007; Valtonen et al. 2005), supporting the common clinical sense that suicidal behavior even in BD patients is linked to depressive symptomatology (Angst et al. 2005; Tondo et al. 2007; Rihmer 2007b; Valtonen et al. 2005).

Although BD, in general, carries the highest risk of suicide (Tondo et al. 2003, 2007; Rihmer 2005; Akiskal 2007; Nordentoft et al. 2011), several studies have shown that BD-II patients have even higher risk than BD-I subjects (Rihmer 2005; Akiskal 2007; Sani et al. 2011; Dennehy et al. 2011; Hantouche et al. 1998; Sato et al. 2005; Undurraga et al. 2012; Bulik et al. 1990; Rihmer et al. 1990; Balazs et al. 2003). However, other studies have found that suicide risk did not seem to vary significantly according to whether the patient had BD-I or BD-II (Hawton et al. 2005; Angst et al. 2005; Schaffer et al. 2015; Tondo et al. 2007; Valtonen et al. 2005; Leverich et al. 2003).

BD shows a high frequency of psychiatric and medical comorbidities (Rihmer and Angst 2009; Leverich et al. 2003; Miller et al. 2014), and it is well documented that comorbid anxiety/anxiety disorders (Hawton et al. 2005; Rihmer 2005, 2007b; Schaffer et al. 2015; Chen and Dilsaver 1996; Balazs et al. 2003; Leverich et al. 2003; Simon et al. 2007), substance-use disorders (Hawton et al. 2005; Rihmer 2005; Schaffer et al. 2015; MacKinnon et al. 2005; Balazs et al. 2003; Leverich et al. 2003; Simon et al. 2007; Clements et al. 2013), personality disorders (mainly borderline personality) (Rihmer 2005; Schaffer et al. 2015; Valtonen et al. 2005; Sanchez-Gistau et al. 2009; Leverich et al. 2003; Clements et al. 2013), attention deficit hyperactivity disorder (ADHD) (Lan et al. 2015), and serious medical illnesses (Hawton et al. 2005; Leverich et al. 2003), particularly in the case of multiple comorbidities, also increase the risk of all forms of suicidal behavior. As successful acute and long-term treatment of BD substantially reduces the risk of both completed and attempted suicides (Tondo et al. 2003; Rihmer 2005, 2007a; Angst et al. 2005; Akiskal 2007; Rihmer and Gonda 2013), lack of medical and family support and the first few days of the therapy, when antidepressants usually do not work (or, rarely, can worsen the depression) (Isometsa et al. 2014; Rihmer and Akiskal 2006; Takeshima and Oka 2013), should also be considered as suicide risk factors.

As for suicide risk factors related to the prior course of BD, previous suicide attempt(s), particularly in the case of violent or more lethal methods, is the most powerful single predictor of future attempts and fatal suicide (Hawton et al. 2005; Tondo et al. 2003; Rihmer 2005, 2007b; Holma et al. 2014; Valtonen et al. 2005; MacKinnon et al. 2005). BD patients in general (Zalsman et al. 2006), and BD-II patients in particular (Vieta et al. 1997; Tondo et al. 2007), use more violent and more lethal suicide methods than patients with MDD or BD-I patients, respectively, particularly males. Therefore, higher rates of suicidal behavior (mainly completed suicide) in patients with BD compared to MDD may be due to a specific effect of BD on males, resulting in more dangerous suicidal behaviors (Zalsman et al. 2006).

Other historical variables, like early onset and early stage of BD (Goodwin and Jamison 2007; Tondo et al. 2003, 2007; Angst et al. 2005; Schaffer et al. 2015; MacKinnon et al. 2005; Simon et al. 2007; Pompili et al. 2013; Gonda et al. 2012), as well as rapid cycling course, predominant depressive polarity, and multiple (i.e., great number of prior) admissions to inpatient care (Hawton et al. 2005; Tondo et al. 2003; Angst et al. 2005; Valtonen et al. 2005; Leverich et al. 2003; Clements et al. 2013; Gonda et al. 2012; Carvalho et al. 2014; Hoyer et al. 2004) have also been shown to increase the chance of both attempted and completed suicides. In accordance with the finding that early stages of the disease are associated with a further increase of suicide risk, a recent study demonstrated that the contribution of suicide to life years lost was the highest in young age bands (Kessing et al. 2015).

Longer periods from onset of illness to the proper diagnosis of BD and/or delay in the provision of appropriate treatment were also found to be risk factors for suicidal behavior in both prospective and retrospective studies (Undurraga et al. 2012; Gonda et al. 2012; Nery-Fernandes et al. 2012; Altamura et al. 2010). Some results also indicated that the depressive polarity of the first affective episode (which is typical for BD-II) is also positively associated with suicidality in BD (Schaffer et al. 2015; Gonda et al. 2012; Chaudhury et al. 2007; Neves et al. 2009).

It is well known that suicide risk is extremely high soon after hospital discharge in psychiatric patients in general and in affective disorder patients in particular (including BD) (Isometsa et al. 2014; Hoyer et al. 2004; Bostwick and Pankratz 2000; Hawton and van Heeringen 2009; Popovic et al. 2014). According to the results of Isometsä and colleagues (Isometsa et al. 2014), the degree of post-discharge risk is the highest for depressive episodes followed by mixed and then by manic episodes; at the same time, post-discharge risk declines with time more slowly for mixed than for depressive episodes. Some investigations have also revealed that among psychiatric patients (including those with BD) suicide risk is highly elevated during the period immediately after admission (Clements et al. 2013; Hoyer et al. 2004, 2009; Hawton and van Heeringen 2009; Popovic et al. 2014).

Personality characteristics also play a significant role in the development and particularly in the manifestation of suicidal behavior. The voluminous literature on this subject consistently shows that aggressive–impulsive personality traits (Valtonen et al. 2005; MacKinnon et al. 2005; Zalsman et al. 2006; Mann et al. 1999; Oquendo et al. 2004; Swann et al. 2007) especially in combination with high levels of current hopelessness and pessimism (Holma et al. 2014; Mann et al. 1999; Oquendo et al. 2004) markedly increase the risk of suicidal behavior in patients with BD and other psychiatric disorders. In BD-I and BD-II depressed patients, the level of impulsivity and the rate of prior suicide attempts increased with increasing number of intra-depressive hypomanic symptoms (Swann et al. 2007), supporting the strong relationship between the bipolar nature of depression and impulsive behavior (Benazzi 2003). Irritable mood (a core symptom of mania and hypomania) and anger attacks (inappropriate, sudden spells of anger associated with autonomic arousal symptoms and behavioral outbursts) are closely linked, and anger attacks are much more common in bipolar depression than in MDD (Benazzi 2003, 2006). Moreover, if anger attacks occur during MDD, the bipolar nature of these "unipolar" depressions was supported by their association with most important validating variables of BD (early onset, atypical or mixed features of depressive episode, family history of BD) (Benazzi 2003).

Cyclothymia/cyclothymic affective temperament—that is, the attenuated form and frequently the precursor of major BD—seems to also be a predisposing factor for suicidal behavior. In patients experiencing a major depressive episode, cyclothymic personality was significantly related to lifetime and current suicidal behavior (ideation and attempts) both in adults and in a pediatric sample (Akiskal et al. 2003; Kochman et al. 2005; Pompili et al. 2012).

The interaction between personality features and illness characteristics in suicidal behavior was best formulated by Mann and colleagues (1999) in their "stress–diathesis model," suggesting that suicidal behavior in psychiatric patients is determined not only by the stressor (acute major psychiatric illness) but also by a diathesis (impulsive, aggressive, pessimistic personality traits).

In spite of the fact that the vast majority of suicide victims in the general population are males and the opposite is true for suicide attempters (Goodwin and Jamison 2007; Harris and Barraclough 1997; Rihmer 2007b; Rihmer and Akiskal 2006), these differences are smaller among suicidal patients with BD (Hawton et al. 2005; Tondo et al. 2003; Angst et al. 2005; Schaffer et al. 2015; Hayes et al. 2015; MacKinnon et al. 2005; Simon et al. 2007), suggesting that in individual cases gender is not a significant predictor for committed and attempted suicide in this otherwise high-risk population. Nevertheless, although gender differences are not so pronounced among BD subjects as among members of the general population, in their recently published meta-analysis Schaffer and colleagues (2015) identified female gender as a risk factor for suicide attempts and male gender as a risk factor for suicide deaths in BD subjects. Lesbian, gay, bisexual, and transgender individuals are at elevated risk for suicidal behavior, particularly if BD and other suicide risk factors are also present (Fitzpatrick et al. 2005).

With regard to suicide risk factors related to personal history, early negative life events (e.g., parental loss, isolation, emotional, physical, and sexual abuse) (Hawton et al. 2005; Rihmer 2005, 2007b; Leverich et al. 2003; Mann et al. 1999), permanent adverse life situations (e.g., unemployment, isolation) (Hawton et al. 2005; Rihmer 2005, 2007b), and acute psychosocial stressors (e.g., loss events, financial disasters) (Rihmer 2005, 2007b; Costa Lda et al. 2015; Leverich et al. 2003; Isometsa et al. 1995) are the most important and clinically useful indicators of possible suicidality, primarily if other risk factors are also present. However, acute psychosocial stressors are commonly dependent on the victim's own behavior, particularly in the case of BD-I (Isometsa et al. 1995). It may be that hypomanic and manic periods can easily lead to aggressive–impulsive behavior, financial extravagance, or episodic promiscuity, thus generating several interpersonal conflicts, marital breakdown, and new negative life events, all of which have a negative impact on the further course of the illness.

Family history of suicidal behavior and/or major mood disorders in first- and second-degree relatives is also a strong risk factor for both attempted and completed suicides in psychiatric patients in general and in BD patients in particular (Hawton et al. 2005; Rihmer 2005, 2007b; Schaffer et al. 2015; Sanchez-Gistau et al. 2009; MacKinnon et al. 2005; Leverich et al. 2003; Mann et al. 1999). However, the familial component of suicidal behavior seems to be partly independent of psychiatric disorders, as relatives of suicide victims are over 10 times more likely than relatives of comparison subjects to attempt or complete suicide even after controlling for psychopathology (Kim et al. 2005). Results of a recent longitudinal study also confirmed that suicidal behavior of parents with mood disorders conveyed elevated odds of suicidal behavior among offspring, even after controlling for some other offspring risk factors for suicidality, for instance, previous suicide attempt and history of mood disorder (both assessed at baseline) or mood disorder at the time point before the suicide attempt (Brent et al. 2015).

The clinically explorable suicide risk factors in BD are listed in Table 4.1. As suicidal behavior in BD patients is very rare in the absence of major mood episodes, suicide risk factors related to these conditions are the most powerful predictors, particularly if other risk factors (and high lethality suicide methods) are also present. Suicide risk factors are additive: the higher the number of risk factors, the higher the chance of suicidal behavior.

4.4   Suicide Protective Factors in Bipolar Disorder

In contrast to several suicide risk factors, only few circumstances are known to have a protective effect against suicidal behavior. Good family and social support, pregnancy and the postpartum period, having a great number of children, holding strong religious beliefs, and restricting lethal suicide methods (e.g., reducing domestic and car exhaust gas toxicity and introducing stricter laws on gun control) whenever possible seem to have some protective effect (Rihmer 2005; Goodwin and Jamison 2007; Marzuk et al. 1997; Dervic et al. 2004; Driver and Abed 2004; Rihmer et al. 2004). However, the most extensively studied suicide protective factor in major mood disorders is acute and long-term pharmacological treatment (Tondo et al. 2003; Rihmer 2005, 2007a, b; Angst et al. 2005; Akiskal 2007; Rihmer and Gonda 2013; Rihmer and Akiskal 2006).

Although suicide is a rare event in the community, it is very frequent among patients with BD, most of whom contact different levels of healthcare some weeks or months before their death (Tondo et al. 2003; Rihmer 2005, 2007a, b; Rihmer et al. 1990). The high rate of BD and recent medical contact before suicidal behavior underline the priority role of healthcare workers in suicide prevention. Unfortunately, less than one-third of BD suicide victims and suicide attempters

Table 4.1 Clinically detectable suicide risk factors in bipolar disorders

1. Risk factors related to acute mood episodes
(a) Severe major depressive episode
– Current suicide attempt, plan, ideation
– Hopelessness, guilt, few reasons for living
– Agitation, depressive mixed state (DSM-5 major depressive episode with mixed features), insomnia
– Psychotic and atypical features
– Bipolar II diagnosis
– Comorbid Axis I (especially anxiety and substance use disorders), Axis II (especially borderline personality disorder), or serious Axis III disorders
– Lack of medical treatment and family/social support
– First few days of the treatment (particularly if appropriate care and co-medication is lacking)
– First few weeks (months) after hospital discharge (especially if hospital admission was due to a depressive or a mixed episode)
– The period immediately after the hospital admission
(b) Manic or hypomanic episode with mixed features (DSM-5)
2. Risk factors related to prior course of the illness
– Previous suicide attempt/ideation (particularly violent/highly lethal methods)
– Early onset/early stage of the illness
– Rapid cycling course
– Depressive polarity of the first and/or the most recent episode; predominantly depressive polarity during the prior course
– High number of previous admissions to inpatient care
– Delay in proper diagnosis and/or treatment; lack of treatment
3. Risk factors related to personality features
– Aggressive/impulsive personality traits
– Cyclothymic, irritable, depressive temperament
– Same-sex orientation, bisexuality
4. Risk factors related to personal history
– Early negative life events (separation, emotional, physical, and sexual abuse)
– Permanent adverse life situations (unemployment, isolation)
– Acute psychosocial stressors (loss events, financial catastrophe)
5. Risk factors related to family history
– Family history of mood disorders (first- and second-degree relatives)
– Family history of suicide and/or suicide attempt (first- and second-degree relatives)

were receiving appropriate pharmacotherapy at the time of the suicide event (Rihmer et al. 1990; Balazs et al. 2003; Isometsa et al. 1994). As about two-thirds of suicide victims die by their first attempt (Goodwin and Jamison 2007; Rihmer 2007b; Isometsa and Lonnqvist 1998), the risk of suicide in BD patients is very high even if the patient never attempted suicide before. The careful estimation of all suicide risks and protective factors (see Table 4.1) is helpful in detecting suicide risk as early as possible—ideally even before the first suicidal act—and intervening prior to the patient making the first suicide act. Successful long-term prophylaxis of these patients results in "hidden suicide prevention."

4.5 Prevention of Suicidal Behavior in Bipolar Patients

Several open clinical studies and randomized controlled trials have consistently found that acute and long-term treatment with lithium and other mood stabilizers (sometimes in combination with antidepressants and antipsychotics) markedly reduces the risk of attempted and completed suicide in BD-I and BD-II, as well as in MDD patients, by an average of 80 % (Tondo et al. 2003; Rihmer 2005; Angst et al. 2005; Rihmer and Gonda 2013; Rihmer and Akiskal 2006). It has been also reported that the suicide rate of BD patients decreased progressively with increasing numbers of prescription of mood stabilizers (Sondergard et al. 2008). This strong antisuicidal effect of lithium seems to be more than the simple result of its episode-prophylactic effect, as it has been reported that during the long-term lithium prophylaxis of 167 recurrent BD or MDD patients with at least one prior suicide attempt, a significant reduction in the number of suicide attempts was found not only in excellent responders to lithium (92 %) but also in moderate (78 %) and poor responders (70 %) (Ahrens and Muller-Oerlinghausen 2001). The clinical message of this finding is that in the case of lithium nonresponse, when the patient has one or more suicide risk factors, instead of switching lithium to another mood stabilizer, the clinician should retain lithium (even on a lower dose) and combine it with another mood stabilizer. BD patients with long-term pharmacotherapy sometimes need (and more frequently receive) antidepressants and/or antipsychotics in addition to their mood stabilizers for shorter or longer periods of time. However, there is a growing body of evidence suggesting that antidepressant monotherapy can worsen the cross-sectional picture as well as the long-term course and outcome of BD and, as a result, can induce suicidal behavior indirectly. Antidepressant monotherapy without the concomitant use of mood stabilizers or atypical antipsychotics in threshold and subthreshold bipolar depressives—who typically have early onset of illness and are young—can result in high rates of antidepressant resistance and can sometimes worsen the cross-sectional picture of depression. This can occur particularly in adolescents and young adults not only by causing (hypo) manic switch but also by inducing or aggravating depressive mixed state/agitation, called also as "activation syndrome," which is the major substrate of suicidal behavior (Dunner 2003; Rihmer 2007a; Akiskal 2007; Rihmer and Gonda 2013; Benazzi 2005; Rihmer and Akiskal 2006; Musil et al. 2013; Takeshima and Oka 2013; Rihmer et al. 2007).

The most recent naturalistic, prospective chart review analysis of 405 BD-I and BD-II patients also showed that mood stabilizer monotherapy markedly reduced the risk of suicidal behavior. The frequency of suicidal behavior was highest in patients with antidepressant and antipsychotic monotherapy, lowest in patients with mood stabilizer monotherapy, and that the risk of patients receiving combination therapy (mood stabilizers + antidepressants or antipsychotics) showed an intermediate position, suggesting that a combination of antidepressants or antipsychotics with mood stabilizers markedly reduced (but did not fully eliminate) this increased suicide risk (Yerevanian et al. 2007a–c). These results support and extend prior and most recent findings (Rihmer 2007a; Akiskal 2007; Rihmer and Akiskal 2006; Takeshima and Oka 2013) and suggest that not only antidepressants but also antipsychotics can worsen the cross-sectional and longitudinal course of BD; this worsening is most clearly reflected in the elevated rate of suicidality. However, some results (e.g., Leon et al. 2014) suggest that antidepressant therapy may be rather protective in regard to suicidal behavior in patients with BD (Leon et al. 2014). Moreover, the recent task force report of the International Society for Bipolar Disorders also concluded that the evidence of an association between either decreased or increased risk of suicidal behavior with antidepressant medication in BD patients is poor (Pacchiarotti et al. 2013). Despite the somewhat inconclusive results, in our opinion, doctors should keep their BD patients on these supplementary medications (e.g., antidepressants) as short a time as possible and the main component of long-term treatment should be the mood stabilizer pharmacotherapy. However, since the risk reduction for suicidal behavior among BD patients treated with mood stabilizers, although clinically significant (Tondo et al. 2003; Rihmer 2005; Angst et al. 2005; Rihmer and Gonda 2013; Rihmer and Akiskal 2006), is still incomplete, development of alternative or supplementary treatment strategies is also required.

Recently, several effective psychosocial interventions for BD have been developed (psychoeducation, cognitive behavioral therapy, interpersonal and social rhythm therapy, etc.), particularly for noncompliant, drug intolerant, and drug-nonresponsive patients (Rihmer 2005; Michalak et al. 2004; Rucci et al. 2002). Since they are designed for relapse/recurrence prevention, they might have efficacy in suicide prevention as well, and a combination of these methods with long-term pharmacotherapy may substantially improve results (Rucci et al. 2002).

References

Ahrens B, Muller-Oerlinghausen B (2001) Does lithium exert an independent antisuicidal effect? Pharmacopsychiatry 34(4):132–136

Akiskal H (2007) Targeting suicide prevention to modifiable risk factors: has bipolar II been overlooked? Acta Psychiatr Scand 116(6):395–402

Akiskal HS, Hantouche EG, Allilaire JF (2003) Bipolar II with and without cyclothymic temperament: "dark" and "sunny" expressions of soft bipolarity. J Affect Disord 73(1–2):49–57

Akiskal HS, Benazzi F, Perugi G, Rihmer Z (2005) Agitated "unipolar" depression re-conceptualized as a depressive mixed state: implications for the antidepressant-suicide controversy. J Affect Disord 85(3):245–258

Altamura AC, Dell'Osso B, Berlin HA, Buoli M, Bassetti R, Mundo E (2010) Duration of untreated illness and suicide in bipolar disorder: a naturalistic study. Eur Arch Psychiatry Clin Neurosci 260(5):385–391

Angst J, Angst F, Gerber-Werder R, Gamma A (2005) Suicide in 406 mood-disorder patients with and without long-term medication: a 40 to 44 years' follow-up. Arch Suicide Res 9(3):279–300

Angst J, Cui L, Swendsen J, Rothen S, Cravchik A, Kessler RC, Merikangas KR (2010) Major depressive disorder with subthreshold bipolarity in the National Comorbidity Survey Replication. Am J Psychiatry 167(10):1194–1201

Balazs J, Lecrubier Y, Csiszer N, Kosztak J, Bitter I (2003) Prevalence and comorbidity of affective disorders in persons making suicide attempts in Hungary: importance of the first depressive episodes and of bipolar II diagnoses. J Affect Disord 76(1–3):113–119

Balazs J, Benazzi F, Rihmer Z, Rihmer A, Akiskal KK, Akiskal HS (2006) The close link between suicide attempts and mixed (bipolar) depression: implications for suicide prevention. J Affect Disord 91(2–3):133–138

Benazzi F (2003) Major depressive disorder with anger: a bipolar spectrum disorder? Psychother Psychosom 72(6):300–306

Benazzi F (2005) Suicidal ideation and depressive mixed states. Psychother Psychosom 74 (1):61–62

Benazzi F (2006) Mood patterns and classification in bipolar disorder. Curr Opin Psychiatry 19 (1):1–8

Bostwick JM, Pankratz VS (2000) Affective disorders and suicide risk: a reexamination. Am J Psychiatry 157(12):1925–1932

Brent DA, Melhem NM, Oquendo M, Burke A, Birmaher B, Stanley B, Biernesser C, Keilp J, Kolko D, Ellis S, Porta G, Zelazny J, Iyengar S, Mann JJ (2015) Familial pathways to early-onset suicide attempt: a 5.6-year prospective study. JAMA Psychiatry 72(2):160–168

Bulik CM, Carpenter LL, Kupfer DJ, Frank E (1990) Features associated with suicide attempts in recurrent major depression. J Affect Disord 18(1):29–37

Carvalho AF, Dimellis D, Gonda X, Vieta E, McLntyre RS, Fountoulakis KN (2014) Rapid cycling in bipolar disorder: a systematic review. J Clin Psychiatry 75(6):e578–e586

Chang CK, Hayes RD, Perera G, Broadbent MT, Fernandes AC, Lee WE, Hotopf M, Stewart R (2011) Life expectancy at birth for people with serious mental illness and other major disorders from a secondary mental health care case register in London. PLoS One 6(5), e19590

Chaudhury SR, Grunebaum MF, Galfalvy HC, Burke AK, Sher L, Parsey RV, Everett B, Mann JJ, Oquendo MA (2007) Does first episode polarity predict risk for suicide attempt in bipolar disorder? J Affect Disord 104(1–3):245–250

Chen YW, Dilsaver SC (1996) Lifetime rates of suicide attempts among subjects with bipolar and unipolar disorders relative to subjects with other Axis I disorders. Biol Psychiatry 39 (10):896–899

Chesney E, Goodwin GM, Fazel S (2014) Risks of all-cause and suicide mortality in mental disorders: a meta-review. World Psychiatry 13(2):153–160

Clements C, Morriss R, Jones S, Peters S, Roberts C, Kapur N (2013) Suicide in bipolar disorder in a national English sample, 1996–2009: frequency, trends and characteristics. Psychol Med 43 (12):2593–2602

Costa Lda S, Alencar AP, Nascimento Neto PJ, dos Santos Mdo S, da Silva CG, Pinheiro Sde F, Silveira RT, Bianco BA, Pinheiro RF Jr, de Lima MA, Reis AO, Rolim Neto ML (2015) Risk factors for suicide in bipolar disorder: a systematic review. J Affect Disord 170:237–254

Crump C, Sundquist K, Winkleby MA, Sundquist J (2013) Comorbidities and mortality in bipolar disorder: a Swedish national cohort study. JAMA Psychiatry 70(9):931–939

Dennehy EB, Marangell LB, Allen MH, Chessick C, Wisniewski SR, Thase ME (2011) Suicide and suicide attempts in the Systematic Treatment Enhancement Program for Bipolar Disorder (STEP-BD). J Affect Disord 133(3):423–427

Dervic K, Oquendo MA, Grunebaum MF, Ellis S, Burke AK, Mann JJ (2004) Religious affiliation and suicide attempt. Am J Psychiatry 161(12):2303–2308

Driver K, Abed RT (2004) Does having offspring reduce the risk of suicide in women? Int J Psychiatry Clin Pract 8(1):25–29

Dunner DL (2003) Clinical consequences of under-recognized bipolar spectrum disorder. Bipolar Disord 5(6):456–463

Fiedorowicz JG, Jancic D, Potash JB, Butcher B, Coryell WH (2014) Vascular mortality in participants of a bipolar genomics study. Psychosomatics 55(5):485–490

Fitzpatrick KK, Euton SJ, Jones JN, Schmidt NB (2005) Gender role, sexual orientation and suicide risk. J Affect Disord 87(1):35–42

Goldberg JF, Harrow M, Whiteside JE (2001) Risk for bipolar illness in patients initially hospitalized for unipolar depression. Am J Psychiatry 158(8):1265–1270

Gonda X, Pompili M, Serafini G, Montebovi F, Campi S, Dome P, Duleba T, Girardi P, Rihmer Z (2012) Suicidal behavior in bipolar disorder: epidemiology, characteristics and major risk factors. J Affect Disord 143(1–3):16–26

Goodwin F, Jamison KR (2007) Manic-depressive illness: bipolar disorders and recurrent depression. Oxford University Press, New York, NY

Hantouche EG, Akiskal HS, Lancrenon S, Allilaire JF, Sechter D, Azorin JM, Bourgeois M, Fraud JP, Chatenet-Duchene L (1998) Systematic clinical methodology for validating bipolar-II disorder: data in mid-stream from a French national multi-site study (EPIDEP). J Affect Disord 50(2–3):163–173

Harris EC, Barraclough B (1997) Suicide as an outcome for mental disorders. A meta-analysis. Br J Psychiatry 170:205–228

Hawton K, van Heeringen K (2009) Suicide. Lancet 373(9672):1372–1381

Hawton K, Sutton L, Haw C, Sinclair J, Harriss L (2005) Suicide and attempted suicide in bipolar disorder: a systematic review of risk factors. J Clin Psychiatry 66(6):693–704

Hayes JF, Miles J, Walters K, King M, Osborn DP (2015) A systematic review and meta-analysis of premature mortality in bipolar affective disorder. Acta Psychiatr Scand 131(6):417–425

Holma KM, Haukka J, Suominen K, Valtonen HM, Mantere O, Melartin TK, Sokero TP, Oquendo MA, Isometsa ET (2014) Differences in incidence of suicide attempts between bipolar I and II disorders and major depressive disorder. Bipolar Disord 16(6):652–661

Hoyer EH, Olesen AV, Mortensen PB (2004) Suicide risk in patients hospitalised because of an affective disorder: a follow-up study, 1973–1993. J Affect Disord 78(3):209–217

Hoyer EH, Licht RW, Mortensen PB (2009) Risk factors of suicide in inpatients and recently discharged patients with affective disorders. A case-control study. Eur Psychiatry 24 (5):317–321

Isometsa ET, Lonnqvist JK (1998) Suicide attempts preceding completed suicide. Br J Psychiatry 173:531–535

Isometsa ET, Henriksson MM, Aro HM, Lonnqvist JK (1994) Suicide in bipolar disorder in Finland. Am J Psychiatry 151(7):1020–1024

Isometsa E, Heikkinen M, Henriksson M, Aro H, Lonnqvist J (1995) Recent life events and completed suicide in bipolar affective disorder. A comparison with major depressive suicides. J Affect Disord 33(2):99–106

Isometsa E, Sund R, Pirkola S (2014) Post-discharge suicides of inpatients with bipolar disorder in Finland. Bipolar Disord 16(8):867–874

Judd LL, Schettler PJ, Akiskal H, Coryell W, Fawcett J, Fiedorowicz JG, Solomon DA, Keller MB (2012) Prevalence and clinical significance of subsyndromal manic symptoms, including irritability and psychomotor agitation, during bipolar major depressive episodes. J Affect Disord 138(3):440–448

Kessing LV, Vradi E, McIntyre RS, Andersen PK (2015) Causes of decreased life expectancy over the life span in bipolar disorder. J Affect Disord 180:142–147

Kessler RC, Borges G, Walters EE (1999) Prevalence of and risk factors for lifetime suicide attempts in the National Comorbidity Survey. Arch Gen Psychiatry 56(7):617–626

Kim CD, Seguin M, Therrien N, Riopel G, Chawky N, Lesage AD, Turecki G (2005) Familial aggregation of suicidal behavior: a family study of male suicide completers from the general population. Am J Psychiatry 162(5):1017–1019

Kochman FJ, Hantouche EG, Ferrari P, Lancrenon S, Bayart D, Akiskal HS (2005) Cyclothymic temperament as a prospective predictor of bipolarity and suicidality in children and adolescents with major depressive disorder. J Affect Disord 85(1–2):181–189

Lan WH, Bai YM, Hsu JW, Huang KL, Su TP, Li CT, Yang AC, Lin WC, Chang WH, Chen TJ, Tsai SJ, Chen MH (2015) Comorbidity of ADHD and suicide attempts among adolescents and young adults with bipolar disorder: a nationwide longitudinal study. J Affect Disord 176:171–175

Leon AC, Fiedorowicz JG, Solomon DA, Li C, Coryell WH, Endicott J, Fawcett J, Keller MB (2014) Risk of suicidal behavior with antidepressants in bipolar and unipolar disorders. J Clin Psychiatry 75(7):720–727

Leverich GS, Altshuler LL, Frye MA, Suppes T, Keck PE Jr, McElroy SL, Denicoff KD, Obrocea G, Nolen WA, Kupka R, Walden J, Grunze H, Perez S, Luckenbaugh DA, Post RM (2003) Factors associated with suicide attempts in 648 patients with bipolar disorder in the Stanley Foundation Bipolar Network. J Clin Psychiatry 64(5):506–515

MacKinnon DF, Potash JB, McMahon FJ, Simpson SG, Depaulo JR Jr, Zandi PP, National Institutes of Mental Health Bipolar Disorder Genetics I (2005) Rapid mood switching and suicidality in familial bipolar disorder. Bipolar Disord 7(5):441–448

Mann JJ, Waternaux C, Haas GL, Malone KM (1999) Toward a clinical model of suicidal behavior in psychiatric patients. Am J Psychiatry 156(2):181–189

Marzuk PM, Tardiff K, Leon AC, Hirsch CS, Portera L, Hartwell N, Iqbal MI (1997) Lower risk of suicide during pregnancy. Am J Psychiatry 154(1):122–123

Michalak EE, Yatham LN, Lam RW (2004) The role of psychoeducation in the treatment of bipolar disorder: a clinical perspective. Clin Appr Bipolar Disord 3(1):5–11

Miller S, Dell'Osso B, Ketter TA (2014) The prevalence and burden of bipolar depression. J Affect Disord 169(Suppl 1):S3–S11

Musil R, Zill P, Seemuller F, Bondy B, Meyer S, Spellmann I, Bender W, Adli M, Heuser I, Fisher R, Gaebel W, Maier W, Rietschel M, Rujescu D, Schennach R, Moller HJ, Riedel M (2013) Genetics of emergent suicidality during antidepressive treatment–data from a naturalistic study on a large sample of inpatients with a major depressive episode. Eur Neuropsychopharmacol 23(7):663–674

Nery-Fernandes F, Quarantini LC, Guimaraes JL, de Oliveira IR, Koenen KC, Kapczinski F, Miranda-Scippa A (2012) Is there an association between suicide attempt and delay of initiation of mood stabilizers in bipolar I disorder? J Affect Disord 136(3):1082–1087

Neves FS, Malloy-Diniz LF, Barbosa IG, Brasil PM, Correa H (2009) Bipolar disorder first episode and suicidal behavior: are there differences according to type of suicide attempt? Rev Bras Psiquiatr 31(2):114–118

Nordentoft M, Mortensen PB, Pedersen CB (2011) Absolute risk of suicide after first hospital contact in mental disorder. Arch Gen Psychiatry 68(10):1058–1064

Oquendo MA, Galfalvy H, Russo S, Ellis SP, Grunebaum MF, Burke A, Mann JJ (2004) Prospective study of clinical predictors of suicidal acts after a major depressive episode in patients with major depressive disorder or bipolar disorder. Am J Psychiatry 161(8):1433–1441

Pacchiarotti I, Bond DJ, Baldessarini RJ, Nolen WA, Grunze H, Licht RW, Post RM, Berk M, Goodwin GM, Sachs GS, Tondo L, Findling RL, Youngstrom EA, Tohen M, Undurraga J, Gonzalez-Pinto A, Goldberg JF, Yildiz A, Altshuler LL, Calabrese JR, Mitchell PB, Thase ME, Koukopoulos A, Colom F, Frye MA, Malhi GS, Fountoulakis KN, Vazquez G, Perlis RH, Ketter TA, Cassidy F, Akiskal H, Azorin JM, Valenti M, Mazzei DH, Lafer B, Kato T, Mazzarini L, Martinez-Aran A, Parker G, Souery D, Ozerdem A, McElroy SL, Girardi P, Bauer M, Yatham LN, Zarate CA, Nierenberg AA, Birmaher B, Kanba S, El-Mallakh RS, Serretti A, Rihmer Z, Young AH, Kotzalidis GD, MacQueen GM, Bowden CL, Ghaemi SN, Lopez-Jaramillo C, Rybakowski J, Ha K, Perugi G, Kasper S, Amsterdam JD, Hirschfeld RM, Kapczinski F, Vieta E (2013) The International Society for Bipolar Disorders (ISBD) task force report on antidepressant use in bipolar disorders. Am J Psychiatry 170(11):1249–1262

Pompili M, Rihmer Z, Innamorati M, Lester D, Girardi P, Tatarelli R (2009) Assessment and treatment of suicide risk in bipolar disorders. Expert Rev Neurother 9(1):109–136

Pompili M, Innamorati M, Rihmer Z, Gonda X, Serafini G, Akiskal H, Amore M, Niolu C, Sher L, Tatarelli R, Perugi G, Girardi P (2012) Cyclothymic-depressive-anxious temperament pattern is related to suicide risk in 346 patients with major mood disorders. J Affect Disord 136 (3):405–411

Pompili M, Gonda X, Serafini G, Innamorati M, Sher L, Amore M, Rihmer Z, Girardi P (2013) Epidemiology of suicide in bipolar disorders: a systematic review of the literature. Bipolar Disord 15(5):457–490

Popovic D, Benabarre A, Crespo JM, Goikolea JM, Gonzalez-Pinto A, Gutierrez-Rojas L, Montes JM, Vieta E (2014) Risk factors for suicide in schizophrenia: systematic review and clinical recommendations. Acta Psychiatr Scand 130(6):418–426

Radomsky ED, Haas GL, Mann JJ, Sweeney JA (1999) Suicidal behavior in patients with schizophrenia and other psychotic disorders. Am J Psychiatry 156(10):1590–1595

Rihmer Z (2005) Prediction and prevention of suicide in bipolar disorder. Clin Neuropsychiatry 2:48–54

Rihmer Z (2007a) Pharmacological prevention of suicide in bipolar patients – a realizable target. J Affect Disord 103(1–3):1–3

Rihmer Z (2007b) Suicide risk in mood disorders. Curr Opin Psychiatry 20(1):17–22

Rihmer Z, Akiskal H (2006) Do antidepressants t(h)reat(en) depressives? Toward a clinically judicious formulation of the antidepressant-suicidality FDA advisory in light of declining national suicide statistics from many countries. J Affect Disord 94(1–3):3–13

Rihmer Z, Angst J (2009) Mood disorders – epidemiology. In: Sadock B, Sadock V, Ruiz P (eds) Kaplan and Sadock's comprehensive textbook of psychiatry, 9th edn. Lippincott Williams and Wilkins, Philadelphia, PA

Rihmer Z, Gonda X (2013) Pharmacological prevention of suicide in patients with major mood disorders. Neurosci Biobehav Rev 37(10 Pt 1):2398–2403

Rihmer Z, Barsi J, Arato M, Demeter E (1990) Suicide in subtypes of primary major depression. J Affect Disord 18(3):221–225

Rihmer Z, Kantor Z, Rihmer A, Seregi K (2004) Suicide prevention strategies–a brief review. Neuropsychopharmacol Hung 6(4):195–199

Rihmer Z, Benazzi F, Gonda X (2007) Suicidal behavior in unipolar depression: focus on mixed states. In: Tatarelli R, Pompili M, Girardi P (eds) Suicide in psychiatric disorders. Nova Science, New York, NY, pp 223–235

Rucci P, Frank E, Kostelnik B, Fagiolini A, Mallinger AG, Swartz HA, Thase ME, Siegel L, Wilson D, Kupfer DJ (2002) Suicide attempts in patients with bipolar I disorder during acute and maintenance phases of intensive treatment with pharmacotherapy and adjunctive psychotherapy. Am J Psychiatry 159(7):1160–1164

Sanchez-Gistau V, Colom F, Mane A, Romero S, Sugranyes G, Vieta E (2009) Atypical depression is associated with suicide attempt in bipolar disorder. Acta Psychiatr Scand 120(1):30–36

Sani G, Tondo L, Koukopoulos A, Reginaldi D, Kotzalidis GD, Koukopoulos AE, Manfredi G, Mazzarini L, Pacchiarotti I, Simonetti A, Ambrosi E, Angeletti G, Girardi P, Tatarelli R (2011) Suicide in a large population of former psychiatric inpatients. Psychiatry Clin Neurosci 65 (3):286–295

Sato T, Bottlender R, Kleindienst N, Moller HJ (2005) Irritable psychomotor elation in depressed inpatients: a factor validation of mixed depression. J Affect Disord 84(2–3):187–196

Schaffer A, Isometsa ET, Tondo L, Moreno DH, Turecki G, Reis C, Cassidy F, Sinyor M, Azorin JM, Kessing LV, Ha K, Goldstein T, Weizman A, Beautrais A, Chou YH, Diazgranados N, Levitt AJ, Zarate CA Jr, Rihmer Z, Yatham LN (2015) International Society for Bipolar Disorders Task Force on Suicide: meta-analyses and meta-regression of correlates of suicide attempts and suicide deaths in bipolar disorder. Bipolar Disord 17(1):1–16

Simon GE, Hunkeler E, Fireman B, Lee JY, Savarino J (2007) Risk of suicide attempt and suicide death in patients treated for bipolar disorder. Bipolar Disord 9(5):526–530

Sokero P, Eerola M, Rytsala H, Melartin T, Leskela U, Lestela-Mielonen P, Isometsa E (2006) Decline in suicidal ideation among patients with MDD is preceded by decline in depression and hopelessness. J Affect Disord 95(1–3):95–102

Sondergard L, Lopez AG, Andersen PK, Kessing LV (2008) Mood-stabilizing pharmacological treatment in bipolar disorders and risk of suicide. Bipolar Disord 10(1):87–94

Swann AC, Moeller FG, Steinberg JL, Schneider L, Barratt ES, Dougherty DM (2007) Manic symptoms and impulsivity during bipolar depressive episodes. Bipolar Disord 9(3):206–212

Szadoczky E, Vitrai J, Rihmer Z, Furedi J (2000) Suicide attempts in the Hungarian adult population. Their relation with DIS/DSM-III-R affective and anxiety disorders. Eur Psychiatry 15(6):343–347

Takeshima M, Oka T (2013) Association between the so-called "activation syndrome" and bipolar II disorder, a related disorder, and bipolar suggestive features in outpatients with depression. J Affect Disord 151(1):196–202

Tidemalm D, Haglund A, Karanti A, Landen M, Runeson B (2014) Attempted suicide in bipolar disorder: risk factors in a cohort of 6086 patients. PLoS One 9(4), e94097

Tondo L, Isacsson G, Baldessarini R (2003) Suicidal behaviour in bipolar disorder: risk and prevention. CNS Drugs 17(7):491–511

Tondo L, Lepri B, Baldessarini RJ (2007) Suicidal risks among 2826 Sardinian major affective disorder patients. Acta Psychiatr Scand 116(6):419–428

Undurraga J, Baldessarini RJ, Valenti M, Pacchiarotti I, Vieta E (2012) Suicidal risk factors in bipolar I and II disorder patients. J Clin Psychiatry 73(6):778–782

Valtonen H, Suominen K, Mantere O, Leppamaki S, Arvilommi P, Isometsa ET (2005) Suicidal ideation and attempts in bipolar I and II disorders. J Clin Psychiatry 66(11):1456–1462

Vieta E, Benabarre A, Colom F, Gasto C, Nieto E, Otero A, Vallejo J (1997) Suicidal behavior in bipolar I and bipolar II disorder. J Nerv Ment Dis 185(6):407–409

Weiner M, Warren L, Fiedorowicz JG (2011) Cardiovascular morbidity and mortality in bipolar disorder. Ann Clin Psychiatry 23(1):40–47

Yerevanian BI, Koek RJ, Mintz J (2007a) Bipolar pharmacotherapy and suicidal behavior. Part 3: impact of antipsychotics. J Affect Disord 103(1–3):23–28

Yerevanian BI, Koek RJ, Mintz J (2007b) Bipolar pharmacotherapy and suicidal behavior. Part I: Lithium, divalproex and carbamazepine. J Affect Disord 103(1–3):5–11

Yerevanian BI, Koek RJ, Mintz J, Akiskal HS (2007c) Bipolar pharmacotherapy and suicidal behavior. Part 2: The impact of antidepressants. J Affect Disord 103(1–3):13–21

Zalsman G, Braun M, Arendt M, Grunebaum MF, Sher L, Burke AK, Brent DA, Chaudhury SR, Mann JJ, Oquendo MA (2006) A comparison of the medical lethality of suicide attempts in bipolar and major depressive disorders. Bipolar Disord 8(5 Pt 2):558–565

Part II
Bipolar Depression: Neurobiology

Chapter 5
The Genetic Basis of Bipolar Disorder

Liping Hou and Francis J. McMahon

Abstract The high heritability of bipolar disorder (BD) means that a full under-standing of etiology must account for the strong influence of inherited genetic variation. An energetic search for specific genetic risk factors over the past 25 years has begun to bear fruit, but most of the genetic risk for BD remains unexplained. In this chapter, we will review the genetic epidemiology of BD, studies aimed at identifying genes that confer risk for the illness, functional genomic studies that seek to elucidate the pathophysiology of BD from the level of genes and gene networks, and pharmacogenomic studies that aim to identify genetic markers of treatment outcome. We conclude with a view to future directions that may ultimately lead to a more complete understanding of this common, severe mental illness and better approaches to diagnosis and treatment.

Keywords GWAS • CACNA1C • ANK3 • TRANK1 • CNV • 16p11 dup • Mania • Depression

5.1   Introduction

Bipolar disorder (BD) is a highly heritable illness, and inherited genetic factors account for most of the individual variation in risk. Despite this fact, the identifi-cation of specific genetic risk factors has proven quite challenging. Recent years have seen a revolution in our understanding of the genetic basis of common, complex disorders like BD owing to innovative study designs such as genome-wide association studies (GWAS). GWAS have now identified a number of repro-ducible genetic markers for BD, but these so far account for only a small portion of

L. Hou, PhD
Human Genetics Branch, Intramural Research Program, National Institutes of Health, Bethesda, MD, USA

F.J. McMahon, MD (✉)
Human Genetics Branch, National Institutes of Health, Bldg. 35, Rm 1A-202, 35 Convent Drive, MSC 3719, Bethesda, MD 20892-3719, USA
e-mail: mcmahonf@mail.nih.gov

© Springer International Publishing Switzerland 2016   73
C.A. Zarate Jr., H.K. Manji (eds.), *Bipolar Depression: Molecular Neurobiology,*
Clinical Diagnosis, and Pharmacotherapy, Milestones in Drug Therapy,
DOI 10.1007/978-3-319-31689-5_5

the heritability. Ongoing studies that examine rarer forms of genetic variation, such as copy number variants (CNVs) and rare functional variants in DNA sequence, may shed light on at least some of the remaining heritability.

5.2   Genetic Epidemiology

5.2.1   Family Studies

Observational and systematic studies conducted in the early twentieth century demonstrated that BD aggregates in families (reviewed in Schulze et al. 2004; Propping 2005). While the lifetime prevalence of BD in the general population is around 1–2 %, multiple studies have reported that the lifetime risk for BD in first-degree relatives of a patient with the illness is increased on the order of five- to 20-fold.

Family studies are also an excellent way to define the range of clinical manifestations of underlying risk genotypes. Classically, probands are ascertained and diagnosed without regard to family history, and their first- and second-degree relatives are then systematically assessed for the presence of traits of interest. The best-known such studies in BD were published in the 1980s (Gershon et al. 1982; Weissman et al. 1984). These studies showed that not only BD but also major depressive disorder (MDD, also known as unipolar depression), dysthymia, cyclothymia, schizoaffective BD, alcoholism, and anxiety disorders were all increased among the first-degree relatives of probands with BD. On the other hand, schizophrenia and other non-affective psychotic illnesses were not found to be increased by most studies. These data were generally interpreted as supporting the existence of a spectrum of bipolar-related conditions distinct from schizophrenia, although some dissenting views persisted (Berrettini 2000).

The availability of large-scale, population-based registries that can be linked to electronic medical records (EMRs) has ushered in a new era of family studies. Population-based studies have several advantages over traditional family study designs, including large sample size, more complete ascertainment, reduced risk of misreporting and recall bias, and the ability to link distant (second degree and beyond) relatives. On the other hand, diagnostic information may be less detailed and precise than what can be obtained from direct, research-grade evaluations blind to proband diagnosis.

Population-based family studies have challenged the view that BD and schizophrenia are distinct at the family level. The largest such study (Lichtenstein et al. 2009), based in Sweden, showed an increased risk of BD and other mood disorder diagnoses among first-degree relatives of individuals with BD comparable to that reported in the earlier studies. However, this study also showed a substantially increased rate of schizophrenia diagnoses among first-degree relatives that contrasted with earlier studies. It is possible that modest sample sizes and incomplete ascertainment led the earlier studies to underestimate the risk of schizophrenia among relatives of people with BD.

One prominent feature of family studies of BD is the lack of a clear Mendelian pattern of inheritance, a cardinal feature of complex disorders. While some segregation analyses have supported a single major locus (Mendelian) model, most studies have been unable to exclude polygenic or multifactorial models. This probably reflects inherent limitations associated with segregation analysis methods to handle heterogeneity and complex modes of inheritance. This is further complicated by evidence that families of probands with BD are characterized by assortative mating (Mathews and Reus 2001), genetic anticipation (McInnis et al. 1993), and parent of origin effects that may reflect genomic imprinting (Stine et al. 1995), mitochondrial inheritance (McMahon et al. 1995), or other factors. Still, most people with BD have a first-degree relative with a diagnosis of BD or MDD (Gershon et al. 1982; Weissman et al. 1984). This contrasts with disorders such as schizophrenia or autism, which typically occur without affected first-degree relatives. This observation fits with the apparently much greater role that de novo genetic changes (CNVs or point mutations) play in schizophrenia and autism than in BD.

5.2.2   Twin Studies

BD is a highly heritable illness. This means that the majority of the individual variation in risk can be explained by inherited genetic variation. This has been a very consistent finding over more than a century of research, despite differences in study populations, case definitions, and analysis methods (Reus and Freimer 1997).

Twin studies published between 1930 and 2004 (Luxenburger 1930; Rosanoff et al. 1935; Bertelsen et al. 1977; McGuffin et al. 2003; Kieseppa et al. 2004) have consistently supported the high heritability of BD, with values on the order of 80 and 90 %. This is remarkable considering how much has changed in terms of diagnostic assessments, twin study methods, and societal trends in the nearly 100-year span of these studies.

Some recent research has suggested that twin studies may overestimate heritability under certain conditions that violate basic assumptions of the twin method. For example, when there is significant gene–environment correlation such that genetic risk factors also increase the likelihood of exposure to environmental risk factors, true heritability can be overestimated (Visscher et al. 2008). An example of this would be alcoholism, where alcoholic parents may pass on risk alleles to their offspring but may also keep more alcohol at home and model excessive drinking behavior for their offspring. Twin studies may also overestimate heritability when identical twins share a more similar environment than dizygotic twins (a violation of the "equal environments" assumption (Eaves 1978)). De novo variation that occurs early in the twinning process may be shared by monozygotic twins but will never be shared by dizygotic twins. To the extent that this de novo event contributes to a trait, heritability estimates based in part on the difference in trait concordance between monozygotic and dizygotic twins may be inflated (Gratten et al. 2013).

5.2.3   Adoption Studies

Two systematic adoption studies in BD (Mendlewicz and Rainer 1977; Wender et al. 1986), while performed in comparatively small samples, further support the notion that genetic factors contribute substantially more to the etiology of BD than environmental factors.

Some population-based registry studies have been published in recent years that largely agree with the results of the classic adoption studies (Lichtenstein et al. 2009; Song et al. 2015). However, the larger sample sizes of the population-based studies also detect significantly increased risks of BD in biologically unrelated adoptee/adoptive parent and adoptive parent/adoptee pairs (OR 1.8–5.4), although the sample sizes were still small (Song et al. 2015). This suggests some nongenetic familial contributions to BD.

5.3   Gene Identification

Genetic epidemiology has provided compelling evidence that genetic factors play a major role in the etiology of BD and has laid the foundation for future studies. However, the methods of genetic epidemiology are not capable of identifying the genes involved or pinpointing the genetic variation to explain the high heritability of BD. This task required a method of genetic mapping, and the advent of genetic linkage studies in humans seemed to fit the bill. When the genetic complexity of BD proved too much for linkage studies, candidate gene association studies came into vogue, but it was not until the advent of GWAS that specific genetic loci involved in BD could be consistently identified. Even then, we did not find exactly what we thought we were looking for.

5.3.1   Linkage Studies

The phenomenon of genetic linkage was first described by Thomas Hunt Morgan in his studies of fruit flies. Morgan realized that traits tended to be inherited together when they were influenced by genes that were near one another on the same chromosome. At first, the application of Morgan's principles to humans proved challenging. In 1980, Botstein, Lander, and White proposed that genetic markers could be used to map genes by linkage even in the human genome (Botstein et al. 1980). Their prediction was soon proven true, with strong findings of genetic linkage in cystic fibrosis (Knowlton et al. 1985; Wainwright et al. 1985), Huntington's Disease (Gusella et al. 1983), and BD (Egeland et al. 1987).

The former two genetic linkages were quickly confirmed and disease-causing mutations identified. But the latter was not and was soon retracted (Kelsoe et al. 1989). This false start foretold much of what would follow in the ensuing decade of genetic linkage studies for BD. It also embodied many of the difficulties—genetic heterogeneity, small effect sizes, and non-Mendelian inheritance patterns—that would by the 1990s come to be seen as the key characteristics of all complex genetic conditions like BD, type 2 diabetes, and cardiovascular disease.

Three major meta-analyses of linkage studies in BD were undertaken in an attempt to increase power by increasing sample size (Badner and Gershon 2002; Segurado et al. 2003; McQueen et al. 2005). While providing support for loci on chromosomes 6q, 8q, 13q, 18q, and 22q, each study tended to highlight nonoverlapping sets of loci. Linkage analysis is now no longer seen as a powerful tool to pinpoint susceptibility genes for complex traits and diseases. While some of the reported linkage regions may ultimately be found to harbor risk alleles, the genetic (or locus) heterogeneity of BD may be so high that it defeats the ability of linkage analyses to separate true findings from a large number of false positives, even with large sample sizes. As noted below, linkage methods may find a new role in family-based sequencing studies.

5.3.2 Candidate Gene Studies

Genetic association studies were the second major molecular approach taken to unlock the genetic mysteries of BD. Instead of searching for genetic markers that were inherited together in families, genetic association studies sought to directly measure genetic differences between groups of unrelated individuals who differed for a trait such as BD. These genetic differences were measured with molecular tools known as genetic markers. The early genetic association studies were only able to examine a few genetic markers at a time, so researchers tried to focus on genes whose known functions seemed to "make sense" as candidates for a given disease. Thus, the era of candidate gene association studies was born.

Candidate gene association studies of BD did make some headway, but with little consensus around the main findings. Several studies, in particular systematic studies of many markers in linkage regions and large-scale candidate gene studies, identified potential susceptibility genes (Craddock and Forty 2006; Detera-Wadleigh and McMahon 2006; Hayden and Nurnberger 2006; Farmer et al. 2007; Kato 2007), but results remained inconsistent across studies (Ioannidis 2007). We now understand that there were at least two major reasons for this inconsistency.

First, with close to 20,000 genes—most of whose functions are unknown—the selection of candidate genes is actually much more difficult than people thought. So the prior probability that any particular marker would be associated with a disease was actually very low. In addition, since individual markers exert only a very small effect on a trait (typically <1 % of the variance, as we now know), the power to detect true association was low in the kinds of sample sizes (in the tens or hundreds) that were typical of a candidate gene study. Taken together, low prior probability and low power to detect true findings meant that most association findings declared significant at the usual p-value thresholds of 5 % or even 1 % would most likely represent false positives (Ioannidis 2003).

The second major reason for inconsistency arose from the fact that there was no commonly accepted catalog of genetic variants from which to select markers for study. Thus, it was difficult to compare studies, since different markers were often used in studies of the same gene. Large efforts to develop standard maps of common genetic variation, such as the HapMap Project (www.hapmap.org; (International HapMap Consortium 2003)), solved this problem by generating large lists of markers and defining "tag" single nucleotide polymorphisms (SNPs) that efficiently interrogate large regions of the genome, known as haplotype blocks.

5.3.3   Genome-Wide Association Studies

Like genetic research in other complex disorders, BD genetics entered the twenty-first century against the backdrop of important developments in human genetics. These included the publication of the sequence of the human genome (McPherson et al. 2001; Venter et al. 2001), the completion of the HapMap, and the advent of DNA microchip technology. These landmark developments have greatly diminished one major impediment in complex genetic research: technical feasibility. Now, financial resources allowing, several thousand samples can be genotyped with several hundred thousand genetic markers or sequenced over hundreds of megabases in a small fraction of the time such an endeavor would have taken just a few years ago. The GWAS, large-scale surveys of copy number variation, and large-scale resequencing studies are all early products of these technological developments. We cannot foresee all that will follow, but it is already clear that the GWAS have opened new windows into the genetic architecture of common complex disorders such as BD.

There are probably about 10 million common SNPs in the human genome. Previous case–control association studies could assay only a fraction of this important source of genetic variation. In GWAS, several hundred thousand or million SNPs are rapidly scanned across the complete genome of a large number of case and control individuals (or, less commonly, case-parent trios). SNPs are selected on the basis of informativeness, without a specific prior hypothesis of etiological involvement in disease, so GWAS are commonly referred to as "hypothesis-free" studies.

GWAS have now been performed for several hundred complex phenotypes. In psychiatry, GWAS have now been published for schizophrenia, obsessive–compulsive disorder, MDD, autism, attention deficit hyperactivity disorder (ADHD), and BD. Association signals have been found and replicated for many of these

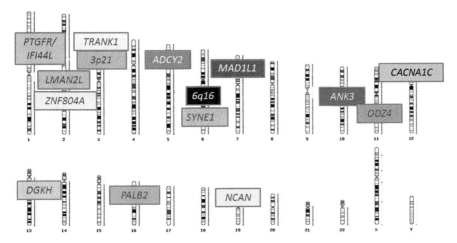

Fig. 5.1 Thanks to the efforts of many groups that have collected and genotyped large case–control samples; about 16 genome-wide significant loci have been identified so far in BD. Loci are indicated here by the name of the nearest gene(s), except for the chromosome 3p21 locus, where many genes are packed into a small region. Each individual locus confers only very small risk, and not every locus has been consistently replicated

disorders (Collins and Sullivan 2013). The early GWAS in BD were based on relatively small samples (Wellcome Trust Case Control Consortium 2007; Baum et al. 2008; Sklar et al. 2008), and it was not initially clear if consistent findings would emerge. But as sample sizes have grown and studies have been combined with sophisticated methods of meta-analysis, we can now confidently point to several loci with reproducible effects (Fig. 5.1). The top SNPs implicate the genes *CACNA1C*, *ANK3*, and *TRANK1*, as well as a region on chromosome 3p that contains many genes. The number of implicated genes is still too small for confident statements about etiologic pathways, but initial data seem to implicate calcium signaling, neuronal transmission, and neuronal development.

These findings support several general conclusions, most of which came as a surprise. Despite the high heritability of BD, there are no common genetic markers that have a large effect on risk. Individual markers typically confer a 1–3 % increased risk in carriers. Taken together, all common markers may account for about 2 % of the individual variance in risk (Wray et al. 2014). GWAS have revealed that many complex disorders, while highly heritable, are actually the product of many genes of small individual effect that add together to increase risk of disease, close to the classic polygenic threshold model (Falconer 1981). Although genetic markers do not necessarily implicate specific genes, none of those that have emerged to date were considered candidate genes by the field and none of the previously reported candidate gene association results have been replicated by GWAS. GWAS are still underpowered to confidently exclude any gene, but it is safe to say that most of the implicated genes have been a complete surprise and that the functions of many of these genes remain largely unknown.

Another surprising finding is that much of the common genetic risk for BD overlaps with schizophrenia and, to a lesser extent, autism (Cross-Disorder Group of the Psychiatric Genomics Consortium 2013a, b). This discovery challenges the prevailing nosology of mental illnesses and suggests that the widely accepted diagnostic categories—while reliable—do not correspond to distinct biological etiologies. The next major challenge for the field will be to confidently assign each replicated marker to a gene and begin to develop gene-based biological hypotheses for further study. This is addressed further below.

5.3.4   CNVs and Other Rare Risk Alleles

Despite the progress made with GWAS, much of the genetic risk for BD remains uncharacterized. In addition to the many common risk alleles, there may also be rarer alleles that confer larger risk, shape the clinical picture, or influence treatment outcomes. This is the pattern for schizophrenia, for example, where common variants account for about one-third of the risk, but many individually rare variants—for instance, large CNVs (Walsh et al. 2008) and de novo point mutations (Xu et al. 2008, 2011, 2012; Fromer et al. 2014)—confer substantial risk in a minority of cases. There has been less progress in identifying rarer alleles that exert large effects in BD, but a few studies have suggested that CNVs do play a role, at least in BD. Several other studies have also reported de novo CNVs in BD, especially chromosome 16p11 duplications (Green et al. 2016). Large-scale sequencing studies of BD are underway, and these may implicate additional rare alleles with larger effects on risk than have been observed so far by GWASs.

Ongoing studies of disorders such as autism (Sebat et al. 2007; Morrow et al. 2008; Sanders et al. 2011, 2012) and schizophrenia (Walsh et al. 2008; Xu et al. 2008, 2011, 2012; Guha et al. 2013) suggest that rare alleles may indeed have some explanatory power in the genetic basis of mental illness, but compelling rare-allele findings have not yet appeared for BD. As noted, existing findings in BD seem to point toward many alleles of small effect. But if this were the whole story, it would not explain the relatively high recurrence rates in first-degree relatives (on the order of 10–15 % (Gershon et al. 1982)). The true genetic architecture of common diseases like BD probably encompasses both common alleles that alone confer small risk (odds ratios 1.1–1.5) and uncommon or rare alleles (including CNVs) that confer larger risks in a few people or families.

While it can be difficult to disprove, the polygenic threshold model (Fig. 5.2) still offers the best overall fit to the existing family, linkage, and association findings for BD (Moser et al. 2015). Polygenic disorders cluster in families and may be highly heritable, but in contrast to monogenic disorders do not show simple inheritance patterns. Classically, risk for disease is spread over many dozens—or hundreds—of distinct genes, each of which confers only a small part of the total risk for disease. Each person's disease risk is influenced by the total burden of risk alleles they carry, with fewer alleles conferring lower and more alleles conferring

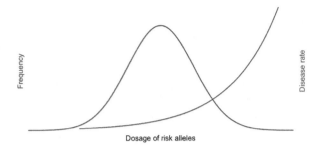

Fig. 5.2 Polygenic threshold model of common, complex disorders. Many alleles are assumed to play a role in risk. Individuals at the *right* of the risk allele distribution (*blue*) are at the highest risk for disease (*red line*). Nongenetic (environmental factors) may also contribute, shifting the disease risk function (*red line*) to the *left*

greater risk. Disease occurs when the allele burden crosses some threshold, although the exact disease threshold for a given person may be influenced by nongenetic factors.

5.4 Functional Genomics

GWAS have identified several risk variants involved in major mental illnesses, but the functional consequences of most variants remain undefined. Genetic variation, both common and rare, may affect coding regions, but most variation lies in regions that play a regulatory role. Thanks to the ENCODE project (Consortium 2012), we have learned much more about these regions in recent years. We can now annotate gene regions based on their interaction with histones, which exert a large influence on gene expression. We are also recognizing large numbers of new genes that produce regulatory RNAs (microRNAs, long-noncoding RNAs, etc.) that appear to play an important role in fine-tuning gene expression over development and in response to environmental stressors (Batista and Chang 2013).

5.4.1 Gene Expression Studies

Genetic variants can cause a range of functional differences that can trigger widespread perturbations in brain development and function. These perturbations may lead to differences in the expression of many genes and may manifest clinically as individual differences in disease risk.

Many common complex traits are mediated by cis-acting regulatory polymorphisms (Bryois et al. 2014) that influence gene expression in a tissue-specific manner (Hernandez et al. 2012). However, direct analysis of the impact of risk variants on gene expression in the brain has been challenging due to the limitations of postmortem studies and the difficulty in accessing appropriate cells from affected patients.

Despite these challenges, gene expression studies in postmortem brain have begun to identify large sets of genes whose expression is consistently perturbed in BD (for a review, see Seifuddin et al. 2013). As sample sizes grow (http://commonmind.org/WP/) and methods improve (Akula et al. 2014), it seems likely that a confident set of differentially expressed genes will be identified. The main limitation of postmortem studies remains the difficulty in distinguishing between gene expression changes that play an etiologic role and those that arise as a consequence of BD or its treatment.

The advent of induced pluripotent stem cell (iPSC) technology (Takahashi and Yamanaka 2006) and new tools that enable the genetic manipulation of patient-derived neuronal cells in vitro (Cong et al. 2013) offer opportunities for new discoveries. iPSCs are particularly valuable for the investigation of polygenic disorders, since the cells capture the full complement of risk alleles within the same genetic background as the individual from whom they derive. Gene expression studies in iPSC-derived neural cells may be a useful strategy for bringing GWAS findings into a biological context (Brennand et al. 2011).

5.4.2   Epigenetics

Epigenetic studies focus on relatively enduring changes in gene expression that are not attributable to DNA sequence (Goldberg et al. 2007). Epigenetic marks include methylation of cytosine residues, which tends to reduce the expression of nearby genes, and acetylation of histones, which tends to have opposing effects on gene function.

Preclinical studies have demonstrated that stressful life experiences, drug exposure, and other postconception events can influence gene expression through epigenetic influences (Lester et al. 2011). These studies raise the exciting possibility that the contribution of environmental factors to the onset of disorders such as BD may ultimately be explainable through epigenetic mechanisms. These mechanisms may prove to be excellent therapeutic targets, perhaps more amenable to curative treatments than risk factors that reside in the inherited sequence of DNA.

5.4.3   Other Studies

The roles of other nontraditional forms of genetic variation have been addressed in several studies, but remain to be fully elucidated. These include telomere shortening, an age-dependent phenomenon that may influence risk for depression (Verhoeven et al. 2014); mitochondrial genetics, which reflect variation in the mitochondrial genome and influence cellular energetics (Anglin et al. 2012); and somatic mutations, which may be particularly common in brain cells and could exert an important influence on individual differences in disease risk or expression (Insel 2014). Much more research is needed to elucidate the potential role of these mechanisms in the genetic basis of BD.

5.5 Pharmacogenomics and Precision Medicine

The treatment of psychiatric disorders is complex, with many different strategies and highly variable outcomes. Pharmacotherapy is a cornerstone of modern psychiatric care, but trial and error to identify the optimal medication regimen often contributes to treatment withdrawal, delayed recovery, or treatment-associated adverse events. The identification of treatment biomarkers could increase the efficiency of pharmacotherapy and help reduce the burden of mental illness.

Pharmacogenetic studies offer the advantage of a potentially shorter path from bench to bedside. This is because genetic discoveries may point to new drug targets or tests that predict treatment outcome without the need to fully elucidate pathophysiology from the gene up (Laje and McMahon 2007). Alleles involved in pharmacogenetic traits may escape natural selection and could thus be common (Pritchard 2001), even though they have a large effect on treatment outcomes. Several such alleles have already been found by GWAS (Higashi et al. 2002; Chung et al. 2004; Shuldiner et al. 2009; Goldstein et al. 2014), suggesting that they may be prevalent.

5.5.1 Mood Stabilizers

Lithium remains a mainstay in the long-term treatment of BD, but response is variable. About 35 % of patients treated with lithium have fewer illness episodes over time, while about 20 % have no response (Calkin and Alda 2012). Three GWAS of lithium response have been published so far, with varying definitions of response (Perlis et al. 2009; Chen et al. 2014; Hou et al. 2016). No replicated results have emerged so far. Genetic markers of lithium response could be valuable for treatment planning and could provide insights into the biological mechanism of lithium action. With regard to BD, lithium response may be considered a prime phenotype (Winokur 1975; Alda 2003). Sample size, however, will be the crucial issue. So far, pharmacogenetic studies of lithium response have been characterized by a lack of statistical power (Dmitrzak-Weglarz et al. 2005; Rybakowski et al. 2005; Bremer et al. 2007). Larger sample sizes may be achieved through international consortia (Schulze et al. 2010), but standardized phenotype characterization across centers is clearly needed.

Carbamazepine is a less widely used anticonvulsant and mood stabilizer that has been associated with severe cutaneous adverse events such as Stevens–Johnson syndrome (SJS). Studies in patients of Asian ancestry have consistently shown that the major histocompatibility allele HLA B*1502 strongly increases risk of SJS and other severe cutaneous events in people taking carbamazepine (Chung et al. 2004). Genetic testing for this allele is now recommended by the U.S. Food and Drug Administration (FDA) in individuals of Asian ancestry. It is unclear if this finding is relevant for other anticonvulsants more commonly used for mood stabilization, such as lamotrigine.

5.5.2 Other Agents

Many psychotropic drugs are metabolized to varying degrees by the cytochrome P450 enzyme system. The genes CYP2D6 and CYP2C19 harbor a number of common functional polymorphisms that affect drug metabolism (pharmacokinetics). Genetic association studies have sought to identify clinically relevant treatment biomarkers in these genes (Gaedigk and Leeder 2014).

Some studies have implicated genetic variation in cytochrome P450 enzyme activity in the efficacy and tolerability of antipsychotics (Tanaka and Hisawa 1999; Basile et al. 2000; Schillevoort et al. 2002). While consistent results have not emerged for the antipsychotics most widely used in BD, the FDA does recommend reduced dosing of some antipsychotics in individuals whose CYP2D genotypes indicate that they are "slow metabolizers" who could suffer increased adverse events at typical dosages.

A few candidate–gene associations with antipsychotics have been replicated. These include associations between antipsychotic efficacy and the dopamine D2 receptor (DRD2) gene (Yamanouchi et al. 2003; Lencz et al. 2006), antipsychotic-induced weight gain and the serotonin 2C receptor (HTR2C) (De Luca et al. 2007; Opgen-Rhein et al. 2010) as well as melanocortin 4 receptor (MC4R) genes (Malhotra et al. 2012; Chowdhury et al. 2013), and between clozapine-induced agranulocytosis and the major histocompatibility gene HLA-DQB1 (Yunis et al. 1995; Goldstein et al. 2014). Sensitivities and specificities for these markers have either not been studied or remain too low to guide clinical practice.

No studies have examined pharmacogenetic predictors of mood stabilization by atypical antipsychotics, although in recent years these drugs have been increasingly used in patients with BD.

5.6 Conclusions and Future Directions

BD, as currently defined in the standard diagnostic manuals (DSM-IV/DSM-5 and ICD 10), is a highly reliable but clinically variable entity. Most of the genetic work in BD has focused on genes that contribute to broader phenotypes, but it is not clear that such genes, if found, will be able to explain the clinical variability in terms of age at onset, symptoms, chronicity, comorbidity, and treatment response that is a hallmark of the BD diagnosis. Without an operational diagnosis, large collaborative studies of BD would not have been possible, since case definitions would have varied too much from center to center. Diagnostic entities in psychiatry are still mainly constructs without a well-defined shared biology. Correspondence between genotype and phenotype, when it finally emerges in psychiatry, is unlikely to show a close resemblance to our current diagnostic systems. It is possible that alleles exist that increase risk for psychopathology in a fairly general way, with other alleles or nongenetic factors influencing precise clinical presentation.

As we begin to accumulate data that may influence genetic risk factors for BD, questions have begun to accumulate. How much of a role will epigenetic factors play? How does SNP or CNV variation affect gene function? How best do we translate from a genetic association finding to a clinically relevant test? Addressing these questions is one of the top priorities for the field in the coming years.

There have been many false starts in genetic studies of BD, but we can now finally be certain that we are on the right path. We now have the necessary tools to address BD and other psychiatric illnesses as genetically influenced common disorders like diabetes, heart disease, and cancer. The relative influence of common and rare alleles varies across disorders and populations, but these risk alleles and the genes they influence provide a solid basis for additional studies. The burgeoning understanding of gene expression and regulation, combined with new tools that enable the genetic manipulation of patient-derived neuronal cells in vitro, means that we can now begin to use what is known about genetic markers to design functional genomic studies aimed at an improved understanding of etiology, pathophysiology and, ultimately, diagnosis and treatment. These are long-term projects, but genomics provides a sound basis for progress.

Disclosures Supported by the NIMH Intramural Research Program. Dr. Hou received additional support from the Brain & Behavior Research Foundation.

References

Akula N, Barb J, Jiang X, Wendland JR, Choi KH, Sen SK, Hou L, Chen DT, Laje G, Johnson K, Lipska BK, Kleinman JE, Corrada-Bravo H, Detera-Wadleigh S, Munson PJ, McMahon FJ (2014) RNA-sequencing of the brain transcriptome implicates dysregulation of neuroplasticity, circadian rhythms and GTPase binding in bipolar disorder. Mol Psychiatry 19(11):1179–1185

Alda M (2003) Pharmacogenetic aspects of bipolar disorder. Pharmacogenomics 4(1):35–40

Anglin RE, Mazurek MF, Tarnopolsky MA, Rosebush PI (2012) The mitochondrial genome and psychiatric illness. Am J Med Genet B Neuropsychiatr Genet 159B(7):749–759

Badner JA, Gershon ES (2002) Meta-analysis of whole-genome linkage scans of bipolar disorder and schizophrenia. Mol Psychiatry 7(4):405–411

Basile VS, Ozdemir V, Masellis M, Walker ML, Meltzer HY, Lieberman JA, Potkin SG, Alva G, Kalow W, Macciardi FM, Kennedy JL (2000) A functional polymorphism of the cytochrome P450 1A2 (CYP1A2) gene: association with tardive dyskinesia in schizophrenia. Mol Psychiatry 5(4):410–417

Batista PJ, Chang HY (2013) Long noncoding RNAs: cellular address codes in development and disease. Cell 152(6):1298–1307

Baum AE, Akula N, Cabanero M, Cardona I, Corona W, Klemens B, Schulze TG, Cichon S, Rietschel M, Nothen MM, Georgi A, Schumacher J, Schwarz M, Abou Jamra R, Hofels S, Propping P, Satagopan J, Detera-Wadleigh SD, Hardy J, McMahon FJ (2008) A genome-wide association study implicates diacylglycerol kinase eta (DGKH) and several other genes in the etiology of bipolar disorder. Mol Psychiatry 13(2):197–207

Berrettini WH (2000) Are schizophrenic and bipolar disorders related? A review of family and molecular studies. Biol Psychiatry 48(6):531–538

Bertelsen A, Harvald B, Hauge M (1977) A Danish twin study of manic-depressive disorders. Br J Psychiatry 130:330–351

Botstein D, White RL, Skolnick M, Davis RW (1980) Construction of a genetic linkage map in man using restriction fragment length polymorphisms. Am J Human Genet 32(3):314–331

Bremer T, Diamond C, McKinney R, Shehktman T, Barrett TB, Herold C, Kelsoe JR (2007) The pharmacogenetics of lithium response depends upon clinical co-morbidity. Mol Diagn Ther 11 (3):161–170

Brennand KJ, Simone A, Jou J, Gelboin-Burkhart C, Tran N, Sangar S, Li Y, Mu Y, Chen G, Yu D, McCarthy S, Sebat J, Gage FH (2011) Modelling schizophrenia using human induced pluripotent stem cells. Nature 473(7346):221–225

Bryois J, Buil A, Evans DM, Kemp JP, Montgomery SB, Conrad DF, Ho KM, Ring S, Hurles M, Deloukas P, Davey Smith G, Dermitzakis ET (2014) Cis and trans effects of human genomic variants on gene expression. PLoS Genet 10(7), e1004461

Calkin C, Alda M (2012) Beyond the guidelines for bipolar disorder: practical issues in long-term treatment with lithium. Can J Psychiatry 57(7):437–445

Chen CH, Lee CS, Lee MT, Ouyang WC, Chen CC, Chong MY, Wu JY, Tan HK, Lee YC, Chuo LJ, Chiu NY, Tsang HY, Chang TJ, Lung FW, Chiu CH, Chang CH, Chen YS, Hou YM, Chen CC, Lai TJ, Tung CL, Chen CY, Lane HY, Su TP, Feng J, Lin JJ, Chang CJ, Teng PR, Liu CY, Chen CK, Liu IC, Chen JJ, Lu T, Fan CC, Wu CK, Li CF, Wang KH, Wu LS, Peng HL, Chang CP, Lu LS, Chen YT, Cheng AT, Taiwan Bipolar C (2014) Variant GADL1 and response to lithium therapy in bipolar I disorder. N Engl J Med 370(2):119–128

Chowdhury NI, Tiwari AK, Souza RP, Zai CC, Shaikh SA, Chen S, Liu F, Lieberman JA, Meltzer HY, Malhotra AK, Kennedy JL, Muller DJ (2013) Genetic association study between antipsychotic-induced weight gain and the melanocortin-4 receptor gene. Pharmacogenomics 13(3):272–279

Chung WH, Hung SI, Hong HS, Hsih MS, Yang LC, Ho HC, Wu JY, Chen YT (2004) Medical genetics: a marker for Stevens-Johnson syndrome. Nature 428(6982):486

Collins AL, Sullivan PF (2013) Genome-wide association studies in psychiatry: what have we learned? Br J Psychiatry 202(1):1–4, http://commonmind.org/WP/. Accessed 15 Oct 2015

Cong L, Ran FA, Cox D, Lin S, Barretto R, Habib N, Hsu PD, Wu X, Jiang W, Marraffini LA, Zhang F (2013) Multiplex genome engineering using CRISPR/Cas systems. Science 339 (6121):819–823

Consortium EP (2012) An integrated encyclopedia of DNA elements in the human genome. Nature 489(7414):57–74

Craddock N, Forty L (2006) Genetics of affective (mood) disorders. Eur J Human Genet 14 (6):660–668

Cross-Disorder Group of the Psychiatric Genomics Consortium (2013a) Identification of risk loci with shared effects on five major psychiatric disorders: a genome-wide analysis. Lancet 381 (9875):1371–1379

Cross-Disorder Group of the Psychiatric Genomics Consortium (2013b) Genetic relationship between five psychiatric disorders estimated from genome-wide SNPs. Nat Genet 45 (9):984–994

De Luca V, Muller DJ, Hwang R, Lieberman JA, Volavka J, Meltzer HY, Kennedy JL (2007) HTR2C haplotypes and antipsychotics-induced weight gain: X-linked multimarker analysis. Hum Psychopharmacol 22(7):463–467

Detera-Wadleigh SD, McMahon FJ (2006) G72/G30 in schizophrenia and bipolar disorder: review and meta-analysis. Biol Psychiatry 60(2):106–114

Dmitrzak-Weglarz M, Rybakowski JK, Suwalska A, Slopien A, Czerski PM, Leszczynska-Rodziewicz A, Hauser J (2005) Association studies of 5-HT2A and 5-HT2C serotonin receptor gene polymorphisms with prophylactic lithium response in bipolar patients. Pharmacol Rep 57 (6):761–765

Eaves LJ (1978) Twins as a basis for the causal analysis of human personality. Prog Clin Biol Res 24A:151–174

Egeland JA, Gerhard DS, Pauls DL, Sussex JN, Kidd KK, Allen CR, Hostetter AM, Housman DE (1987) Bipolar affective disorders linked to DNA markers on chromosome 11. Nature 325 (6107):783–787

Falconer DS (1981) Introduction to quantitative genetics. Longmans Green, London

Farmer A, Elkin A, McGuffin P (2007) The genetics of bipolar affective disorder. Curr Opin Psychiatry 20(1):8–12

Fromer M, Pocklington AJ, Kavanagh DH, Williams HJ, Dwyer S, Gormley P, Georgieva L, Rees E, Palta P, Ruderfer DM, Carrera N, Humphreys I, Johnson JS, Roussos P, Barker DD, Banks E, Milanova V, Grant SG, Hannon E, Rose SA, Chambert K, Mahajan M, Scolnick EM, Moran JL, Kirov G, Palotie A, McCarroll SA, Holmans P, Sklar P, Owen MJ, Purcell SM, O'Donovan MC (2014) De novo mutations in schizophrenia implicate synaptic networks. Nature 506(7487):179–184

Gaedigk A, Leeder JS (2014) CYP2D6 and pharmacogenomics: where does future research need to focus? Part 1: Technical aspects. Pharmacogenomics 15(4):407–410

Gershon ES, Hamovit J, Guroff JJ, Dibble E, Leckman JF, Sceery W, Targum SD, Nurnberger JI Jr, Goldin LR, Bunney WE Jr (1982) A family study of schizoaffective, bipolar I, bipolar II, unipolar, and normal control probands. Arch Gen Psychiatry 39(10):1157–1167

Goldberg AD, Allis CD, Bernstein E (2007) Epigenetics: a landscape takes shape. Cell 128 (4):635–638

Goldstein JI, Jarskog LF, Hilliard C, Alfirevic A, Duncan L, Fourches D, Huang H, Lek M, Neale BM, Ripke S, Shianna K, Szatkiewicz JP, Tropsha A, van den Oord EJ, Cascorbi I, Dettling M, Gazit E, Goff DC, Holden AL, Kelly DL, Malhotra AK, Nielsen J, Pirmohamed M, Rujescu D, Werge T, Levy DL, Josiassen RC, Kennedy JL, Lieberman JA, Daly MJ, Sullivan PF (2014) Clozapine-induced agranulocytosis is associated with rare HLA-DQB1 and HLA-B alleles. Nat Commun 5:4757

Gratten J, Visscher PM, Mowry BJ, Wray NR (2013) Interpreting the role of de novo protein-coding mutations in neuropsychiatric disease. Nat Genet 45(3):234–238

Green EK, Rees E, Walters JT, Smith KG, Forty L, Grozeva D, Moran JL, Sklar P, Ripke S, Chambert KD, Genovese G, McCarroll SA, Jones I, Jones L, Owen MJ, O'Donovan MC, Craddock N, Kirov G (2016) Copy number variation in bipolar disorder. Mol Psychiatry 21 (1):89–93

Guha S, Rees E, Darvasi A, Ivanov D, Ikeda M, Bergen SE, Magnusson PK, Cormican P, Morris D, Gill M, Cichon S, Rosenfeld JA, Lee A, Gregersen PK, Kane JM, Malhotra AK, Rietschel M, Nothen MM, Degenhardt F, Priebe L, Breuer R, Strohmaier J, Ruderfer DM, Moran JL, Chambert KD, Sanders AR, Shi J, Kendler K, Riley B, O'Neill T, Walsh D, Malhotra D, Corvin A, Purcell S, Sklar P, Iwata N, Hultman CM, Sullivan PF, Sebat J, McCarthy S, Gejman PV, Levinson DF, Owen MJ, O'Donovan MC, Lencz T, Kirov G, Molecular Genetics of Schizophrenia C, Wellcome Trust Case Control C (2013) Implication of a rare deletion at distal 16p11.2 in schizophrenia. JAMA Psychiatry 70(3):253–260

Gusella JF, Wexler NS, Conneally PM, Naylor SL, Anderson MA, Tanzi RE, Watkins PC, Ottina K, Wallace MR, Sakaguchi AY et al (1983) A polymorphic DNA marker genetically linked to Huntington's disease. Nature 306(5940):234–238

Hayden EP, Nurnberger JI Jr (2006) Molecular genetics of bipolar disorder. Genes Brain Behav 5 (1):85–95

Hernandez DG, Nalls MA, Moore M, Chong S, Dillman A, Trabzuni D, Gibbs JR, Ryten M, Arepalli S, Weale ME, Zonderman AB, Troncoso J, O'Brien R, Walker R, Smith C, Bandinelli S, Traynor BJ, Hardy J, Singleton AB, Cookson MR (2012) Integration of GWAS SNPs and tissue specific expression profiling reveal discrete eQTLs for human traits in blood and brain. Neurobiol Dis 47(1):20–28

Higashi MK, Veenstra DL, Kondo LM, Wittkowsky AK, Srinouanprachanh SL, Farin FM, Rettie AE (2002) Association between CYP2C9 genetic variants and anticoagulation-related outcomes during warfarin therapy. JAMA 287(13):1690–1698

Hou L, Heilbronner U, Degenhardt F, Adli M, Akiyama K, Akula N et al (2016) Genetic variants associated with response to lithium treatment in bipolar disorder: a genome-wide association study. Lancet 387(10023):1085–1093

Insel TR (2014) Brain somatic mutations: the dark matter of psychiatric genetics? Mol Psychiatry 19(2):156–158

International HapMap Consortium (2003) The International HapMap Project. Nature 426 (6968):789–796

Ioannidis JP (2003) Genetic associations: false or true? Trends Mol Med 9(4):135–138

Ioannidis JP (2007) Non-replication and inconsistency in the genome-wide association setting. Hum Hered 64(4):203–213

Kato T (2007) Molecular genetics of bipolar disorder and depression. Psychiatry Clin Neurosci 61 (1):3–19

Kelsoe JR, Ginns EI, Egeland JA, Gerhard DS, Goldstein AM, Bale SJ, Pauls DL, Long RT, Kidd KK, Conte G et al (1989) Re-evaluation of the linkage relationship between chromosome 11p loci and the gene for bipolar affective disorder in the Old Order Amish. Nature 342 (6247):238–243

Kieseppa T, Partonen T, Haukka J, Kaprio J, Lonnqvist J (2004) High concordance of bipolar I disorder in a nationwide sample of twins. Am J Psychiatry 161(10):1814–1821

Knowlton RG, Cohen-Haguenauer O, Van Cong N, Frezal J, Brown VA, Barker D, Braman JC, Schumm JW, Tsui LC, Buchwald M et al (1985) A polymorphic DNA marker linked to cystic fibrosis is located on chromosome 7. Nature 318(6044):380–382

Laje G, McMahon FJ (2007) The pharmacogenetics of major depression: past, present, and future. Biol Psychiatry 62(11):1205–1207

Lencz T, Robinson DG, Xu K, Ekholm J, Sevy S, Gunduz-Bruce H, Woerner MG, Kane JM, Goldman D, Malhotra AK (2006) DRD2 promoter region variation as a predictor of sustained response to antipsychotic medication in first-episode schizophrenia patients. Am J Psychiatry 163(3):529–531

Lester BM, Tronick E, Nestler E, Abel T, Kosofsky B, Kuzawa CW, Marsit CJ, Maze I, Meaney MJ, Monteggia LM, Reul JM, Skuse DH, Sweatt JD, Wood MA (2011) Behavioral epigenetics. Ann NY Acad Sci 1226:14–33

Lichtenstein P, Yip BH, Bjork C, Pawitan Y, Cannon TD, Sullivan PF, Hultman CM (2009) Common genetic determinants of schizophrenia and bipolar disorder in Swedish families: a population-based study. Lancet 373(9659):234–239

Luxenburger H (1930) Psychiatrisch-neurologische Zwillings pathologie. Zbl Ges Neurol Psychiat 56:145–180

Malhotra AK, Correll CU, Chowdhury NI, Muller DJ, Gregersen PK, Lee AT, Tiwari AK, Kane JM, Fleischhacker WW, Kahn RS, Ophoff RA, Meltzer HY, Lencz T, Kennedy JL (2012) Association between common variants near the melanocortin 4 receptor gene and severe antipsychotic drug-induced weight gain. Arch Gen Psychiatry 69(9):904–912

Mathews CA, Reus VI (2001) Assortative mating in the affective disorders: a systematic review and meta-analysis. Compr Psychiatry 42(4):257–262

McGuffin P, Rijsdijk F, Andrew M, Sham P, Katz R, Cardno A (2003) The heritability of bipolar affective disorder and the genetic relationship to unipolar depression. Arch Gen Psychiatry 60 (5):497–502

McInnis MG, McMahon FJ, Chase GA, Simpson SG, Ross CA, DePaulo JR Jr (1993) Anticipation in bipolar affective disorder. Am J Hum Genet 53(2):385–390

McMahon FJ, Stine OC, Meyers DA, Simpson SG, DePaulo JR (1995) Patterns of maternal transmission in bipolar affective disorder. Am J Hum Genet 56(6):1277–1286

McPherson JD, Marra M, Hillier L, Waterston RH, Chinwalla A, Wallis J, Sekhon M, Wylie K, Mardis ER, Wilson RK, Fulton R, Kucaba TA, Wagner-McPherson C, Barbazuk WB, Gregory SG, Humphray SJ, French L, Evans RS, Bethel G, Whittaker A, Holden JL, McCann OT, Dunham A, Soderlund C, Scott CE, Bentley DR, Schuler G, Chen HC, Jang W, Green ED, Idol JR, Maduro VV, Montgomery KT, Lee E, Miller A, Emerling S, Kucherlapati RS, Kucherlapati R, Scherer S, Gorrell JH, Sodergren E, Clerc-Blankenburg K, Tabor P, Naylor S, Garcia D, de Jong PJ, Catanese JJ, Nowak N, Osoegawa K, Qin S, Rowen L, Madan A, Dors M, Hood L, Trask B, Friedman C, Massa H, Cheung VG, Kirsch IR, Reid T, Yonescu R, Weissenbach J, Bruls T, Heilig R, Branscomb E, Olsen A, Doggett N, Cheng JF, Hawkins T, Myers RM, Shang J, Ramirez L, Schmutz J, Velasquez O, Dixon K, Stone NE, Cox DR, Haussler D, Kent WJ, Furey T, Rogic S, Kennedy S, Jones S, Rosenthal A, Wen G, Schilhabel M, Gloeckner G, Nyakatura G, Siebert R, Schlegelberger B, Korenberg J, Chen XN, Fujiyama A, Hattori M, Toyoda A, Yada T, Park HS, Sakaki Y, Shimizu N, Asakawa S, Kawasaki K, Sasaki T, Shintani A, Shimizu A, Shibuya K, Kudoh J, Minoshima S, Ramser J, Seranski P, Hoff C, Poustka A, Reinhardt R, Lehrach H, International Human Genome Mapping C (2001) A physical map of the human genome. Nature 409(6822):934–941

McQueen MB, Devlin B, Faraone SV, Nimgaonkar VL, Sklar P, Smoller JW, Abou Jamra R, Albus M, Bacanu SA, Baron M, Barrett TB, Berrettini W, Blacker D, Byerley W, Cichon S, Coryell W, Craddock N, Daly MJ, Depaulo JR, Edenberg HJ, Foroud T, Gill M, Gilliam TC, Hamshere M, Jones I, Jones L, Juo SH, Kelsoe JR, Lambert D, Lange C, Lerer B, Liu J, Maier W, Mackinnon JD, McInnis MG, McMahon FJ, Murphy DL, Nothen MM, Nurnberger JI, Pato CN, Pato MT, Potash JB, Propping P, Pulver AE, Rice JP, Rietschel M, Scheftner W, Schumacher J, Segurado R, Van Steen K, Xie W, Zandi PP, Laird NM (2005) Combined analysis from eleven linkage studies of bipolar disorder provides strong evidence of suscep-tibility loci on chromosomes 6q and 8q. Am J Hum Genet 77(4):582–595

Mendlewicz J, Rainer JD (1977) Adoption study supporting genetic transmission in manic–depressive illness. Nature 268(5618):327–329

Morrow EM, Yoo SY, Flavell SW, Kim TK, Lin Y, Hill RS, Mukaddes NM, Balkhy S, Gascon G, Hashmi A, Al-Saad S, Ware J, Joseph RM, Greenblatt R, Gleason D, Ertelt JA, Apse KA, Bodell A, Partlow JN, Barry B, Yao H, Markianos K, Ferland RJ, Greenberg ME, Walsh CA (2008) Identifying autism loci and genes by tracing recent shared ancestry. Science 321 (5886):218–223

Moser G, Lee SH, Hayes BJ, Goddard ME, Wray NR, Visscher PM (2015) Simultaneous discovery, estimation and prediction analysis of complex traits using a bayesian mixture model. PLoS Genet 11(4), e1004969

Opgen-Rhein C, Brandl EJ, Muller DJ, Neuhaus AH, Tiwari AK, Sander T, Dettling M (2010) Association of HTR2C, but not LEP or INSIG2, genes with antipsychotic-induced weight gain in a German sample. Pharmacogenomics 11(6):773–780

Perlis RH, Smoller JW, Ferreira MA, McQuillin A, Bass N, Lawrence J, Sachs GS, Nimgaonkar V, Scolnick EM, Gurling H, Sklar P, Purcell S (2009) A genome wide association study of response to lithium for prevention of recurrence in bipolar disorder. Am J Psychiatry 166 (6):718–725

Pritchard JK (2001) Are rare variants responsible for susceptibility to complex diseases? Am J Hum Genet 69(1):124–137

Propping P (2005) The biography of psychiatric genetics: from early achievements to historical burden, from an anxious society to critical geneticists. Am J Med Genet B Neuropsychiatr Genet 136B(1):2–7

Reus VI, Freimer NB (1997) Understanding the genetic basis of mood disorders: where do we stand? Am J Hum Genet 60(6):1283–1288

Rosanoff AJ, Handy LM, Plesset IR (1935) The etiology of manic-depressive syndromes with special reference to their occurrence in twins. Am J Psychiatry 91(4):725–762

Rybakowski JK, Suwalska A, Czerski PM, Dmitrzak-Weglarz M, Leszczynska-Rodziewicz A, Hauser J (2005) Prophylactic effect of lithium in bipolar affective illness may be related to serotonin transporter genotype. Pharmacol Rep 57(1):124–127

Sanders SJ, Ercan-Sencicek AG, Hus V, Luo R, Murtha MT, Moreno-De-Luca D, Chu SH, Moreau MP, Gupta AR, Thomson SA, Mason CE, Bilguvar K, Celestino-Soper PB, Choi M, Crawford EL, Davis L, Wright NR, Dhodapkar RM, DiCola M, DiLullo NM, Fernandez TV, Fielding-Singh V, Fishman DO, Frahm S, Garagaloyan R, Goh GS, Kammela S, Klei L, Lowe JK, Lund SC, McGrew AD, Meyer KA, Moffat WJ, Murdoch JD, O'Roak BJ, Ober GT, Pottenger RS, Raubeson MJ, Song Y, Wang Q, Yaspan BL, Yu TW, Yurkiewicz IR, Beaudet AL, Cantor RM, Curland M, Grice DE, Gunel M, Lifton RP, Mane SM, Martin DM, Shaw CA, Sheldon M, Tischfield JA, Walsh CA, Morrow EM, Ledbetter DH, Fombonne E, Lord C, Martin CL, Brooks AI, Sutcliffe JS, Cook EH Jr, Geschwind D, Roeder K, Devlin B, State MW (2011) Multiple recurrent de novo CNVs, including duplications of the 7q11.23 Williams syndrome region, are strongly associated with autism. Neuron 70(5):863–885

Sanders SJ, Murtha MT, Gupta AR, Murdoch JD, Raubeson MJ, Willsey AJ, Ercan-Sencicek AG, DiLullo NM, Parikshak NN, Stein JL, Walker MF, Ober GT, Teran NA, Song Y, El-Fishawy P, Murtha RC, Choi M, Overton JD, Bjornson RD, Carriero NJ, Meyer KA, Bilguvar K, Mane SM, Sestan N, Lifton RP, Gunel M, Roeder K, Geschwind DH, Devlin B, State MW (2012) De novo mutations revealed by whole-exome sequencing are strongly associated with autism. Nature 485(7397):237–241

Schillevoort I, de Boer A, van der Weide J, Steijns LS, Roos RA, Jansen PA, Leufkens HG (2002) Antipsychotic-induced extrapyramidal syndromes and cytochrome P450 2D6 genotype: a case-control study. Pharmacogenetics 12(3):235–240

Schulze TG, Fangerau H, Propping P (2004) From degeneration to genetic susceptibility, from eugenics to genetics, from Bezugsziffer to LOD score: the history of psychiatric genetics. Int Rev Psychiatry 16(4):246–259

Schulze TG, Alda M, Adli M, Akula N, Ardau R, Bui ET, Chillotti C, Cichon S, Czerski P, Del Zompo M, Detera-Wadleigh SD, Grof P, Gruber O, Hashimoto R, Hauser J, Hoban R, Iwata N, Kassem L, Kato T, Kittel-Schneider S, Kliwicki S, Kelsoe JR, Kusumi I, Laje G, Leckband SG, Manchia M, Macqueen G, Masui T, Ozaki N, Perlis RH, Pfennig A, Piccardi P, Richardson S, Rouleau G, Reif A, Rybakowski JK, Sasse J, Schumacher J, Severino G, Smoller JW, Squassina A, Turecki G, Young LT, Yoshikawa T, Bauer M, McMahon FJ (2010) The International Consortium on Lithium Genetics (ConLiGen): an initiative by the NIMH and IGSLI to study the genetic basis of response to lithium treatment. Neuropsychobiology 62 (1):72–78

Sebat J, Lakshmi B, Malhotra D, Troge J, Lese-Martin C, Walsh T, Yamrom B, Yoon S, Krasnitz A, Kendall J, Leotta A, Pai D, Zhang R, Lee YH, Hicks J, Spence SJ, Lee AT, Puura K, Lehtimaki T, Ledbetter D, Gregersen PK, Bregman J, Sutcliffe JS, Jobanputra V, Chung W, Warburton D, King MC, Skuse D, Geschwind DH, Gilliam TC, Ye K, Wigler M (2007) Strong association of de novo copy number mutations with autism. Science 316 (5823):445–449

Segurado R, Detera-Wadleigh SD, Levinson DF, Lewis CM, Gill M, Nurnberger JI Jr, Craddock N, DePaulo JR, Baron M, Gershon ES, Ekholm J, Cichon S, Turecki G, Claes S, Kelsoe JR, Schofield PR, Badenhop RF, Morissette J, Coon H, Blackwood D, McInnes LA, Foroud T, Edenberg HJ, Reich T, Rice JP, Goate A, McInnis MG, McMahon FJ, Badner JA, Goldin LR, Bennett P, Willour VL, Zandi PP, Liu J, Gilliam C, Juo SH, Berrettini WH, Yoshikawa T, Peltonen L, Lonnqvist J, Nothen MM, Schumacher J, Windemuth C, Rietschel M, Propping P, Maier W, Alda M, Grof P, Rouleau GA, Del-Favero J, Van Broeckhoven C, Mendlewicz J, Adolfsson R, Spence MA, Luebbert H, Adams LJ, Donald

JA, Mitchell PB, Barden N, Shink E, Byerley W, Muir W, Visscher PM, Macgregor S, Gurling H, Kalsi G, McQuillin A, Escamilla MA, Reus VI, Leon P, Freimer NB, Ewald H, Kruse TA, Mors O, Radhakrishna U, Blouin JL, Antonarakis SE, Akarsu N (2003) Genome scan meta-analysis of schizophrenia and bipolar disorder, Part III: Bipolar disorder. Am J Hum Genet 73(1):49–62

Seifuddin F, Pirooznia M, Judy JT, Goes FS, Potash JB, Zandi PP (2013) Systematic review of genome-wide gene expression studies of bipolar disorder. BMC Psychiatry 13(1):213

Shuldiner AR, O'Connell JR, Bliden KP, Gandhi A, Ryan K, Horenstein RB, Damcott CM, Pakyz R, Tantry US, Gibson Q, Pollin TI, Post W, Parsa A, Mitchell BD, Faraday N, Herzog W, Gurbel PA (2009) Association of cytochrome P450 2C19 genotype with the antiplatelet effect and clinical efficacy of clopidogrel therapy. JAMA 302(8):849–857

Sklar P, Smoller JW, Fan J, Ferreira MA, Perlis RH, Chambert K, Nimgaonkar VL, McQueen MB, Faraone SV, Kirby A, de Bakker PI, Ogdie MN, Thase ME, Sachs GS, Todd-Brown K, Gabriel SB, Sougnez C, Gates C, Blumenstiel B, Defelice M, Ardlie KG, Franklin J, Muir WJ, McGhee KA, MacIntyre DJ, McLean A, VanBeck M, McQuillin A, Bass NJ, Robinson M, Lawrence J, Anjorin A, Curtis D, Scolnick EM, Daly MJ, Blackwood DH, Gurling HM, Purcell SM (2008) Whole-genome association study of bipolar disorder. Mol Psychiatry 13(6):558–569

Song J, Bergen SE, Kuja-Halkola R, Larsson H, Landen M, Lichtenstein P (2015) Bipolar disorder and its relation to major psychiatric disorders: a family-based study in the Swedish population. Bipolar Disord 17(2):184–193

Stine OC, Xu J, Koskela R, McMahon FJ, Gschwend M, Friddle C, Clark CD, McInnis MG, Simpson SG, Breschel TS, Vishio E, Riskin K, Feilotter H, Chen E, Shen S, Folstein S, Meyers DA, Botstein D, Marr TG, DePaulo JR (1995) Evidence for linkage of bipolar disorder to chromosome 18 with a parent-of-origin effect. Am J Hum Genet 57(6):1384–1394

Takahashi K, Yamanaka S (2006) Induction of pluripotent stem cells from mouse embryonic and adult fibroblast cultures by defined factors. Cell 126(4):663–676

Tanaka E, Hisawa S (1999) Clinically significant pharmacokinetic drug interactions with psychoactive drugs: antidepressants and antipsychotics and the cytochrome P450 system. J Clin Pharm Ther 24(1):7–16

Venter JC, Adams MD, Myers EW, Li PW, Mural RJ, Sutton GG, Smith HO, Yandell M, Evans CA, Holt RA, Gocayne JD, Amanatides P, Ballew RM, Huson DH, Wortman JR, Zhang Q, Kodira CD, Zheng XH, Chen L, Skupski M, Subramanian G, Thomas PD, Zhang J, Gabor Miklos GL, Nelson C, Broder S, Clark AG, Nadeau J, McKusick VA, Zinder N, Levine AJ, Roberts RJ, Simon M, Slayman C, Hunkapiller M, Bolanos R, Delcher A, Dew I, Fasulo D, Flanigan M, Florea L, Halpern A, Hannenhalli S, Kravitz S, Levy S, Mobarry C, Reinert K, Remington K, Abu-Threideh J, Beasley E, Biddick K, Bonazzi V, Brandon R, Cargill M, Chandramouliswaran I, Charlab R, Chaturvedi K, Deng Z, Di Francesco V, Dunn P, Eilbeck K, Evangelista C, Gabrielian AE, Gan W, Ge W, Gong F, Gu Z, Guan P, Heiman TJ, Higgins ME, Ji RR, Ke Z, Ketchum KA, Lai Z, Lei Y, Li Z, Li J, Liang Y, Lin X, Lu F, Merkulov GV, Milshina N, Moore HM, Naik AK, Narayan VA, Neelam B, Nusskern D, Rusch DB, Salzberg S, Shao W, Shue B, Sun J, Wang Z, Wang A, Wang X, Wang J, Wei M, Wides R, Xiao C, Yan C, Yao A, Ye J, Zhan M, Zhang W, Zhang H, Zhao Q, Zheng L, Zhong F, Zhong W, Zhu S, Zhao S, Gilbert D, Baumhueter S, Spier G, Carter C, Cravchik A, Woodage T, Ali F, An H, Awe A, Baldwin D, Baden H, Barnstead M, Barrow I, Beeson K, Busam D, Carver A, Center A, Cheng ML, Curry L, Danaher S, Davenport L, Desilets R, Dietz S, Dodson K, Doup L, Ferriera S, Garg N, Gluecksmann A, Hart B, Haynes J, Haynes C, Heiner C, Hladun S, Hostin D, Houck J, Howland T, Ibegwam C, Johnson J, Kalush F, Kline L, Koduru S, Love A, Mann F, May D, McCawley S, McIntosh T, McMullen I, Moy M, Moy L, Murphy B, Nelson K, Pfannkoch C, Pratts E, Puri V, Qureshi H, Reardon M, Rodriguez R, Rogers YH, Romblad D, Ruhfel B, Scott R, Sitter C, Smallwood M, Stewart E, Strong R, Suh E, Thomas R, Tint NN, Tse S, Vech C, Wang G, Wetter J, Williams S, Williams M, Windsor S, Winn-Deen E, Wolfe K, Zaveri J, Zaveri K, Abril JF, Guigo R, Campbell MJ, Sjolander KV, Karlak B, Kejariwal A, Mi H, Lazareva B, Hatton T, Narechania A, Diemer K,

Muruganujan A, Guo N, Sato S, Bafna V, Istrail S, Lippert R, Schwartz R, Walenz B, Yooseph S, Allen D, Basu A, Baxendale J, Blick L, Caminha M, Carnes-Stine J, Caulk P, Chiang YH, Coyne M, Dahlke C, Mays A, Dombroski M, Donnelly M, Ely D, Esparham S, Fosler C, Gire H, Glanowski S, Glasser K, Glodek A, Gorokhov M, Graham K, Gropman B, Harris M, Heil J, Henderson S, Hoover J, Jennings D, Jordan C, Jordan J, Kasha J, Kagan L, Kraft C, Levitsky A, Lewis M, Liu X, Lopez J, Ma D, Majoros W, McDaniel J, Murphy S, Newman M, Nguyen T, Nguyen N, Nodell M, Pan S, Peck J, Peterson M, Rowe W, Sanders R, Scott J, Simpson M, Smith T, Sprague A, Stockwell T, Turner R, Venter E, Wang M, Wen M, Wu D, Wu M, Xia A, Zandieh A, Zhu X (2001) The sequence of the human genome. Science 291(5507):1304–1351

Verhoeven JE, Revesz D, Epel ES, Lin J, Wolkowitz OM, Penninx BW (2014) Major depressive disorder and accelerated cellular aging: results from a large psychiatric cohort study. Mol Psychiatry 19(8):895–901

Visscher PM, Hill WG, Wray NR (2008) Heritability in the genomics era–concepts and misconceptions. Nat Rev Genet 9(4):255–266

Wainwright BJ, Scambler PJ, Schmidtke J, Watson EA, Law HY, Farrall M, Cooke HJ, Eiberg H, Williamson R (1985) Localization of cystic fibrosis locus to human chromosome 7cen-q22. Nature 318(6044):384–385

Walsh T, McClellan JM, McCarthy SE, Addington AM, Pierce SB, Cooper GM, Nord AS, Kusenda M, Malhotra D, Bhandari A, Stray SM, Rippey CF, Roccanova P, Makarov V, Lakshmi B, Findling RL, Sikich L, Stromberg T, Merriman B, Gogtay N, Butler P, Eckstrand K, Noory L, Gochman P, Long R, Chen Z, Davis S, Baker C, Eichler EE, Meltzer PS, Nelson SF, Singleton AB, Lee MK, Rapoport JL, King MC, Sebat J (2008) Rare structural variants disrupt multiple genes in neurodevelopmental pathways in schizophrenia. Science 320 (5875):539–543

Weissman MM, Gershon ES, Kidd KK, Prusoff BA, Leckman JF, Dibble E, Hamovit J, Thompson WD, Pauls DL, Guroff JJ (1984) Psychiatric disorders in the relatives of probands with affective disorders. The Yale University–National Institute of Mental Health Collaborative Study. Arch Gen Psychiatry 41(1):13–21

Wellcome Trust Case Control Consortium (2007) Genome-wide association study of 14,000 cases of seven common diseases and 3,000 shared controls. Nature 447(7145):661–678

Wender PH, Kety SS, Rosenthal D, Schulsinger F, Ortmann J, Lunde I (1986) Psychiatric disorders in the biological and adoptive families of adopted individuals with affective disorders. Arch Gen Psychiatry 43(10):923–929

Winokur G (1975) The Iowa 500: heterogeneity and course in manic-depressive illness (bipolar). Compr Psychiatry 16(2):125–131

Wray NR, Lee SH, Mehta D, Vinkhuyzen AA, Dudbridge F, Middeldorp CM (2014) Research review: polygenic methods and their application to psychiatric traits. J Child Psychol Psychiatry 55(10):1068–1087

Xu B, Roos JL, Levy S, van Rensburg EJ, Gogos JA, Karayiorgou M (2008) Strong association of de novo copy number mutations with sporadic schizophrenia. Nat Genet 40(7):880–885

Xu B, Roos JL, Dexheimer P, Boone B, Plummer B, Levy S, Gogos JA, Karayiorgou M (2011) Exome sequencing supports a de novo mutational paradigm for schizophrenia. Nat Genet 43 (9):864–868

Xu B, Ionita-Laza I, Roos JL, Boone B, Woodrick S, Sun Y, Levy S, Gogos JA, Karayiorgou M (2012) De novo gene mutations highlight patterns of genetic and neural complexity in schizophrenia. Nat Genet 44(12):1365–1369

Yamanouchi Y, Iwata N, Suzuki T, Kitajima T, Ikeda M, Ozaki N (2003) Effect of DRD2, 5-HT2A, and COMT genes on antipsychotic response to risperidone. Pharmacogenomics 3 (6):356–361

Yunis JJ, Corzo D, Salazar M, Lieberman JA, Howard A, Yunis EJ (1995) HLA associations in clozapine-induced agranulocytosis. Blood 86(3):1177–1183

Chapter 6
Understanding the Neurobiology of Bipolar Depression

Araba F. Chintoh and L. Trevor Young

Abstract Despite the extensive research in the field, the precise etiology of bipolar disorder (BD) is not clear; neither, then, is our understanding of the pathogenesis of bipolar depression. What we do know is largely gleaned from investigations of patients with BD irrespective of their mood state. The most consistent neuropathological findings include structural, cellular, and functional changes in cortical and limbic regions, and most explanations for the illness involve pathways that ultimately result in cerebral atrophy and cell loss. Though many theories have been proposed, here, the focus is on the current leading hypotheses of mitochondrial involvement, oxidative stress, the role of inflammation, and neurotrophic factors. Future research is required with a specific focus on the depressive phase of BD in addition to the study of biomarkers to aid clinicians in the diagnosis and targeted treatment of bipolar depression.

Keywords Bipolar disorder • Depression • Neuroplasticity • Mitochondrial dysfunction • Oxidative stress • Inflammation

6.1 Understanding the Neurobiology of Bipolar Depression

Despite extensive research in the field, the precise etiology of bipolar disorder (BD) is not clear. Neither, then, do we have a fulsome understanding of the neurobiology of bipolar depression. Some insights have been gained from knowledge of the mechanisms associated with major depressive disorder (MDD); however, it is largely believed that bipolar depression is unique because of its differential response to conventional antidepressant treatment as well as its divergent prognosis and course of illness. Our understanding of the pathogenesis of

A.F. Chintoh, MD, PhD
Department of Psychiatry, University of Toronto, Toronto, ON, Canada

L.T. Young, MD, PhD (✉)
Department of Psychiatry, University of Toronto, 1 King's College Circle, Room 2109,
Toronto, ON, Canada, M5S 1A8
e-mail: ltrevor.young@utoronto.ca

© Springer International Publishing Switzerland 2016
C.A. Zarate Jr., H.K. Manji (eds.), *Bipolar Depression: Molecular Neurobiology,
Clinical Diagnosis, and Pharmacotherapy*, Milestones in Drug Therapy,
DOI 10.1007/978-3-319-31689-5_6

bipolar depression comes from the investigation of patients with BD irrespective of their mood state. Few studies focus exclusively on bipolar depression. Subsequently, the existing literature exploring the neurobiology of BD is robust but can be piecemeal with respect to the depressive state. This chapter will feature key areas of research into the pathophysiology of BD, taking the opportunity to highlight depression-specific findings where available.

Though many theories have been proposed, here, the focus is on the current leading hypotheses of mitochondrial involvement, oxidative stress, the role of inflammation, and neurotrophic factors. Other hypotheses have implicated the serotonergic system and neurotransmitter dysfunction (see Wang and Young (2009) for review) but will not be discussed here in detail. The preceding chapter provides an in-depth discussion of genetic factors implicated in bipolar depression and so will not be reviewed again here.

We begin with a review of the neuropathological features of BD since most explanations for the illness involve pathways that ultimately result in cerebral atrophy and cell loss. Finally, we attend to the future of research in this field, highlighting the need for biomarkers to guide diagnosis and treatment of this complex illness.

6.2 Neuropathological Studies Reveal Cellular Loss in Cerebral Cortical and Limbic Regions

Structural, cellular, and functional changes have long been observed in the brains of patients with BD. These changes are often classified based on region, allowing researchers to postulate relationships between neuroanatomy and behavioral correlates of BD. The neural basis of affective states includes the subcortical limbic system as well as cortical regions like the anterior cingulate cortex (ACC) and dorsolateral prefrontal cortex (dlPFC). The limbic system comprises primitive structures (e.g., amygdala, hippocampus, thalamus) essential to one's ability to perceive, process, and create memories about emotions and experiences with emotional valence (Janak and Tye 2015). The cerebral cortex is implicated in emotion through its extensive connections to the limbic system and other brain regions. In the cortex, complex sensory information is processed and, through its connections, the cortex is believed to modulate one's experience of emotion (Salzman and Fusi 2010). In BD, neuroanatomical abnormalities are observed in both limbic and cortical areas, and these changes are fundamental to understanding the neurobiology of the illness.

The amygdala is widely understood to be the seat of our emotions and is necessarily implicated in BD, with its prominent affective states. Using MRI to estimate amygdala volume, Hartberg and colleagues (2015) recently reported a decrease in left amygdala volume in patients with BD not treated with lithium. These researchers note that lithium-treated patients displayed larger amygdalar volumes, suggesting a neuroprotective effect for lithium. This study corroborates other evidence linking amygdala structure to BD (Foland-Ross et al. 2012; Inal-Emiroglu et al. 2015; Kittel-Schneider et al. 2015). Previous reports investigating volumes and neurons in amygdala of postmortem specimens highlight a decrease in the size of neuronal cell bodies as well as decreased density of neurons in certain nuclei of the amygdala.

The hippocampus is known for its role in memory, and the thalamus is known as a relay station between cortical and subcortical regions; however, both regions play a key role in regulating emotion and have been implicated in the neurobiology of BD. Postmortem brain samples from patients with BD have smaller CA1 pyramidal hippocampal neurons compared to a control population (Liu et al. 2007). Some reports support a reduction in thalamic volume and support cells in BD patients (Bielau et al. 2005; Byne et al. 2008). In contrast, there is compelling evidence for decreased hippocampal volume in MDD coming from both imaging and postmortem brain studies (Campbell and Macqueen 2004). This potentially highlights different pathophysiological mechanisms in MDD vs. BD.

The ACC is thought to be a hub in the network between our cortical cognitive capacity and our subcortical limbic emotional experiences, and the subgenual region of the ACC (sgACC) is particularly implicated in emotion regulation (Ghaznavi and Deckersbach 2012). In patients with bipolar depression, gray matter volume, blood flow, and metabolism in the sgACC are decreased. Early investigation of postmortem ACC specimens from BD patients revealed a deficit in both the number and density of glia cells, i.e., the nonneuronal cells that play a supportive function in the central nervous system facilitating neuronal growth and differentiation (Wake et al. 2013). Notably, the structural changes observed in the sgACC were present in patients with a family history of BD at the outset of their illness (Hirayasu et al. 1999). Further, recent imaging analyses highlight the functional connectivity between brain regions and note that the children of patients with BD are more likely to exhibit atypical connection patterns between cortical neurons (including the sgACC) and subcortical areas (Singh et al. 2014). Taken together, these findings suggest that pathophysiologic changes in patients with this disorder are not entirely due to the damaging effect of repeated mood episodes but can be present at the outset of the illness. Here again, lithium treatment is associated with a reversal in the sgACC gray matter volume deficits observed in patients with BD (Moore et al. 2009). In addition, response to ketamine—whose novel therapeutic use shows efficacy in treatment-resistant depression—is linked to sgACC activity in patients with bipolar depression (Nugent et al. 2014) (Fig. 6.1).

The dlPFC is implicated in attention and executive functioning—patients with BD have long been known to have decreased performance on tasks that require these cognitive capacities (Rubinsztein et al. 2006; Taylor and Abrams 1987). The dlPFC has also been identified as a central locus in the cortical emotional processing network that plays a role in cognitive reappraisals of emotionally salient events (Morawetz et al. 2016). Thus, it is expected that the dlPFC is a region fundamental to the pathophysiology of BD where maladaptive affect regulation and extreme appraisals of affective states are key deficits (Palmier-Claus et al. 2015).

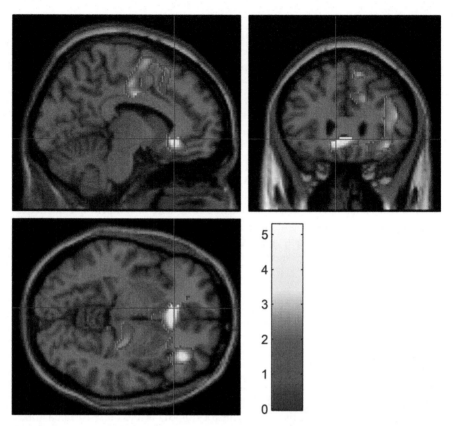

Fig. 6.1 Areas where regional metabolic rate of glucose (rMRGlu) following placebo infusion was significantly correlated with percent improvement in Montgomery–Åsberg Depression Rating Scale (MADRS) score following ketamine administration. Crosshair is centered on the finding in the subgenual anterior cingulate cortex (sgACC), but the cluster with the peak voxel is in the dorsal cingulate. Image threshold at $p < 0.05$ uncorrected. The extent threshold was set such that only the clusters remaining significant after correction for multiple comparisons ($p_{corrected} < 0.01$) are shown (Nugent et al. 2014). Reprinted with permission from Nugent et al. (2014), Fig. 2

Indeed, patients with BD display deficient connections between the dlPFC and subcortical limbic regions, suggesting a possible mechanism for emotion dysregulation in BD (Radaelli et al. 2015).

Examination of the cellular architecture of the dlPFC found decreased density of glial cells in certain cortical layers of brain specimens of BD patients. Studies also observed a significant reduction in the number of pyramidal neurons with a trend towards decreased neuron size. Subsequent studies inspecting postmortem samples corroborated these results. MRI quantifying volume of gray matter showed decreased dlPFC volumes in children and adolescents with BD (Adleman et al. 2012), suggesting that these structural changes may underlie the pathogenesis of BD.

These above-noted structural abnormalities are widely accepted as fundamental to the pathology of BD, though the mechanisms that underlie these changes remain unclear. In the next section, we describe some of the potential models that lead to cell loss including work from our own laboratory.

6.3 Proposed Mechanisms Underlying Cortical and Limbic Cell Loss

The proposed mechanisms thought to underlie these structural brain changes fall loosely into two categories: (1) processes that increase destruction of neuronal cells or (2) processes with dysfunctional neuroprotective mechanisms. We begin by examining the role of mitochondrial dysfunction and the destructive impact of oxidative stress and then explore the role of inflammation before concluding with a discussion of neurotrophic factors.

6.3.1 The Role of Mitochondrial Dysfunction

The past 20 years have seen a growing literature supporting the concept of dysfunctional mitochondrial energy metabolism in the pathogenesis of BD. Mitochondria are ubiquitous organelles providing cellular energy in the form of adenosine triphosphate (ATP). Energy is created via the process of oxidative phosphorylation, i.e., the transfer of electrons via mitochondrial membrane proteins (the mitochondrial electron transport chain (mETC)) and creation of a proton gradient. The mETC is a series of protein complexes (I–V) situated in the inner mitochondrial membrane. Protons are shuttled across these complexes and the resulting electrochemical gradient allows for the transfer of electrons to an oxygen molecule (i.e., reduction of oxygen) and the creation of ATP.

Reactive oxygen species (ROS) are formed as a byproduct of oxidative phosphorylation. These compounds have both physiological and pathological functions. In our normal physiological state, we utilize endogenous antioxidants to counteract the pathological or harmful effects of a buildup of ROS. Human brain tissue is specifically susceptible to ROS damage because of its high oxygen content and ROS overproduction; alternately, ROS that overwhelm the endogenous antioxidants can create neuronal damage specifically targeting lipids (key in neuronal membranes), proteins, and DNA (Moniczewski et al. 2015). The pathological effects of ROS, or oxidative stress, are reflected via a number of peripheral and central indicators and are associated with the pathophysiology of BD. ROS levels are elevated in BD, and it has been suggested that this likely reflects deficient energy metabolism resulting from mitochondrial dysfunction (Wang 2007). Many lines of evidence highlight a role for mitochondrial dysfunction in BD; for instance, the mitochondria of patients with BD exhibit morphological changes and distributional abnormalities within cells (Cataldo et al. 2010).

Patients differentially express genes related to energy metabolism with decreased expression of genes encoding for certain mitochondrial subunits, and downregulation of mitochondrial genes has been observed (Kato 2007; Naydenov et al. 2007; Sun et al. 2006; Washizuka et al. 2005). In depressed patients with BD, there is evidence for an increase in the expression of genes encoding for the mitochondrial ETC that is hypothesized to reflect an increase in turnover of mitochondria in this population (Beech et al. 2010).

Increased lactate levels have been found in brain tissue and cerebrospinal fluid from patients (Kato et al. 1992; Regenold et al. 2009). Earlier research by Kato and colleagues identified decreased intracellular pH in patients with BD, while a significant increase in pH was observed in patients during a depressive state (Kato et al. 1992). Low pH is associated with an increase in lactate—the byproduct of anaerobic glycolysis, which is the alternative energy-generating process arising when the function of mitochondria is reduced. The process of oxidative phosphorylation in mitochondria is meant to be the main source of cerebral energy, but in conditions of mitochondrial dysfunction, there can be a shift to anaerobic glycolysis and, subsequently, an accumulation of lactate. There is some evidence to suggest that it is the increase in lactate—following a state of mitochondrial dysfunction—that accounts for the reduced pH in patients with BD. It has been suggested that the increase in intracellular pH observed during the depressive state might reflect the brain's attempt to correct pH imbalance (Kato et al. 1992; Stork and Renshaw 2005).

Phosphocreatine (PCr) is created from creatinine and ATP and functions as an energy store for use during acute neuronal activity (MacDonald et al. 2006). Chronically decreased levels of PCr are indicative of mitochondrial dysfunction as evidenced by low brain PCr levels in other known mitochondrial disorders (Barbiroli et al. 1993; Chaturvedi and Flint Beal 2013; Eleff et al. 1990). Reductions in PCr have long been noted in BD, with research identifying low focal levels of PCr in the left frontal lobe of patients in the depressed state (Kato et al. 1994, 1995; Moore et al. 1997).

In all, there is impressive evidence supporting a role for mitochondrial dysfunction in the pathophysiology of BD. Oxidative stress is implicated as a proposed mechanism generating the brain atrophy and decreased cell density expressed in the population. Though data describing associations of oxidative stress and BD are abundant, the direction of the association remains unknown—does BD create oxidative stress or does oxidative stress create a milieu in which BD can develop?

6.3.2 Mitochondrial Dysfunction Leads to Oxidative Stress

Lipid peroxidation refers to the specific oxidative stress damage to lipids. This damage is key in CNS pathology because of the high content of lipids in white matter in the brain. Thiobarbituric acid reactive substances (TBARS) are byproducts of lipid peroxidation that can be quantified from serum samples. Elevated serum levels of TBARS were quantified from a population of BD patients in various affective states (i.e., mania, depression, and euthymia (Andreazza et al. 2007a; Kunz et al. 2008)). Marked elevations were noted in manic patients, though levels in depressed patients were also increased (Kunz et al. 2008). Bannerjee and colleagues investigated the effects of lithium on TBARS and found that BD patients had elevated serum TBARS and that lithium treatment significantly reduced markers of lipid peroxidation (Banerjee et al. 2012).

4-Hydroxy-2-nonenal (4-HNE), another marker of lipid peroxidation, is significantly increased in the postmortem brain specimens of patients with BD (Wang et al. 2009). Notably, the 4-HNE was sampled from cells of the ACC—a region with noted structural deficits and highlighted as key in the pathogenesis of BD (discussed above in Sect. 6.1.1). This postmortem sample included patients who had received pharmacological treatment for BD; however, subsequent analysis suggest that the differences identified were indicative of the pathological process in BD rather than changes induced by chronic medication (Wang et al. 2009). Malondialdehyde (MDA) is created when ROS degrade cellular lipid components, and peripheral serum levels of MDA are increased in patients with BD compared to nonpsychiatric controls (Can et al. 2011). Interestingly, post hoc analyses that stratified patients by medication subtype found that those on a combination of antidepressant and antipsychotic medication (a popular combination in bipolar depression) exhibited decreased serum MDA values. In addition, the impact of antipsychotics and antidepressants suggest that they modulate antioxidant capacity to combat oxidative stress (Tang and Wang 2013). Recent evidence links lipid peroxidation measured peripherally to changes in white matter tracts, which may underscore the importance of this process to the pathophysiology of BD (Versace et al. 2014). These findings also identify a potential biomarker to be measured in blood and might specifically reflect changes in brain in this disorder (Fig. 6.2).

Similarly, proteins are vulnerable to oxidative stress damage that can be measured by protein carbonylation, and when ROS attack amino acid side chains of proteins, 3-nitrotyrosine (3-NT) is formed. Increased levels of protein carbonylation (i.e., oxidative stress damage of proteins) have been detected in BD patients in a number of studies. Kapczinski and colleagues quantified serum levels of 3-NT and found a positive correlation with BD patients in a depressive state (Kapczinski et al. 2011).

8-Hydroxy-2-deoxyguanosine (8-OHdG) is measured to detect ROS-induced DNA damage. Levels of 8-OHdG are elevated in patients with BD and correlate positively to number of previous manic episodes (Soeiro-de-Souza et al. 2013). Longitudinal observation of 8-OHdG levels across the affective states in BD found elevations in 8-OHdG at baseline, six months, and 12 months, independent of symptomatology (Munkholm et al. 2015). Other studies corroborate the finding of increased oxidative damage to nucleic acids of DNA and RNA (D'Addario et al. 2012; Dell'Osso et al. 2014; Huzayyin et al. 2014).

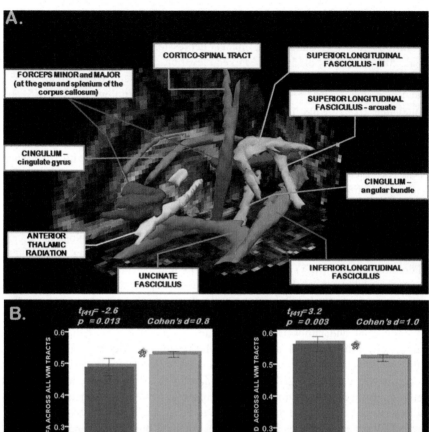

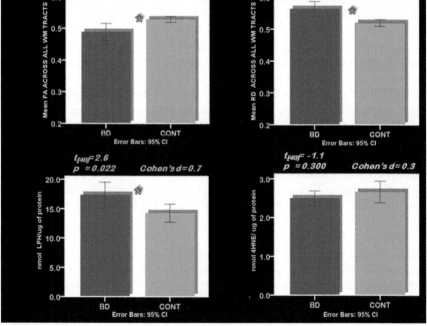

Fig. 6.2 Panel (**a**). The posterior distribution of each white matter (WM) tract is displayed in isosurface mode. The forceps minor and major are represented in *red*; the anterior thalamic radiation in *yellow*; the angular bundle of the cingulum in *light green*; the cingulate gyrus of the cingulum in *emerald green*; the cortico-spinal tract in *purple*; the inferior longitudinal fasciculus in

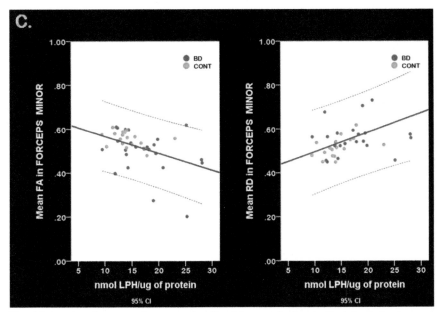

Fig. 6.2 (continued) orange; the arcuate bundle of the superior longitudinal fasciculus in *aquamarine*; the III bundle of the superior longitudinal fasciculus in *gray*; and the uncinate fasciculus in *blue*. All fibers were thresholded at 20 % of their maximum. The background image depicts the fractional anisotropy (FA) image in color-code convention in one of our participants: voxels with a *red* color define left-to-right oriented fibers; voxels with a *blue* color define inferior-to-superior oriented fibers; and voxels with a *green* color define anterior-to-posterior oriented fibers. The isosurface superimposed on the sagittal view of the colored FA image shows the characteristic anterior–posterior alignment of the cingulum (*green voxels*) and the origin of the forceps minor and major (*red voxels* in the genu and splenium of the corpus callosum). Panel (**b**). Error bar graphs depict the between-group differences in central measures (*top*; FA in the *left corner* and radial diffusivity (RD) in the *right corner*) across all WM tracts and in peripheral measures (*bottom*; lipid hydroperoxides (LPH) in the *left corner* and 4-HNE in the *right corner*) in 24 currently euthymic patients with bipolar disorder (BD) and 19 gender-matched healthy controls (CONT). Panel (**c**). Scatter plot graphs represent the linear relationship between mean FA (*left*) and mean RD (*right*) and LPH across all study participants in the forceps minor. Reprinted with permission from Versace et al. (2014), Fig. 1

There is strong evidence to support the notion of increased oxidative damage in BD. Many findings are generally applicable to the illness while others show selective changes for the depressed state.

6.3.3   Decreased Antioxidant Capacity Facilitates Oxidative Stress Damage

Oxidative stress damage can be magnified by deficient antioxidant capacity. Some of our endogenous antioxidants include superoxide dismutase (SOD), catalase, and glutathione (Soczynska et al. 2008), and abnormalities in antioxidant activity have been reported in patients with BD. Though it is reasonable to expect deficient antioxidant activity if oxidative stress damage were to underlie the pathogenesis of bipolar depression, some research demonstrates an increase in antioxidant activity. Andreazza and colleagues identified increases in SOD activity in BD patients during affective states (both depressed and manic) though not during euthymia (Andreazza et al. 2007a). Similar results have been replicated (Kuloglu et al. 2002; Kunz et al. 2008). The authors conceptualize this increase in SOD activity as an overall deficient state of antioxidant capacity responding to the increased oxidative stress during an affective (e.g., depressive) episode. Other studies have reported lower levels of SOD in bipolar depression. Selek and colleagues found that SOD levels were significantly lower in patients with bipolar depression compared to controls, and though these levels improved over the course of pharmacological treatment, they did not reach the baseline levels in the control group (Selek et al. 2008). As the brain has limited levels of SOD and catalase, glutathione plays a primary role as an antioxidant, and literature supports abnormalities of this antioxidant in BD patients (Bengesser et al. 2015; Gawryluk et al. 2011).

It follows, then, that bolstering antioxidant capacity provides a potential treatment option for bipolar depression. N-acetylcysteine (NAC) is a compound that creates the precursor to glutathione, and administration of NAC can be used to increase physiological stores of this antioxidant. In one study, patients with bipolar depression had their maintenance medication augmented with NAC for eight weeks (Berk et al. 2008) and displayed a significant reduction in depressive symptoms. These patients were then followed in a randomized, placebo-controlled trial divided into two groups: one receiving NAC daily for six months and the other placebo. The NAC group had a prolonged reduction in depressive symptoms (Berk et al. 2011). Another small study of BD patients supports these data; higher remission rates for affective symptoms were observed in the NAC-treated patients vs. their control group (Magalhães et al. 2011).

6.3.4 Overactive Inflammatory Pathways Are Implicated in BD

We now turn our focus to the role of inflammation in BD. This theory is salient in bipolar depression specifically because the link between markers of inflammation and clinical states of depression has long been known. There has been extensive research investigating the role of inflammation and cytokines in MDD (see Zunszain et al. (2011) for a review), with more recent interest in BD. It is not yet clear if inflammation has a causative role in the pathogenesis of BD or if manic and depressive episodes chronically activate the immune system, which then gets identified by alterations in central and peripheral inflammatory markers. Regardless, the role of inflammation is receiving considerable attention in the field, as it is linked with affective episodes, oxidative stress, and the medical comorbidities common in patients with BD. Given the growing research indicating the importance of the inflammatory process, we take the opportunity to explore the literature here.

C-reactive protein (CRP) is an acute phase reactant protein created in the liver and released into the bloodstream after acute injury, infection, or other systemic inflammation (Dargel et al. 2015). Clinically, serum CRP is measured to detect an acute inflammatory response or to monitor the condition of someone with a chronic inflammatory illness. Overall, the literature robustly supports elevations of CRP in BD. Most data highlight a positive correlation between serum CRP and manic episodes, and there are reports of a similar association in the depressive phase of the illness (Cunha et al. 2008; Dargel et al. 2015; De Berardis et al. 2008; Dickerson et al. 2007, 2013; Goldstein et al. 2011).

Cytokines are another class of proteins that are implicated in inflammatory states. Every cell has the capacity to generate this heterogeneous group of proteins though specific immune-related cells can enhance their production (e.g., lymphocytes) (Müller and Schwarz 2007). Cytokines can have varied effects depending on their physiological targets, and these potent molecules tend to be classified as proinflammatory (i.e., inducing inflammatory responses) or anti-inflammatory. Alterations in circulating cytokines have been recorded in both manic and depressed patients with BD with literature supporting a positive association between proinflammatory cytokines and mood episodes (Goldstein et al. 2011; Modabbernia et al. 2013; Rao et al. 2010; Su et al. 2011). A comprehensive meta-analysis of research quantifying peripheral levels of proinflammatory markers found evidence for elevated levels of cytokines, including tumor necrosis factor alpha (TNF-α) and interleukin-4 (IL-4) (Munkholm et al. 2013). The studies appraised included manic, depressed, and euthymic patients. Other data support differences between these markers based on affective valence, where CRP and soluble IL-2 receptor (sIL-2R) are consistently elevated in manic episodes, and IL-6 and TNF-α tend to be higher in depressive states (Cunha et al. 2008; De Berardis et al. 2008; Ortiz-Domínguez et al. 2007).

Pharmacological treatments have varied effects on these inflammatory markers. Lithium treatment has been implicated as anti-inflammatory, given that a comparison between treated and untreated patients revealed lower levels of cytokines in the lithium group (Boufidou et al. 2004). Interestingly, in vitro models using cells from a normal control sample revealed a proinflammatory effect of lithium treatment (Knijff et al. 2007; Petersein et al. 2015) though other preclinical models identified a reduction in cytokines with valproic acid (Himmerich et al. 2014).

The role of inflammation in BD is strengthened by acknowledgement of the high prevalence of certain medical conditions in this population. BD is highly comorbid with cardiovascular disease and type 2 diabetes mellitus, at rates exceeding those found in MDD (Angst et al. 2002; Birkenaes et al. 2007). This morbidity, and subsequent increased mortality, is not immediately related to the factors common in severe and persistent mental illness (e.g., suicide, exposure to medications, low income, physical inactivity, obesity, etc.). Inflammation is implicated in the pathogenesis of both cardiovascular disease and insulin resistance and, increasingly, research is linking these altered inflammatory networks to those that might underlie BD (Goldstein and Young 2013).

In terms of mechanism, recent research links the overactivation of inflammatory pathways to oxidative stress. Studies suggest that the overproduction of ROS by the dysfunctional mitochondrial membrane protein, complex I, can stimulate inflammatory networks and increase levels of proinflammatory cytokines. The nod-like receptor pyrin domain containing 3 (NLRP3) inflammasome is hypothesized as one mechanism for this activation. The NLRP3 inflammasome is a protein complex that can gauge oxidative species. In the presence of ROS, NLRP3 activates downstream inflammatory processes via the enzyme, caspase-1 (López-Armada et al. 2013). When ROS production is blocked, there is no activation of the NLRP3 inflammasome (Zhou et al. 2011). New research from our lab identified increases in NLRP3 components specific to mitochondrial samples from patients with BD but not in other brain or cytoplasmic samples (Kim, HK et al., in preparation). These findings implicate both oxidative stress and increased activation of the inflammasome in the pathophysiology of BD.

6.3.5   Oxidative Stress and Inflammation Can Accelerate Apoptosis

Growing literature highlights the process of apoptosis as central to the morphologic changes in the brains of patients with BD and oxidative stress. Its subsequent activation of inflammation has a significant impact on this critical cellular pathway. Apoptosis is the programmed process of cell death involving an ordered process with many steps, including cell shrinkage, dissociation of nuclear components, labeling the cell for destruction, and removal by phagocytosis. Dysregulated apoptotic pathways have been postulated as the mechanism leading to the cell loss and brain atrophy seen in the brains of patients with BD (Gigante et al. 2011). Reductions in nuclear size and DNA fragmentation have been observed at higher rates in patient samples than control populations (Buttner et al. 2007; Munkholm et al. 2015; Uranova et al. 2001). Further, the amount of DNA damage observed was positively correlated with depressive symptoms (Andreazza et al. 2007b).

A number of proteins regulate apoptosis. B-cell lymphoma 2 (Bcl-2) denotes one family of these proteins, and Bcl-2 proteins are considered antiapoptotic—thus protective of neuronal cells. Brain-derived neurotrophic factor (BDNF), a neurotrophin, is also considered an antiapoptotic factor. Proapoptotic compounds include caspase-3, caspase-9, Bcl-2-associated death promotor (BAD), and Bcl-2-associated X protein (BAX). Analysis of prefrontal cortical samples from brains of patients with BD revealed abnormal apoptotic pathways (Kim et al. 2010); specifically, the authors identified a decrease in protective, antiapoptotic factors (i.e., Bcl-2 and BDNF) and an increase in the expression of proapoptotic factors (i.e., BAX, BAD, caspase-3, and 9).

The link between BD and medical comorbidities was introduced in the section above. Here, we highlight the association because of the role of apoptosis in cellular aging. Telomere length is one indication of cellular aging as telomeres are regions on genetic material that shorten with each cell division in the life cycle; abnormal apoptotic pathways, as well as oxidative stress, are known to induce pathological shortening of telomeres (Lindqvist et al. 2015). BD patients have notably shortened telomeres (Lima et al. 2015), and one study provides evidence for a negative correlation between telomere length and number of depressive episodes in BD-II patients (Elvsåshagen et al. 2011). The concept of apoptosis-induced cellular aging synthesizes the morphological changes, mitochondrial dysfunction, inflammatory associations, and clinical comorbidities seen in BD. Future research will have to investigate and integrate knowledge about the impact of affective states within this model.

6.3.6 Growth Factors Are Implicated in Both Oxidative Stress and Inflammatory Models of BD

Dysfunctional neurotrophic factors are another potential mechanism by which cellular atrophy may occur. In general, neurotrophic factors are proteins responsible for the growth and maintenance of neurons and their support cells. They act by binding to tyrosine kinase receptors and initiating a cascade of events that inhibits apoptotic pathways. This suggests an alternative or potentially synergistic mechanism underlying the cell loss and brain atrophy observed in the brains of patients with BD. The neurotrophins of the nerve growth family are heterogenous and include BDNF, neurotrophin-3 (NT-3), and NT-4/5. Glial cell line-derived neurotrophic factor (GDNF) and vascular endothelial growth factor (VEGF) are other growth factors connected to the progression of mood disorders. To date, the bulk of the research has focused on BDNF in BD, though evidence does implicate a role for other growth factors.

BDNF is the most widely studied neurotrophin, and is implicated in the neurobiology of BD via a number of proposed mechanisms. BDNF is a key protein in regulating the survival of neurons in both the central and peripheral nervous system. As a growth factor, BDNF plays a role in promoting the development and differentiation of neurons and synapses. It is also negatively correlated with measures of oxidative stress (Kapczinski et al. 2008). The concept of neuronal growth and survival is fundamental and underlies the idea that low levels of BDNF create an inability to protect the survival or promote the growth of neurons in the brain areas implicated in BD (Hashimoto 2010). Patients with the BDNF polymorphism gene (resulting in deficient BDNF activity) had the smallest hippocampal volumes compared to patients homozygous for the gene and nonpsychiatric controls (Chepenik et al. 2009). Other data demonstrate decreased volume of the ACC in patients carrying the polymorphism gene vs. homozygotes (Matsuo et al. 2009). It has also been suggested that dysfunctional BDNF activity prevents neuroplasticity and malformation of specific cortical and subcortical regions, given that the BDNF polymorphism gene is associated with childhood, adolescent, and early onset BD (Inal-Emiroglu et al. 2015). Decreased peripheral BDNF levels have been observed during both episodes of depression and mania (de Oliveira et al. 2009; Fernandes et al. 2009, 2011; Machado-Vieira et al. 2007). Finally, there exist data to support the notion that BDNF levels improve with pharmacotherapy (Dwivedi and Zhang 2015; Yasuda et al. 2009). Of interest, ECT—one of the few known effective options for treatment resistant depression in BD—increases BDNF in multiple brain areas and, further, chronic convulsive stimuli can generate regrowth of neurons.

Levels of NT-3 and NT-4/5 are increased in manic and depressive episodes (Loch et al. 2015; Rybakowski et al. 2013; Walz et al. 2007). Loch and colleagues specifically investigated depressed patients with BD who were medication free at the time of study entry. Their baseline levels of NT-3 and NT-4/5 were elevated compared to healthy controls, and these levels did not change significantly after six weeks of lithium treatment (Loch et al. 2015).

GDNF has been investigated in BD with mixed results. Some studies reported a decrease of this neurotrophin in BD patients during manic and depressive phases (Takebayashi et al. 2006; Zhang et al. 2010), while others reported an increase or no difference compared to euthymic patients or healthy controls (Otsuki et al. 2008; Rosa et al. 2006; Rybakowski et al. 2013; Tunca et al. 2015).

VEGF is not a neurotrophic factor, but a growth factor involved in the development and proliferation of blood vessels. It is believed that VEGF is implicated in CNS diseases through its impact on cerebral vasculature (Scola and Andreazza 2015). In BD, there is evidence to support increased peripheral levels of VEGF (Lee and Kim 2012; Shibata et al. 2013) and a role for VEGF in response to treatments specific to bipolar depression. The association of VEGF and BD can be linked to our earlier discussion of inflammation and cardiovascular comorbidity. Vasculogenesis and endothelial dysfunction are implicated in the pathophysiology of cardiovascular disease (with a role for both VEGF and BDNF), and the hope is that future research will elucidate how these changes in growth factors are involved in the pathogenesis of both these physical and mental illnesses (Goldstein and Young 2013).

6.4   Neurobiology of BD: The Past, Present, and Future

Over 20 years ago, Post proposed the "kindling" hypothesis of BD (Post 1992). He acknowledged that psychosocial stressors are important as triggers for mood episodes in early BD and theorized that subsequent neuropathological and genetic changes could sensitize the brain, resulting in autonomously triggered mood episodes later in the illness. This work integrated biopsychosocial evidence available at the time and presented a comprehensive theory for the neurobiology of BD.

Stage	Clinical features	Biomarkers	Cognition	Maintenance treatment	Prognosis
Latent	At risk for developing BD, positive family history, mood or anxiety symptoms without criteria for threshold BD	Polymorphisms that confer susceptibility to BD	No impairment	↓ Exposure to pathogens	Good prognosis when protected from pathogens
I	Well-defined periods of euthymia without overt psychiatric symptoms	↑ TNF-α ↑ 3-nytrotyrosine	No impairment	Mood stabilizer monotherapy; psychoeducation	Good prognosis with careful prophylaxis
II	Symptoms in interepisodic periods related to comorbidities	↑ TNF-α ↓ BDNF ↑ 3-nytrotyrosine	Transient impairment	Combined treatment (pharmacotherapy + psychotherapy; focus on the treatment of comorbidities)	Prognosis depends on how well comorbidities can be managed. Worse than stage I
III	Marked impairment in cognition and functioning	Morphometric changes in brain may be present ↑ TNF-α ↓ BDNF ↑ 3-nytrotyrosine	Severe cognitive impairment associated with functioning impairment (unable to work or very impaired performance)	Complex regimens usually required; consider innovative strategies	Reserved prognosis; rescue therapy required
IV	Unable to live autonomously owing to cognitive and functional impairment	Ventricular enlargement and/or white matter hyperintensities ↑ TNF-α ↓ BDNF ↑ Glutathione reductase and transferase ↑ 3-nytrotyrosine	Cognitive impairment prevents patients from living independently	Palliative; daycare center	Poor prognosis

Table 1. Clinical staging in bipolar disorder.

BD: Bipolar disorder; BDNF: Brain-derived neurotrophic factor.

Fig. 6.3 Reprinted with permission from Kapczinski et al. (2009), Table 1

In the intervening years, however, advances in research techniques both in psychiatry and other fields have given rise to the staging model of BD (Kapczinski et al. 2014). In a similar fashion, this model aims to integrate the many dimensions of BD and outlines a series of progressive stages from a latent phase to the significantly symptomatic Stage IV, where patients have substantial functional, clinical, and cognitive impairments (Fig. 6.3). The model is presented as a means to target and predict response to treatment. Indeed, one common criticism in the field of psychiatry is that our diagnostic structure relies largely on patients' accounts of their subjective experience and collateral interpretations of behavior change, as well as our own clinical observations. This diagnostic process can lead to misdiagnosis and inefficient or inappropriate treatment. Undertreating bipolar depression has a significant impact on both patients and their support networks as sustained depressive symptoms can be incredibly debilitating.

Peripheral biomarkers are fundamental to the staging model, as the hope is that they can be used to identify the neuroprogression at different stages of BD. The concept of biomarkers is at the forefront of research now with a global acknowledgement of a need for biologically valid markers that can aid in early identification, diagnosis, and treatment of those with mental illness. The neurobiological pathways described in this chapter highlight some of the most promising areas for future development. Markers of oxidative stress, lipid peroxidation, protein carbonylation, inflammation, apoptosis, and neurotrophin status are all potential areas for investigation.

6.5 Summary

Knowledge about the neurobiology of bipolar depression is gleaned largely from what is known about the pathophysiology of BD in general. Cell loss and brain atrophy in cortical and limbic areas are consistent findings and specific to BD. Though the mechanisms underlying these morphological changes are not fully understood, promising hypotheses include oxidative stress induced by mitochondrial dysfunction and dysregulated inflammatory processes. These potential mechanisms are thought to lead to an increase in apoptosis. Neurotrophins are also implicated in the pathophysiology of BD with links to both oxidative stress and inflammation. The future of this field will depend on research with a specific focus on the depressive phase of BD as well as the study of biomarkers to aid clinicians in the diagnosis and targeted treatment of bipolar depression.

References

Adleman NE, Fromm SJ, Razdan V, Kayser R, Dickstein DP, Brotman MA, Pine DS, Leibenluft E (2012) Cross-sectional and longitudinal abnormalities in brain structure in children with severe mood dysregulation or bipolar disorder. J Child Psychol Psychiatry 53(11):1149–1156

Andreazza AC, Cassini C, Rosa AR, Leite MC, de Almeida LMV, Nardin P, Cunha ABN, Cereser KM, Santin A, Gottfried C, Salvador M, Kapczinski F, Goncalves CA (2007a) Serum S100B and antioxidant enzymes in bipolar patients. J Psychiatr Res 41(6):523–529

Andreazza AC, Noronha Frey B, Erdtmann B, Salvador M, Rombaldi F, Santin A, Goncalves CA, Kapczinski F (2007b) DNA damage in bipolar disorder. Psychiatry Res 153(1):27–32

Angst F, Stassen HH, Clayton PJ, Angst J (2002) Mortality of patients with mood disorders: follow-up over 34–38 years. J Affect Dis 68(2–3):167–181

Banerjee U, Dasgupta A, Rout JK, Singh OP (2012) Effects of lithium therapy on Na+–K+–ATPase activity and lipid peroxidation in bipolar disorder. Prog Neuropsychopharmacol Biol Psychiatry 37(1):56–61

Barbiroli B, Montagna P, Martinelli P, Lodi R, Iotti S, Cortelli P, Funicello R, Zaniol P (1993) Defective brain energy metabolism shown by in vivo 31P MR spectroscopy in 28 patients with mitochondrial cytopathies. J Cereb Blood Flow Metab 13(3):469–474

Beech RD, Lowthert L, Leffert JJ, Mason PN, Taylor MM, Umlauf S, Lin A, Lee JY, Maloney K, Muralidharan A, Lorberg B, Zhao H, Newton SS, Mane S, Epperson CN, Sinha R, Blumberg H, Bhagwagar Z (2010) Increased peripheral blood expression of electron transport chain genes in bipolar depression. Bipolar Disord 12(8):813–824

Bengesser SA, Lackner N, Birner A, Fellendorf FT (2015) Peripheral markers of oxidative stress and antioxidative defense in euthymia of bipolar disorder—Gender and obesity effects. J Affect Disord 172:367–374

Berk M, Copolov DL, Dean O, Lu K, Jeavons S, Schapkaitz I, Anderson-Hunt M, Bush AI (2008) N-acetyl cysteine for depressive symptoms in bipolar disorder—a double-blind randomized placebo-controlled trial. Biol Psychiatry 64(6):468–475

Berk M, Dean O, Cotton SM, Gama CS, Kapczinski F, Fernandes BS, Kohlmann K, Jeavons S, Hewitt K, Allwang C, Cobb H, Bush AI, Schapkaitz I, Dodd S, Malhi GS (2011) The efficacy of N-acetylcysteine as an adjunctive treatment in bipolar depression: an open label trial. J Affect Disord 135(1–3):389–394

Bielau H, Trübner K, Krell D, Agelink MW, Bernstein HG, Stauch R, Mawrin C, Danos P, Gerhard L, Bogerts B, Baumann B (2005) Volume deficits of subcortical nuclei in mood disorders. Eur Arch Psychiatry Clin Neurosci 255(6):401–412

Birkenaes AB, Opjordsmoen S, Brunborg C, Engh JA, Jonsdottir H, Ringen PA, Simonsen C, Vaskinn A, Birkeland KI, Friis S, Sundet K, Andreassen OA (2007) The level of cardiovascular risk factors in bipolar disorder equals that of schizophrenia: a comparative study. J Clin Psychiatry 68(6):917–923

Boufidou F, Nikolaou C, Alevizos B, Liappas IA, Christodoulou GN (2004) Cytokine production in bipolar affective disorder patients under lithium treatment. J Affect Disord 82(2):309–313

Buttner N, Bhattacharyya S, Walsh J, Benes FM (2007) DNA fragmentation is increased in non-GABAergic neurons in bipolar disorder but not in schizophrenia. Schizophr Res 93 (1–3):33–41

Byne W, Tatusov A, Yiannoulos G, Vong GS, Marcus S (2008) Effects of mental illness and aging in two thalamic nuclei. Schizophr Res 106(2–3):172–181

Campbell S, Macqueen G (2004) The role of the hippocampus in the pathophysiology of major depression. J Psychiatry Neurosci 29(6):417–426

Can M, Güven B, Atik L, Konuk N (2011) Lipid peroxidation and serum antioxidant enzymes activity in patients with bipolar and major depressive disorders. J Mood Disord 1(1):14

Cataldo AM, McPhie DL, Lange NT, Punzell S, Elmiligy S, Ye NZ, Froimowitz MP, Hassinger LC, Menesale EB, Sargent LW, Logan DJ, Carpenter AE, Cohen BM (2010) Abnormalities in mitochondrial structure in cells from patients with bipolar disorder. Am J Pathol 177 (2):575–585

Chaturvedi RK, Flint Beal M (2013) Mitochondrial diseases of the brain. Free Radic Biol Med 63:1–29

Chepenik LG, Fredericks C, Papademetris X, Spencer L, Lacadie C, Wang F, Pittman B, Duncan JS, Staib LH, Duman RS, Gelernter J, Blumberg HP (2009) Effects of the brain-derived neurotrophic growth factor val66met variation on hippocampus morphology in bipolar disorder. Neuropsychopharmacology 34(4):944–951

Cunha ÂB, Andreazza AC, Gomes FA, Frey BN, Silveira LE, Gonçalves CA, Kapczinski F (2008) Investigation of serum high-sensitive C-reactive protein levels across all mood states in bipolar disorder. Eur Arch Psychiatry Clin Neurosci 258(5):300–304

D'Addario C, Dell'Osso B, Palazzo MC, Benatti B, Lietti L, Cattaneo E, Galimberti D, Fenoglio C, Cortini F, Scarpini E, Arosio B, Di Francesco A, Di Benedetto M, Romualdi P, Candeletti S, Mari D, Bergamaschini L, Bresolin N, Maccarrone M, Altamura AC (2012) Selective DNA methylation of BDNF promoter in bipolar disorder: differences among patients with BDI and BDII. Neuropsychopharmacology 37(7):1647–1655

Dargel AA, Godin O, Kapczinski F, Kupfer DJ, Leboyer M (2015) C-reactive protein alterations in bipolar disorder: a meta-analysis. J Clin Psychiatry 76(2):142–150

De Berardis D, Conti CM, Campanella D, Carano A, Scali M, Valchera A, Serroni N, Pizzorno AM, D'Albenzio A, Fulcheri M, Gambi F, La Rovere R, Cotellessa C, Salerno RM, Ferro FM (2008) Evaluation of C-reactive protein and total serum cholesterol in adult patients with bipolar disorder. Int J Immunopathol Pharmacol 21(2):319–324

de Oliveira GS, Ceresér KM, Fernandes BS, Kauer-Sant'Anna M, Fries GR, Stertz L, Aguiar B, Pfaffenseller B, Kapczinski F (2009) Decreased brain-derived neurotrophic factor in medicated and drug-free bipolar patients. J Psychiatr Res 43(14):1171–1174

Dell'Osso B, D'Addario C, Carlotta Palazzo M, Benatti B, Camuri G, Galimberti D, Fenoglio C, Scarpini E, Di Francesco A, Maccarrone M, Altamura AC (2014) Epigenetic modulation of BDNF gene: differences in DNA methylation between unipolar and bipolar patients. J Affect Disord 166:330–333

Dickerson F, Stallings C, Origoni A, Boronow J, Yolken R (2007) Elevated serum levels of C-reactive protein are associated with mania symptoms in outpatients with bipolar disorder. Prog Neuropsychopharmacol Biol Psychiatry 31(4):952–955

Dickerson F, Stallings C, Origoni A, Vaughan C, Khushalani S, Yolken R (2013) Elevated C-reactive protein and cognitive deficits in individuals with bipolar disorder. J Affect Disord 150(2):456–459

Dwivedi T, Zhang H (2015) Lithium-induced neuroprotection is associated with epigenetic modification of specific BDNF gene promoter and altered expression of apoptotic-regulatory proteins. Front Neurosci 8:457

Eleff SM, Barker PB, Blackband SJ, Chatham JC, Lutz NW, Johns DR, Bryan RN, Hurko O (1990) Phosphorus magnetic resonance spectroscopy of patients with mitochondrial cytopathies demonstrates decreased levels of brain phosphocreatine. Ann Neurol 27 (6):626–630

Elvsåshagen T, Vera E, Bøen E, Bratlie J, Andreassen OA, Josefsen D, Malt UF, Blasco MA, Boye B (2011) The load of short telomeres is increased and associated with lifetime number of depressive episodes in bipolar II disorder. J Affect Disord 135(1–3):43–50

Fernandes BS, Gama CS, Kauer-Sant'Anna M, Lobato MI, Belmonte-de-Abreu P, Kapczinski F (2009) Serum brain-derived neurotrophic factor in bipolar and unipolar depression: a potential adjunctive tool for differential diagnosis. J Psychiatr Res 43(15):1200–1204

Fernandes BS, Gama CS, Maria C"eserér K, Yatham LN, Fries GR, Colpo G, de Lucena D, Kunz M, Gomes FA, Kapczinski F (2011) Brain-derived neurotrophic factor as a state-marker of mood episodes in bipolar disorders: a systematic review and meta-regression analysis. J Psychiatr Res 45(8):995–1004

Foland-Ross LC, Brooks JO, Mintz J, Bartzokis G, Townsend J, Thompson PM, Altshuler LL (2012) Mood-state effects on amygdala volume in bipolar disorder. J Affect Disord 139 (3):298–301

Gawryluk JW, Wang J-F, Andreazza AC, Shao L, Young LT (2011) Decreased levels of gluta-thione, the major brain antioxidant, in post-mortem prefrontal cortex from patients with psychiatric disorders. Int J Neuropsychopharmacol 14(1):123–130

Ghaznavi S, Deckersbach T (2012) Rumination in bipolar disorder: evidence for an unquiet mind. Biol Mood Anxiety Disord 2(1):2

Gigante AD, Young LT, Yatham LN, Andreazza AC, Nery FG, Grinberg LT, Heinsen H, Lafer B (2011) Morphometric post-mortem studies in bipolar disorder: possible association with oxidative stress and apoptosis. Int J Neuropsychopharmacol 14(8):1075–1089

Goldstein B, Young LT (2013) Toward clinically applicable biomarkers in bipolar disorder: focus on BDNF, inflammatory markers, and endothelial function. Curr Psychiatry Rep 15(12):1–7

Goldstein BI, Collinger KA, Lotrich F, Marsland AL, Gill M-K, Axelson DA, Birmaher B (2011) Preliminary findings regarding proinflammatory markers and brain-derived neurotrophic factor among adolescents with bipolar spectrum disorders. J Child Adolesc Psychopharmacol 21 (5):479–484

Hartberg CB, Jørgensen KN, Haukvik UK, Westlye LT, Melle I, Andreassen OA, Agartz I (2015) Lithium treatment and hippocampal subfields and amygdala volumes in bipolar disorder. Bipolar Disord 17(5):496–506

Hashimoto K (2010) Brain-derived neurotrophic factor as a biomarker for mood disorders: an historical overview and future direction. Psychiatry Clin Neurosci 64(4):341–357

Himmerich H, Bartsch S, Hamer H, Mergl R, Schonherr J, Petersein C, Munzer A, Kirkby KC, Bauer K, Sack U (2014) Modulation of cytokine production by drugs with antiepileptic or mood stabilizer properties in anti-CD3- and anti-Cd40-stimulated blood in vitro. Oxid Med Cell Longev 2014:806162

Hirayasu Y, Shenton ME, Salisbury DF, Kwon JS et al (1999) Subgenual cingulate cortex volume in first-episode psychosis. Am J Psychiatry 156(7):1091–1093

Huzayyin AA, Andreazza AC, Turecki G, Cruceanu C, Rouleau GA, Alda M, Young LT (2014) Decreased global methylation in patients with bipolar disorder who respond to lithium. Int J Neuropsychopharmacol 17(4):561–569

Inal-Emiroglu FN, Karabay N, Resmi H, Guleryuz H (2015) Correlations between amygdala volumes and serum levels of BDNF and NGF as a neurobiological marker in adolescents with bipolar disorder. J Affect Disord 182:50–56

Janak PH, Tye KM (2015) From circuits to behaviour in the amygdala. Nature 517(7534):284–292

Kapczinski F, Frey BN, Andreazza AC, Kauer-Sant'Anna M, Cunha AB, Post RM (2008) Increased oxidative stress as a mechanism for decreased BDNF levels in acute manic episodes. Rev Bras Psiquiatr 30(3):243–245

Kapczinski F, Dias VV, Kauer-Sant'Anna M, Frey BN, Grassi-Oliveira R, Colom F, Berk M (2009) Clinical implications of a staging model for bipolar disorders. Expert Rev Neurother 9 (7):957–966

Kapczinski F, Dal-Pizzol F, Teixeira AL, Magalhaes PVS, Kauer-Sant'Anna M, Klamt F, Moreira JCF, Augusto de Bittencourt Pasquali M, Fries GR, Quevedo J, Gama CS, Post R (2011) Peripheral biomarkers and illness activity in bipolar disorder. J Psychiatr Res 45(2):156–161

Kapczinski F, Magalhães PVS, Balanzá-Martinez V, Dias VV, Frangou S, Gama CS, Gonzalez-Pinto A, Grande I, Ha K, Kauer-Sant'Anna M, Kunz M, Kupka R, Leboyer M, Lopez-Jaramillo C, Post RM, Rybakowski JK, Scott J, Strejilevitch S, Tohen M, Vazquez G, Yatham L, Vieta E, Berk M (2014) Staging systems in bipolar disorder: an International Society for Bipolar Disorders Task Force Report. Acta Psychiatr Scand 130(5):354–363

Kato T (2007) Mitochondrial dysfunction as the molecular basis of bipolar disorder: therapeutic implications. CNS Drugs 21(1):1–11

Kato T, Takahashi S, Shioiri T, Inubushi T (1992) Brain phosphorous metabolism in depressive disorders detected by phosphorus-31 magnetic resonance spectroscopy. J Affect Disord 26 (4):223–230

Kato T, Takahashi S, Shioiri T, Murashita J, Hamakawa H, Inubushi T (1994) Reduction of brain phosphocreatine in bipolar II disorder detected by phosphorus-31 magnetic resonance spectroscopy. J Affect Disord 31(2):125–133

Kato T, Shioiri T, Murashita J, Hamakawa H, Takahashi Y, Inubushi T, Takahashi S (1995) Lateralized abnormality of high energy phosphate metabolism in the frontal lobes of patients with bipolar disorder detected by phase-encoded 31P-MRS. Psychol Med 25(03):557–566

Kim H-W, Rapoport SI, Rao JS (2010) Altered expression of apoptotic factors and synaptic markers in postmortem brain from bipolar disorder patients. Neurobiol Dis 37(3):596–603

Kittel-Schneider S, Wobrock T, Scherk H, Schneider-Axmann T, Trost S, Zilles D, Wolf C, Schmitt A, Malchow B, Hasan A, Backens M, Reith W, Falkai P, Gruber O, Reif A (2015) Influence of DGKH variants on amygdala volume in patients with bipolar affective disorder and schizophrenia. Eur Arch Psychiatry Clin Neurosci 265(2):127–136

Knijff EM, Nadine Breunis M, Kupka RW, de Wit HJ, Ruwhof C, Akkerhuis GW, Nolen WA, Drexhage HA (2007) An imbalance in the production of IL-1ß and IL-6 by monocytes of bipolar patients: restoration by lithium treatment. Bipolar Disord 9(7):743–753

Kuloglu M, Ustundag B, Atmaca M, Canatan H, Tezcan AE, Cinkilinc N (2002) Lipid peroxidation and antioxidant enzyme levels in patients with schizophrenia and bipolar disorder. Cell Biochem Funct 20(2):171–175

Kunz M, Gama CS, Andreazza AC, Salvador M, Ceresér KM, Gomes FA, Belmonte-de-Abreu PS, Berk M, Kapczinski F (2008) Elevated serum superoxide dismutase and thiobarbituric acid reactive substances in different phases of bipolar disorder and in schizophrenia. Prog Neuropsychopharmacol Biol Psychiatry 32(7):1677–1681

Lee B-H, Kim Y-K (2012) Increased plasma VEGF levels in major depressive or manic episodes in patients with mood disorders. J Affect Disord 136(1–2):181–184

Lima IMM, Barros A, Rosa DV, Albuquerque M (2015) Analysis of telomere attrition in bipolar disorder. J Affect Disord 172:43–47

Lindqvist D, Epel ES, Mellon SH, Penninx BW, Révész D, Verhoeven JE, Reus VI, Lin J, Mahan L, Hough CM, Rosser R, Saverio Bersani F, Blackburn EH, Wolkowitz OM (2015) Psychiatric disorders and leukocyte telomere length: underlying mechanisms linking mental illness with cellular aging. Neurosci Biobehav Rev 55:333–364

Liu L, Schulz SC, Lee S, Reutiman TJ, Fatemi SH (2007) Hippocampal CA1 pyramidal cell size is reduced in bipolar disorder. Cell Mol Neurobiol 27(3):351–358

Loch AA, Zanetti MV, de Sousa RT, Chaim TM, Serpa MH, Gattaz WF, Teixeira AL, Machado-Vieira R (2015) Elevated neurotrophin-3 and neurotrophin 4/5 levels in unmedicated bipolar depression and the effects of lithium. Prog Neuropsychopharmacol Biol Psychiatry 56:243–246

López-Armada MJ, Riveiro-Naveira RR, Vaamonde-García C, Valcárcel-Ares MN (2013) Mitochondrial dysfunction and the inflammatory response. Mitochondrion 13:106–118

MacDonald ML, Naydenov A, Chu M, Matzilevich D, Konradi C (2006) Decrease in creatine kinase messenger RNA expression in the hippocampus and dorsolateral prefrontal cortex in bipolar disorder. Bipolar Disord 8(3):255–264

Machado-Vieira R, Dietrich MO, Leke R, Cereser VH, Zanatto V, Kapczinski F, Souza DO, Portela LV, Gentil V (2007) Decreased plasma brain derived neurotrophic factor levels in unmedicated bipolar patients during manic episode. Biol Psychiatry 61(2):142–144

Magalhães PV, Dean OM, Bush AI, Copolov DL, Malhi GS, Kohlmann K, Jeavons S, Schapkaitz I, Anderson-Hunt M, Berk M (2011) N-acetyl cysteine add-on treatment for bipolar II disorder: a subgroup analysis of a randomized placebo-controlled trial. J Affect Disord 129 (1–3):317–320

Matsuo K, Walss-bass C, Nery FG, Nicoletti MA, Hatch JP, Frey BN, Monkul ES, Zunta-soares GB, Bowden CL, Escamilla MA, Soares JC (2009) Neuronal correlates of brain-derived neurotrophic factor Val66Met polymorphism and morphometric abnormalities in bipolar disorder. Neuropsychopharmacology 34(8):1904–1913

Modabbernia A, Taslimi S, Brietzke E, Ashrafi M (2013) Cytokine alterations in bipolar disorder: a meta-analysis of 30 studies. Biol Psychiatry 74(1):15–25

Moniczewski A, Gawlik M, Smaga I, Niedzielska E, Krzek J, Przegaliński E, Pera J, Filip M (2015) Oxidative stress as an etiological factor and a potential treatment target of psychiatric disorders. Part 1. Chemical aspects and biological sources of oxidative stress in the brain. Pharmacol Rep 67(3):560–568

Moore CM, Christensen JD, Lafer B, Fava M et al (1997) Lower levels of nucleoside triphosphate in the basal ganglia of depressed subjects: a phosphorous-31 magnetic resonance spectroscopy study. Am J Psychiatry 154(1):116–118

Moore GJ, Cortese BM, Glitz DA, Zajac-Benitez C, Quiroz JA, Uhde TW, Drevets WC, Manji HK (2009) A longitudinal study of the effects of lithium treatment on prefrontal and subgenual prefrontal gray matter volume in treatment-responsive bipolar disorder patients. J Clin Psychiatry 70(5):699–705

Morawetz C, Bode S, Baudewig J, Kirilina E, Heekeren HR (2016) Changes in effective connectivity between dorsal and ventral prefrontal regions moderate emotion regulation. Cereb Cortex 26(5):1923–1937

Müller N, Schwarz MJ (2007) The immune-mediated alteration of serotonin and glutamate: towards an integrated view of depression. Mol Psychiatry 12(11):988–1000

Munkholm K, Bräuner JV, Kessing LV, Vinberg M (2013) Cytokines in bipolar disorder vs. healthy control subjects: a systematic review and meta-analysis. J Psychiatr Res 47 (9):1119–1133

Munkholm K, Poulsen HE, Kessing LV, Vinberg M (2015) Elevated levels of urinary markers of oxidatively generated DNA and RNA damage in bipolar disorder. Bipolar Disord 17 (3):257–268

Naydenov AV, MacDonald ML, Ongur D, Konradi C (2007) Differences in lymphocyte electron transport gene expression levels between subjects with bipolar disorder and normal controls in response to glucose deprivation stress. Arch Gen Psychiatry 64(5):555–564

Nugent AC, Diazgranados N, Carlson PJ, Ibrahim L, Luckenbaugh DA, Brutsche N, Herscovitch P, Drevets WC, Zarate CA (2014) Neural correlates of rapid antidepressant response to ketamine in bipolar disorder. Bipolar Disord 16(2):119–128

Ortiz-Domínguez A, Hernández ME, Berlanga C, Gutiérrez-Mora D, Moreno J, Heinze G, Pavón L (2007) Immune variations in bipolar disorder: phasic differences. Bipolar Disord 9 (6):596–602

Otsuki K, Uchida S, Watanuki T, Wakabayashi Y, Fujimoto M, Matsubara T, Funato H, Watanabe Y (2008) Altered expression of neurotrophic factors in patients with major depression. J Psychiatr Res 42(14):1145–1153

Palmier-Claus JE, Dodd A, Tai S, Emsley R, Mansell W (2015) Appraisals to affect: testing the integrative cognitive model of bipolar disorder. Br J Clin Psychol 2015

Petersein C, Sack U, Mergl R, Schonherr J, Schmidt F, Lichtblau N, Kirkby K, Bauer K, Himmerich H (2015) Impact of lithium alone and in combination with antidepressants on cytokine production in vitro. J Neural Transm 122(1):109–122

Post RM (1992) Transduction of psychosocial stress into the neurobiology of recurrent affective disorder. Am J Psychiatry 149(8):999–1010

Radaelli D, Sferrazza Papa G, Vai B, Poletti S, Smeraldi E, Colombo C, Benedetti F (2015) Fronto-limbic disconnection in bipolar disorder. Eur Psychiatry 30(1):82–88

Rao JS, Harry GJ, Rapoport SI, Kim HW (2010) Increased excitotoxicity and neuroinflammatory markers in postmortem frontal cortex from bipolar disorder patients. Mol Psychiatry 15 (4):384–392

Regenold WT, Phatak P, Marano CM, Sassan A, Conley RR, Kling MA (2009) Elevated cerebrospinal fluid lactate concentrations in patients with bipolar disorder and schizophrenia: implications for the mitochondrial dysfunction hypothesis. Biol Psychiatry 65(6):489–494

Rosa AR, Frey BN, Andreazza AC, Cereser KM, Cunha ABM, Quevedo J, Santin A, Gottfried C, Goncalves CA, Vieta E, Kapczinski F (2006) Increased serum glial cell line-derived neurotrophic factor immunocontent during manic and depressive episodes in individuals with bipolar disorder. Neurosci Lett 407(2):146–150

Rubinsztein JS, Michael A, Underwood BR, Tempest M, Sahakian BJ (2006) Impaired cognition and decision-making in bipolar depression but no 'affective bias' evident. Psychol Med 36 (5):629–639

Rybakowski JK, Permoda-Osip A, Skibinska M, Adamski R, Bartkowska-Sniatkowska A (2013) Single ketamine infusion in bipolar depression resistant to antidepressants: are neurotrophins involved? Hum Psychopharmacol 28(1):87–90

Salzman CD, Fusi S (2010) Emotion, cognition, and mental state representation in amygdala and prefrontal cortex. Ann Rev Neurosci 33:173–202

Scola G, Andreazza AC (2015) The role of neurotrophins in bipolar disorder. Prog Neuropsychopharmacol Biol Psychiatry 56:122–128

Selek S, Savas HA, Gergerlioglu HS, Bulbul F, Uz E, Yumru M (2008) The course of nitric oxide and superoxide dismutase during treatment of bipolar depressive episode. J Affect Disord 107 (1–3):89–94

Shibata T, Yamagata H, Uchida S, Otsuki K, Hobara T, Higuchi F, Abe N, Watanabe Y (2013) The alteration of hypoxia inducible factor-1 (HIF-1) and its target genes in mood disorder patients. Prog Neuropsychopharmacol Biol Psychiatry 43:222–229

Singh MK, Chang KD, Kelley RG, Saggar M, Reiss AL, Gotlib IH (2014) Early signs of anomalous neural functional connectivity in healthy offspring of parents with bipolar disorder. Bipolar Disord 16(7):678–689

Soczynska JK, Kennedy SH, Chow CSM, Woldeyohannes HO, Konarski JZ, McIntyre RS (2008) Acetyl-L-carnitine and a-lipoic acid: possible neurotherapeutic agents for mood disorders? Expert Opin Investig Drugs 17(6):827–843

Soeiro-de-Souza MG, Andreazza AC, Carvalho AF, Machado-Vieira R, Young LT, Moreno RA (2013) Number of manic episodes is associated with elevated DNA oxidation in bipolar I disorder. Int J Neuropsychopharmacol 16(7):1505–1512

Stork C, Renshaw PF (2005) Mitochondrial dysfunction in bipolar disorder: evidence from magnetic resonance spectroscopy research. Mol Psychiatry 10(10):900–919

Su SC, Sun MT, Wen MJ, Lin CJ, Chen YC, Hung YJ (2011) Brain-derived neurotrophic factor, adiponectin, and proinflammatory markers in various subtypes of depression in young men. Int J Psychiatry Med 42(3):211–226

Sun X, Jun-Feng W, Tseng M, Young LT (2006) Downregulation in components of the mito-chondrial electron transport chain in the postmortem frontal cortex of subjects with bipolar disorder. J Psychiatr Neurosci 31(3):189–196

Takebayashi M, Hisaoka K, Nishida A, Tsuchioka M, Miyoshi I, Kozuru T, Hikasa S, Okamoto Y, Shinno H, Morinobu S, Yamawaki S (2006) Decreased levels of whole blood glial cell line-derived neurotrophic factor (GDNF) in remitted patients with mood disorders. Int J Neuropsychopharmacol 9(5):607–612

Tang V, Wang J-F (2013) Oxidative stress in bipolar disorder. Biochem Anal Biochem 2(2):1–8

Taylor MA, Abrams R (1987) Cognitive impairment patterns in schizophrenia and affective disorder. J Neurol Neurosurg Psychiatry 50(7):895–899

Tunca Z, Kivircik Akdede B, Özerdem A, Alkin T, Polat S, Ceylan D, Bayin M, Cengizçetin Kocuk N, Simsek S, Resmi H, Akan P (2015) Diverse glial cell line-derived neurotrophic factor (GDNF) support between mania and schizophrenia: a comparative study in four major psychiatric disorders. Eur Psychiatry 30(2):198–204

Uranova N, Orlovskaya D, Vikhreva O, Zimina I, Kolomeets N, Vostrikov V, Rachmanova V (2001) Electron microscopy of oligodendroglia in severe mental illness. Brain Res Bull 55 (5):597–610

Versace A, Andreazza AC, Young LT, Fournier JC, Almeida JR, Stiffler RS, Lockovich JC, Aslam HA, Pollock MH, Park H, Nimgaonkar VL, Kupfer DJ, Phillips ML (2014) Elevated serum measures of lipid peroxidation and abnormal prefrontal white matter in euthymic bipolar adults: toward peripheral biomarkers of bipolar disorder. Mol Psychiatry 19(2):200–208

Wake H, Moorhouse AJ, Miyamoto A, Nabekura J (2013) Microglia: actively surveying and shaping neuronal circuit structure and function. Trends Neurosci 36(4):209–217

Walz JC, Andreazza AC, Frey BN, Cacilhas AA, Cereser KMM, Cunha ABM, Weyne F, Stertz L, Santin A, Goncalves CA, Kapczinski F (2007) Serum neurotrophin-3 is increased during manic and depressive episodes in bipolar disorder. Neurosci Lett 415(1):87–89

Wang JF (2007) Defects of mitochondrial electron transport chain in bipolar disorder: implications for mood-stabilizing treatment. Can J Psychiatry 52(12):753–762

Wang JF, Young LT (2009) Understanding the neurobiology of bipolar depression. In: Zarate CA, Manji HK (eds) Bipolar depression: molecular neurobiology, clinical diagnosis and pharma-cotherapy. Milestones in drug therapy. Birkhäuser Verlag, Basel, pp 77–94

Wang JF, Shao L, Sun X, Young LT (2009) Increased oxidative stress in the anterior cingulate cortex of subjects with bipolar disorder and schizophrenia. Bipolar Disord 11(5):523–529

Washizuka S, Kakiuchi C, Mori K, Tajima O, Akiyama T, Kato T (2005) Expression of mitochondria-related genes in lymphoblastoid cells from patients with bipolar disorder. Bipo-lar Disord 7(2):146–152

Yasuda S, Liang MH, Marinova Z, Yahyavi A, Chuang DM (2009) The mood stabilizers lithium and valproate selectively activate the promoter IV of brain-derived neurotrophic factor in neurons. Mol Psychiatry 14(1):51–59

Zhang X, Zhang Z, Sha W, Xie C, Xi G, Zhou H, Zhang Y (2010) Effect of treatment on serum glial cell line-derived neurotrophic factor in bipolar patients. J Affect Disord 126 (1–2):326–329

Zhou R, Yazdi AS, Menu P, Tschopp J (2011) A role for mitochondria in NLRP3 inflammasome activation. Nature 469(7329):221–225

Zunszain PA, Anacker C, Cattaneo A, Carvalho LA, Pariante CM (2011) Glucocorticoids, cytokines and brain abnormalities in depression. Prog Neuropsychopharmacol Biol Psychiatry 35(3):722–729

Chapter 7
Chronotherapeutics in Bipolar and Major Depressive Disorders: Implications for Novel Therapeutics

Wallace C. Duncan Jr.

Abstract Chronotherapeutic interventions (CTs) produce rapid antidepressant effects and, when applied sequentially, maintain an enduring antidepressant response following the initial CT intervention or in association with traditional drug therapies. Rapid antidepressant effects associated with CTs (sleep deprivation (SD), partial sleep deprivation (PSD), sleep phase advance (SPA)) and with the novel therapeutic ketamine are present in both major depressive disorder (MDD) and bipolar disorder (BD). The effects of the N-methyl-D-aspartate (NMDA) antagonist ketamine and CTs on sleep slow waves (SWS), brain-derived neurotrophic factor (BDNF), cortical excitability, and neuronal plasticity are present in MDD. Whether slow wave effects are also present in drug-free BD or healthy controls requires further investigation. The existing literature suggests that there are important differences in MDD versus BD patients in the regulation of SWS that may underlie diagnostic differences in slow wave response to ketamine. These differences further suggest that mood stabilizers may affect the expression of slow waves and moderation of mood cycles in BD.

Keywords Bipolar disorder • Sleep deprivation • Chronotherapeutics • Ketamine • BDNF

7.1 Introduction

Chronotherapeutic (CT) interventions rapidly activate antidepressant mechanisms that potentially converge with pathways implicated in novel treatments for major depressive (MDD) and bipolar (BD) disorders. Identifying biological mechanisms common to both CT and other novel, rapid-acting treatments is critical for developing more effective antidepressant treatments. Desired benefits of both treatment

W.C. Duncan Jr., PhD (✉)
Experimental Therapeutics & Pathophysiology Branch, Intramural Research Program, National Institute of Mental Health, National Institutes of Health, Bethesda, MD 20892, USA
e-mail: wduncan@mail.nih.gov

© Springer International Publishing Switzerland 2016
C.A. Zarate Jr., H.K. Manji (eds.), *Bipolar Depression: Molecular Neurobiology, Clinical Diagnosis, and Pharmacotherapy*, Milestones in Drug Therapy, DOI 10.1007/978-3-319-31689-5_7

approaches are rapid and enduring relief of depressive symptoms. Since different circadian clock (Partonen 2012) and sleep profiles (de Maertelaer et al. 1987; Duncan et al. 1979; Fossion et al. 1998; Giles et al. 1986) characterize MDD and BD patient groups, one might predict that rapid antidepressant responses to CTs and other novel treatments, and associated biological markers, might similarly distinguish the two groups. Consistent with clinical differences between MDD and BD, the N-methyl-D-aspartate (NMDA) antagonist ketamine has different sleep slow wave (SWS) effects in MDD and BD, although rapidly acting antidepressant effects are present in both patient groups. The possible biological and pharmacological basis of this observation is explored in this chapter.

The fact that CTs alter sleep levels and circadian timing, as well as rapidly elevate mood, suggests links between rapid-acting antidepressant treatments, the sleep homeostat, and the circadian clock. Gene linkage studies indicate that associations between clock genes, gene products, and MDD/BD are complex. Numerous clinical and preclinical studies indicate associations between circadian genes and these disorders (Etain et al. 2014; Novakova et al. 2015; McClung et al. 2005; McClung 2011; Hampp et al. 2008). A recent survey of controlled mood disorder genome-wide association studies (GWAS) (Partonen 2012) identified four clock genes associated with MDD (*RORA, CRY1*) or BD (*RORB, NR1D1*) and two with seasonal affective disorder (*NPAS2, CRY2*). A survey of clock gene networks common to bipolar spectrum illnesses and lithium-responsive genes found enriched association of core clock and clock-controlled genes (McCarthy et al. 2012). In addition, an examination of reward circuitry showed that the NMDA antagonist ketamine—which has been shown to have rapid and robust antidepressant effects in both MDD and BD—activated clock-related genes in the nucleus accumbens (Zhao et al. 2014). Identifying clock and sleep-related pathways related to CT benefits may provide important clues that spur the development of novel treatments. Notably, lithium's effects on the molecular signaling pathways of the circadian system appear to be associated with its mood stabilizing properties, inspiring efforts to identify novel interventions. The continuing discovery of key genetic and molecular elements of the circadian clock and their interaction with the sleep homeostat will clarify mechanisms of established nonpharmacological and drug-based therapies as well as the development of new treatment regimens.

This chapter will highlight the role of sleep homeostatic and circadian factors that form the basis of CT interventions to treat MDD and BD. The key CT interventions—including sleep deprivation (SD), partial sleep deprivation (PSD), sleep phase advance (SPA), bright-light therapy (BLT), and dark therapy (DT)— are briefly summarized. Importantly, the extension of rapid antidepressant effects with repeated (sequenced) chronotherapy is discussed as it applies to prolonging the initial clinical benefits of CTs, traditional antidepressants, or novel agents. This chapter will then summarize the results of studies using the novel therapeutic agent ketamine to examine the interaction between neuroplasticity-associated SWS production and mood in MDD. These MDD studies are compared with results from BD patients receiving mood stabilizers and discussed in the context of EEG sleep findings in unmedicated BD patients. Finally, the role of sleep homeostasis and clock gene effects on ketamine's rapid antidepressant properties is discussed.

7.2 Homeostatic and Circadian Contributions to Rapid Antidepressant Interventions

The S-deficiency hypothesis of depression (Borbely and Wirz-Justice 1982; Borbely 1987) proposes that sleep disturbances and depressive symptoms are both related to disrupted mechanisms of sleep homeostasis. The key components of the two process model (Borbely 1982) of human sleep–wake regulation are Process S, a homeostatic hourglass-like mechanism that tracks the duration of prior sleep and wakefulness, and Process C, a self-sustaining circadian timekeeper that provides internal synchrony to multiple secondary oscillators that comprise the circadian system, as well as synchronizing the internal milieu with the external environment. This two process system regulates the timing, duration, and structure of human sleep. In healthy persons, extended prior wakefulness is associated with a homeostatic regulated increase in SWS/slow wave activity (SWA) during recovery sleep, an important consequence of SD therapy associated with mood response.

In mood disorders, both homeostatic and circadian processes are dysregulated, with major impact on sleep–wake cycles, circadian rhythms of body temperature, hormones, behavior, and mood. As discussed later, patients with typical depression exhibit disturbed sleep characterized by decreased sleep and sleep continuity, low levels of SWS or SWA, short REM latency, and increased and early morning waking. Many of these disturbances are consistent with abnormal function of the sleep homeostat (SWS deficiency), as well as the central clock (temperature and hormone rhythms).

Normalization of deficient levels of S (indirectly measured in EEG sleep studies by SWS or SWA in depressed persons) by extended wakefulness is thought to be linked with next-day rapid antidepressant effects (Borbely and Wirz-Justice 1982). The S-deficiency hypothesis has served as a theoretical framework for manipulating sleep and circadian timing and consequent effects on mood. Relapse after recovery sleep has also been attributed to reset the sleep homeostat, thus reestablishing a depressogenic state. Shortened or phase-shifted sleep interventions, such as PSD or SPA, have been evaluated for their capacity to alter homeostatic mechanisms, as described above, as well as their ability to affect interactions between sleep (Process S) and the circadian system (Process C). The framework of this hypothesis has recently provided structure for investigating potential molecular mechanisms of SD and its rapid antidepressant effects, as well as synaptic homeostasis (Tononi and Cirelli 2006).

7.3 Chronotherapeutics Affect Rapid and Durable Mood Response

As summarized below, CTs have rapid antidepressant effects when applied as a single intervention. When applied sequentially or in combination with conventional antidepressants or subsequent CT interventions, they can extend mood benefits.

7.3.1 Basic CTs

Rapid relief of depressive symptoms is a common feature of CT interventions. Few side effects are associated with CTs with the exception of fatigue and the urge to sleep during extended wakefulness. CT response latencies range from hours (SD, PSD) to several days (SPA, BLT). In addition, DT, related "extended-night" interventions (Wehr et al. 1998; Wirz-Justice et al. 1999; Barbini et al. 2005), or blue-wavelength blocking glasses (van der Lely et al. 2015; Phelps 2008) have all been found to control symptoms of mania, hypomania, and rapid cycling in BD (Wehr et al. 1998; Wirz-Justice et al. 1999; Barbini et al. 2005; van der Lely et al. 2015; Phelps 2008). DT has also been used to treat the symptoms of schizoaffective disorder (Gomez-Bernal 2009).

Numerous experimental interventions have explored the antidepressant mechanism(s) of EEG sleep stage-specific or phase-dependent CT interventions (e.g., selective SWS or REM SD, early versus late PSD, SPA); a detailed discussion of these interventions is beyond the scope of this chapter. Some of these experimental interventions, in addition to clarifying underlying mechanisms, have become useful clinical tools (e.g., PSD, SPA). Also, additional clinician-mediated interventions with chronobiological components have been used to treat depression in MDD and BD; these include Cognitive Behavioral Therapy-Insomnia (CBT-I) (Kaplan and Harvey 2013; Soehner et al. 2013; Morin et al. 2007; Spielman et al. 1987) and Interpersonal Social Rhythm Therapy (IPSRT) (Frank et al. 2000).

7.3.2 Sequential Chronotherapies (Nonpharmacological)

Sequential CT interventions have been used to extend treatment benefits (Benedetti et al. 2001; Berger et al. 1997). Importantly, combined CTs (e.g., total SD followed by SPA and BLT interventions (Sahlem et al. 2014)) rapidly improve mood and reduce suicidality, thus providing rapid relief to this vulnerable patient population.

Sequentially applied CTs capture and extend the initial rapid antidepressant effects of SD, thereby activating or extending the mechanisms that provide rapid and enduring relief. These sequential interventions may activate response pathways that are unique to the durable response, target different symptom clusters (thus amplifying clinical benefits), or may maintain activation of the initial target pathway. Identifying the biological substrates of the durable response is critical to extending the benefits of CTs and novel therapeutics.

The fact that relapses are prevented by repeated SPA (Benedetti et al. 2001; Berger et al. 1997) indicates that sleep is not sufficient for relapse, but is, nevertheless, a major mediator. The specific factors are not clear but may be sleep stage or circadian phase dependent. At the molecular level, the glycogen synthase kinase (GSK) promotor variant (rs334558*C) interacts with the long/short form of the serotonin transporter (5-HTTLPR) 5HT allele to extend a post-SD antidepressant response after recovery sleep (Benedetti et al. 2012), suggesting a molecular target for therapies designed to prolong response. As discussed later, both flumazenil-associated sleep–wake homeostatic effect (Hemmeter et al. 2007) and ketamine-induced brain derived neurotrophic factor (BDNF) release (Haile et al. 2014) are also associated with prolonged response.

7.3.3 Chronotherapeutic Augmentation of Conventional Antidepressant Treatments

CTs augment the effects of antidepressant medications. When a conventional antidepressant and a mood stabilizer (sertraline and lithium) were combined with three CT interventions (SD + BL + SPA), BD patients showed fewer depressive symptoms within 48 hours, with effects persisting for seven weeks (Wu et al. 2009). In a study of 39 mixed BD patients (13 were treatment-resistant and 23 were medicated (14 were receiving lithium and nine were receiving an SSRI)), three SD interventions augmented with morning BLT were associated with a 50 % reduction in Hamilton Depression Rating Scale (HAM-D) scores in 26 patients (Benedetti et al. 2007). In a third study, the use of a mood stabilizer (lithium) combined with SD and BLT alleviated suicidal ideation and mood symptoms (Benedetti et al. 2014). The interactions between CTs with electroconvulsive therapy (ECT) and repetitive transcranial magnetic stimulation (rTMS) have also been explored. One study found that rTMS prolonged the effects of SD (Eichhammer et al. 2002), but a second study found that active rTMS was not superior to sham rTMS in stabilizing the antidepressant effects of SD (Kreuzer et al. 2012).

7.4 Ketamine as a Prototype for Investigating the Chronobiological Mechanisms of Novel Therapeutics

The patterns of disrupted sleep and circadian rhythms in depression, the effects of the mood stabilizer lithium on the circadian clock, and the rapid effects of SD on both sleep homeostasis and mood combine to form a network for exploring the mechanistic links between CTs and the development of novel rapid antidepressant treatments. The exploration of lithium's effects on circadian function, extensively reviewed elsewhere (Gould and Manji 2002; Lenox et al. 2002), illustrates the potential benefits. Lithium's inhibition of GSK3β (Gould and Manji 2005) and subsequent delay of the circadian clock via reduced period (PER) phosphorylation and delayed nuclear entry of PER and cryptochrome (CRY) proteins following lithium treatment are associated with delayed timing of the central clock. The stability of interacting clock gene feedback loops may then contribute to lithium's mood stabilizing properties. The enriched presence of lithium-responsive clock genes within downstream and clock-controlled gene pathways (McCarthy et al. 2012) is consistent with lithium's functional role in stabilizing feedback loops. To further understand the homeostatic process and how it relates to the development of novel therapeutics, recent investigations have focused on molecular and SWS associations with the rapid antidepressant effects of the NMDA antagonist ketamine.

7.4.1 Ketamine and the Molecular Elements of Neuroplasticity and Sleep

The rapid antidepressant properties of ketamine are mediated via NMDA receptor blockade and by altered glutamatergic signaling downstream of the NMDA blockade, thus resulting in increased synaptic strength and plasticity (Duman and Aghajanian 2012). The sequence of events that leads from ketamine-induced NMDA receptor blockade to rapid improvement in mood, cognition, and behavior includes: (a) blockade of NMDA receptor-induced firing of gamma aminobutyric acid (GABA)ergic interneurons, (b) disinhibition of glutamatergic pyramidal cells (Moghaddam et al. 1997), (c) increased glutamate release and activation of α-amino-3-hydroxy-5-methyl-4-isoxazolepropionic acid (AMPA) receptors, and (d) activity-dependent release of BDNF (Li et al. 2010; Maeng and Zarate 2007). Changes in glutamatergic transmission activate the mammalian target of rapamycin (mTOR) signaling pathway and affect downstream changes in dendritic spines and local synaptic protein synthesis, including BDNF (Duman and Aghajanian 2012). BDNF secretion, activation of the tropomyosin-receptor-kinase B (TrkB) receptor, and downstream trafficking lead to further dendritic structural complexity, spine and BDNF synthesis, and synaptic plasticity. Polymorphisms such as BDNF

Val66Met (Chen et al. 2004; Egan et al. 2003) alter the functional effects of BDNF trafficking. Ultimately, changes in critical local neuronal circuits converge via enhanced synaptic plasticity and neuronal synchronization, especially in areas involved in mood and behavior, to produce rapid antidepressant effects (Maeng and Zarate 2007; Zarate et al. 2006). Many steps in this signaling cascade reverse the effects of ketamine in various rodent models of depression. For instance, synaptic spine production associated with ketamine use is blocked by rapamycin, which prevents activation of the mTOR pathway (Li et al. 2010; Maeng and Zarate 2007), and biological and behavioral markers of chronic unpredictable stress (decreased spine density, depressive-like behavior) are known to be reversed with ketamine (Li et al. 2011). Anisomycin-induced inhibition of rapid (30 minutes) BDNF synthesis after ketamine treatment prevented the long-term behavioral effects of ketamine measured with the forced swim test (Autry et al. 2011). It is also possible that blockade of spontaneous (rather than evoked) glutamatergic activation of the NMDA receptor blockade—which is critical to ketamine's mechanism of action (Autry et al. 2011)—inactivates eukaryotic elongation factor 2 (eEF2) kinase, and consequently blunts eFF2 phosphorylation and de-suppresses BDNF (Kavalali and Monteggia 2012), but additional studies are required to clarify these pathways.

Sleep EEG and evoked potentials are interpreted as markers of altered synaptic plasticity in humans (Huber et al. 2004, 2006), consistent with synaptic homeostasis (Tononi and Cirelli 2006). High-density EEG studies have shown that interventions such as rotation learning and high-frequency TMS associated with synaptic potentiation in local cortical circuits lead to local increases in SWA during subsequent sleep (Huber et al. 2004). Interventions such as arm immobilization, which is associated with synaptic depression, lead to a local reduction in SWA (Huber et al. 2006). Computer simulations indicate that sleep SWA directly reflects synaptic strength due to changes in neural synchronization and recruitment (Esser et al. 2007; Vyazovskiy et al. 2007). Several studies that directly examined the effect of BDNF on EEG SWS also noted a close relationship between SWA and BDNF (Faraguna et al. 2008; Huber et al. 2007). These studies found that SWA was increased by intrahemispheric infusion of BDNF and by behavioral interventions that increase central levels of BDNF (Huber et al. 2007), as well as the plasticity-related genes *Arc*, *Homer*, and *NGFI-A* (Huber et al. 2007), but was diminished by BDNF antagonism (Faraguna et al. 2008). Acoustic suppression of SWA activity and its capacity to diminish perceptual learning (Aeschbach et al. 2008) suggest that decreased SWS levels may contribute to cognitive and memory deficits in some depressed patients. Finally, reduced production of SWS is present in human carriers of the BDNF Met allele of the Val66Met polymorphism (Bachmann et al. 2012), thus establishing another link between BDNF, SWA, and mood.

7.5 Clinical Effects of Ketamine on Sleep Slow Waves and Mood

Based on the above-described associations between SWS, cognition, and memory, and ketamine's capacity to increase slow waves (Campbell and Feinberg 1996; Feinberg and Campbell 1993), it would not be unexpected that ketamine's sleep effects could provide a link between mood and cognition. Below, I review ketamine's effects on mood and sleep in MDD and BD.

7.5.1 Major Depressive Disorder

Rapid antidepressant response to ketamine is related to decreased waking, as well as increased total sleep, SWS, SWA, and REM sleep (Fig. 7.1, left panels) (Duncan et al. 2013b). Interestingly, clinical response to ketamine was predicted by a low baseline delta sleep ratio, a measure of deficient early night production of SWS (Duncan et al. 2013a). Normalizing the deficient early night production of SWS by increasing early production of SWS (Fig. 7.1, top left panel) appears to be associated with ketamine's rapid antidepressant effects.

The effects of other novel therapeutics also extend to SWS effects. Similar to ketamine's effects, the effects of rTMS on plasticity (Cohen et al. 2010) are also associated with both sleep–wake and/or circadian-dependent processes. Applying rTMS to the left dorsolateral prefrontal cortex (DLPFC) increases SWA at F3, possibly reflecting locally enhanced synaptic plasticity, similar to the early night effects of ketamine on SWS. Increased production of SWA was more evident during the first half of the sleep period, similar to ketamine's effects (see Fig. 7.1, top left panel), suggesting homeostatic regulation of SWA (Saeki et al. 2013).

Interestingly, studies using ketamine show links between neuroplasticity, SWS, and mood on the one hand and the neurotrophin BDNF on the other hand. Responders to ketamine infusion showed a correlation between ketamine-induced SWS production and BDNF levels (Duncan et al. 2013b), a finding consistent with preclinical studies of BDNF on SWS production (Huber et al. 2007). Increased BDNF levels are also associated with fast-acting antidepressant interventions (Duncan et al. 2013b; Giese et al. 2014; Haile et al. 2014; Gorgulu and Caliyurt 2009).

The relationship between BDNF, mood, and sleep has a genetic basis. A BDNF polymorphism is associated with mood response to ketamine (Laje et al. 2012), as well as with SWS production (Bachmann et al. 2012). Markers of BDNF activity also predict subsequent response to ketamine. BDNF levels at 240 minutes post-ketamine infusion predicted Montgomery–Asberg Depression Rating Scale (MADRS) scores up to 72 hours post-ketamine infusion (Haile et al. 2014). Further, a diurnal rhythm for BDNF is present in SD responders but not in nonresponders (Giese et al. 2014).

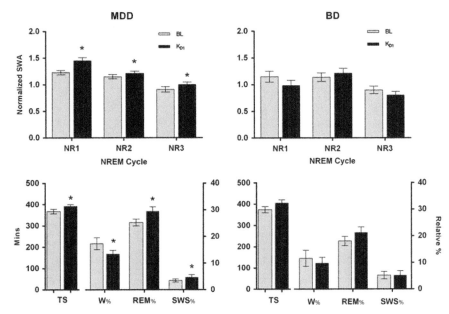

Fig. 7.1 Slow wave activity (SWA, *top*) during non-REM cycles 1–3 (NR1, 2, 3) and selected sleep measures (*bottom*) from patients with major depressive disorder (MDD) (*left panels*, $n = 30$) and bipolar disorder (BD) (*right panels*, $n = 15$) were compared during baseline (BL, *gray bars*) and the first night (D1) after ketamine infusion (K_{D1}, *black bars*) (* denotes $p < .05$ versus the baseline mean). All patients were treatment resistant with severe depression at the time of ketamine treatment. BD patients were receiving maintenance mood stabilizers (Lithium, $n = 11$; valproate, $n = 4$). The *lower left panel* shows that in MDD patients, ketamine significantly improved sleep quality (increased Total Sleep (TS), sleep slow waves (SWS), REM%, and decreased Waking% (W%) on D1 (*left*)). The *lower right panel* shows that with the exception of SWS, which slightly decreased after ketamine, BD patients also showed improved sleep quality as measured by TS, W, and REM%. In MDD (*top left panel*), ketamine increased early SWA during NR1, thus enhancing the nighttime decline in SWA across successive non-REM cycles. Similar effects were not apparent in BD (*top right panel*) due to diagnostic differences in homeostatic regulation or to the effects of mood stabilizers on sleep homeostasis and SWS

7.5.2   Bipolar Disorder

In contrast to MDD, increased production of SWS by ketamine was not observed in a cohort of BD patients maintained on mood stabilizers (Fig. 7.1, top right panel). Although rapid antidepressant effects were observed (Duncan and Zarate 2013), SWA levels in BD declined post-ketamine specifically during non-REM cycles 1 and 3 (Fig. 7.1, top right panel). There are several possible explanations for this finding. First, although some studies suggest that baseline SWS levels are similar between depressed BD and MDD patients (Table 7.1), SWS expression may be differently altered by state-dependent waking history (Linkowski et al. 1986), REM sleep pressure (Kupfer and Ehlers 1989), or genetic factors. Second, mood stabilizers such as lithium affect SWS and may, therefore, alter ketamine-induced slow

Table 7.1 Drug-free EEG sleep studies in bipolar disorder (BD) during depressed (D), manic (M), and remitted (R) states relative to major depressive disorder (MDD), bipolar depressed, or healthy control groups

Mood state D/M/R	Homeostatic marker SWS	REM markers RD	RL	REM (%, m)	Sleep continuity A	SE	TS	CNTRL GRP	Comments Contrasts (BD vs MDD) or (BD vs HC)	Reference
D	↔		↓*	↔	↑*			MDD	BD-I (14), BD-II (14), MDD (14); *trend for BD-I vs MDD and BD-II	Fossion et al. (1998)
D	↔↑	↔	↔↑	↔↑	↔↑	↔↑	↔↑	MDD	REM efficiency: BD < MDD	Duncan et al. (1979)
D	↔	↔	←	↔(m)		↔	←	MDD	(BD-I, BD-II) vs MDD; RL, TS higher in BD-II than MDD	Giles et al. (1986)
D	↔	↔	↔	↔(%)	↔		↔	MDD	BD (11) vs MDD (8); fewer sleep spindles in MDD than BD	de Maertelaer et al. (1987)
D	↔	←	→	↔(%)	↔	→	↔	MDD	BD (10) vs MDD (14)	Lauer et al. (1992)
D	↓*	→	←	→	↔			MDD	BD (5) vs MDD (19) vs HC (35); Adolescent BD conversion from MDD (28); ↓SWS linked to BD conversion, *p < .07	Rao et al. (2002)
D	↔		↔	↔(m)	↔	↔		HC	BD (10) vs HC (10); mild depression	Jernajczyk (1986)
D	↔	↔	↔	↑(m)	↔			HC	BD-I (7), BD-II (19) vs HC (26); anergic BD; REM minutes trended up	Thase et al. (1989)
D	↔↑	↔	↓↑	↔↑(m)	↔↑	↓↑	↔↑	HC	BD (22); HC (36); MDD (36); F ANOVA: MDD × BD × HC, $p < .05$	Duncan et al. (1979)
D	→	↔	↔	↓(m)	←	→	→	HC	BD (60) vs HC (200); age-, sex-matched	Jovanovic (1977)
M	↔	↔	↔	↔	↔	↔	→	MDD	Daytime naps may have affected night sleep	Hudson et al. (1992)

								MDD	Trend for reduced SWS in MDD compared with M and HC	
M	↑*	↑*	↑	↔	↔	↔	↔	↔		Trend for reduced SWS in MDD compared with M and HC	Linkowski et al. (1986)
M	↑*	↔	↔	↔	↔	↔	↔	↔	D	BD: M(6) vs D (6) vs MDD (6) vs HC (6)	Linkowski et al. (1986)
M	↓	↓	↑	↓	↑	↑	↑	↔	HC	M (19), MDD (19), HC (19); age-matched, young populations; Stage 3 SWS%: M > HC	Hudson et al. (1992)
M	↓	↓	↑	↔	↔	↔	↔	↔	HC	M (9) vs HC (9)	Hudson et al. (1988)
R	↔	↓	↔	↓	↔	↔	↔	↔	HC	Remitted BD (14) vs HC (15)	Sitaram et al. (1982)
R	↔	↔	↔	↓	↔	↔	↔	↔	HC	Remitted BD (10) vs HC (10) age-matched; number of arousals ↑ BD	Knowles et al. (1986)

D depressed; *M* mania, hypomania; *E/R* euthymic or remitted; *SWS* slow wave sleep; *SWA* slow wave activity; *RD* REM density; *RL* REM latency; *REM* (%, m = minutes); *A* awake (minutes); *SE* sleep efficiency; *TS* total sleep; *HC* healthy control; *BD* bipolar disorder; *MDD* major depressive disorder

* = trend: $p < 0.1$; (F) = ANOVA

wave production, a possibility discussed in greater detail in the next section. This suggests that the capacity of mood stabilizers to affect homeostatic production of SWS is linked to their clinical benefits as mood stabilizers.

If a blunting of the ketamine-induced SWS is associated with mood stabilizer treatment (as suggested by ketamine's effects in BD), this finding may be in accord with the report that SWS levels correlate with the intensity of subsequent mania (Eidelman et al. 2010) such that low SWS levels are associated with reduced future mania. Thus, a property of mood stabilizers might be their ability to minimize night to night fluctuations in SWS. This property would likely be an effect shared by dark or bed-rest interventions, i.e., CT interventions that impose fixed, extended bed-rest schedules, thus controlling the duration of prior wakefulness and minimizing night-to-night fluctuations in SWS. In the following section, drug-free EEG sleep studies in BD depression and mania are compared with MDD and healthy control groups while focusing on differences in sleep continuity, REM sleep, and SWS differences.

7.6 Disrupted Circadian and Sleep Wake Patterns in Bipolar Disorder

BD is characterized by episodic mood cycles as well as disrupted circadian patterns in mood, sleep, energy, and appetite that present both research and treatment challenges. While mood cycles vary in duration, the presence of abnormal circadian patterns of body temperature, sleep–wake, and hormones provides evidence of abnormal regulation of core and associated circadian clock mechanisms. Treatment of BD is complicated by emerging mood cycles, by the limited success of traditional antidepressant medications in treating the disorder, and by treatment resistance, relapse, or emerging mania that result in dynamic changes in sleep during treatment. The observations that episodes of mania are associated with reduced need for sleep and amount of sleep, and that episodes of depression exhibit complex patterns of sleep disruption, complicate research efforts to accurately measure SWS and functional aspects of the homeostat in BD. Further, pathological function of the sleep homeostat itself might directly contribute to the pathology of the illness.

SWS, REM sleep, and sleep continuity exhibit increased variability in both MDD and BD (Harvey 2008; Soreca 2014). Unraveling the sources of this variability in BD is challenging. Relatively few drug-free EEG sleep studies are available to inform the dynamic relationships that exist between sleep homeostasis and change of affective state in BD (Table 7.1). SWS markers of sleep homeostasis are confounded by daytime sleep. Findings in a large sample of manic BD patients indicated that 10 % of waking clinical EEG samples exhibited micro-sleep episodes (Small et al. 1999). Gender differences as well as differences in underlying opponent circadian processes may also contribute to the pattern of SWS expression (Frey et al. 2012a, b; Goldschmied et al. 2014).

Table 7.1—which lists several potential markers of sleep homeostasis—summarizes the drug-free, controlled EEG sleep studies conducted in depressed, manic, and remitted BD patients in order to explore possible relationships between selected sleep measures, mood, and diagnostic groups. Control populations within each study (either healthy control, mood state control, or MDD control) are indicated.

7.6.1 Bipolar Depression Versus Controls

Four of 10 EEG sleep studies of bipolar depression reported decreased sleep continuity (either low sleep efficiency and total sleep or increased time awake) relative to control populations (two versus MDD controls and three versus healthy controls); five reported no difference and one found increased total sleep versus an MDD group (Table 7.1). Five of 10 studies reported abnormal REM measures (decreased REM latency, time spent in REM sleep, or REM density) relative to controls (four versus MDD controls and one versus healthy controls); the remaining five studies found no difference. Finally, two out of 10 studies identified decreased SWS in BD depression compared with healthy and MDD control groups. Taken together, the results suggest that relative to healthy control or MDD controls, disturbances of sleep continuity and REM sleep are more common in BD depression and remission than abnormal SWS levels.

7.6.2 Bipolar Mania Versus Controls

Five studies compared EEG sleep during mania relative to either MDD, bipolar depression, or a healthy control group. Decreased sleep continuity was present in three of five studies, consistent with clinical observations. REM sleep measures often distinguished mania from depression in BD. Three studies found decreased REM latency (two versus an MDD control group), and three studies found increased REM density (similar to a case study (Gillin et al. 1977)) in manic patients versus an MDD control group.

SWS differences between BD depression and mania are variable and remarkable in that they are rather minor. One report of six subjects studied in both depressed and manic phases found no difference in SWS levels between states (Linkowski et al. 1986). A second report found few SWS differences relative to healthy controls (Hudson et al. 1988). A third age-controlled study found similar SWS in young manic patients relative to MDD. In contrast, Stage 3 sleep was elevated in individuals with MDD versus healthy controls (Hudson et al. 1992). In remitted patients relative to healthy controls, there was evidence for persisting REM sleep differences (elevated REM density and REM %) as well as increased waking.

In contrast to the studies listed in Table 7.1 in which EEG sleep was measured during brief, mood state-specific samples, two studies that examined dynamic state changes in BD implicated SWS in the progression of mania. In a study of latent BD, low SWS was present in adolescent MDD patients who later converted to BD (Rao et al. 2002), suggesting that low SWS is a trait-like feature of the illness. In a second study conducted in medicated patients, SWS and REM density measures correlated with the severity of future manic and depressive symptoms, respectively (Eidelman et al. 2010). SWS levels correlated with the future severity of manic symptoms, such that low SWS levels were followed by less severe symptoms, thus associating with the clinical course of BD to later affect state-like features. Whether a restorative function of SWS contributes to the level of mania in BD is unknown. Mood modulation is clearer in the case of MDD, when increased slow wave production after SD is associated with improved mood. As discussed previously in the context of ketamine's effects on slow wave production, drugs or CT interventions (such as dark therapy) that buffer day-to-day variations in SWS also contribute toward stabilizing mood. Such mechanisms may be related to the capacity of mood stabilizers to buffer ketamine-induced generation of SWS in BD.

7.7 Association of Clock Gene and Sleep Homeostat Function with Rapid and Durable Antidepressant Response

Identification of biomarkers of a rapid and durable response to CTs, as well as existing novel therapeutics, is important for developing new and more effective treatments. The CTs associated with rapid antidepressant mechanisms implicate molecular elements associated with the two process model, such as SWS-associated neuroplasticity, as well as circadian clock genes that govern the regulation of the circadian system. The fact that sleep homeostasis and clock genes interact at the molecular level (Franken 2013; Franken and Dijk 2009) allows for the possibility that this interaction might affect clinical response.

Clock genes and their associated molecules (Bunney and Bunney 2012, 2013), as well as SWS-associated neural plasticity (Duncan et al. 2013b), have been linked to ketamine's rapid antidepressant effects. The temporal relationship between specific clock genes versus sleep homeostatic mechanisms can inform our understanding of how these processes contribute to rapid antidepressant response. However, linking chronobiological mechanisms of effective rapid treatments requires examination of the day-to-day relationship between mood change on the one hand and changes in clock genes and sleep homeostasis on the other. If sleep homeostasis and the circadian clock are associated with the (rapid) antidepressant effects of CTs and novel antidepressants, markers associated with these processes would be closely linked in time with rapid response. In the case of sleep homeostasis, SWS-related changes in neural plasticity and rapid antidepressant response occur within 12 hours post-ketamine infusion (Duncan et al. 2013b; Duncan and Zarate 2013).

While conventional antidepressant drug treatments (e.g., SSRIs) require weeks for clinical benefit, preclinical studies of these drugs nevertheless indicate that they rapidly alter clock genes as well as markers of clock input, suggesting that such clock effects are not sufficient for rapid mood effects. In contrast to conventional antidepressant treatments, CT interventions and ketamine rapidly increase SWS production, indicating that this rapid increase in SWS is an essential part of the rapid mood response. This suggests a key role of SWS-associated processes, such as neurotrophin release and neuronal plasticity, with the rapid-acting effects of SD and novel therapeutics such as ketamine. Subsequent to the initial events that trigger the response, it is possible that the interaction between homeostatic processes and clock gene-associated events is important for sustaining antidepressant response.

Consistent with BDNF's association with effective rapid antidepressant interventions, ketamine's effects on BDNF are associated with rapid mood response (Duncan et al. 2013b; Duncan and Zarate 2013), and the magnitude of BDNF and SWS effects is correlated in ketamine responders (Duncan et al. 2013b). Further, BDNF levels four hours post-ketamine infusion predicted mood response three days post-infusion (Haile et al. 2014). The presence of a diurnal rhythm of BDNF in PSD responders links the circadian system to other rapidly acting treatment interventions (Giese et al. 2014). Similar to SD and the effects of sleep homeostasis on clock gene expression (for a review, see Franken and Dijk (2009)), other novel therapeutics (such as rTMS) are also linked to clock and circadian function. Plasticity associated with rTMS was found to correlate with cortisol awakening response, a circadian biomarker possibly regulated by peripheral circadian *CLOCK* genes (Clow et al. 2014). Like other conventional antidepressants, rTMS increased REM latency, a circadian phase marker suggesting a link between rTMS antidepressant effects and circadian rhythms (Cohrs et al. 1998).

While their role in mediating rapid response and relapse are not fully understood, naps and recovery sleep appear to both diminish rapid antidepressant effects and promote relapse; in contrast, extending prior wakefulness is a necessary factor for inducing and maintaining rapid-acting effects. Despite the capacity for naps to blunt rapid antidepressant response (Dallaspezia and Benedetti 2011), pharmacological treatments that reduce microsleeps/naps (i.e., modafinil, caffeine, flumazenil) do not substantially enhance the rapid antidepressant response associated with SD (Beck et al. 2010; Hemmeter et al. 2007). Still, a modafinil case report found clinical benefit by preventing daytime naps (Even et al. 2005), and the benzodiazepine receptor antagonist flumazenil prolonged rapid antidepressant response to SD on the day after recovery sleep (Hemmeter et al. 2007). In addition, while few responders (5–10 %) remain euthymic after recovery sleep (Benedetti et al. 1999; Benedetti and Colombo 2011), the fact that a rapid SD response can be extended by sequential CT interventions that allow sleep (PSD, SPA, BLT) (Berger et al. 1997) indicates that recovery sleep per se is not sufficient for relapse. Rather, the use of altered sleep schedules and BLT to affect circadian timekeeping and to prolong prior waking facilitates both the rapid and durable features of the antidepressant response. Because PSD, SPA, and BLT alter sleep timing and the duration of prior wakefulness, a durable response may be related to an interaction between sleep–wake and circadian system networks. A potential mechanism of rapid and durable response is the association of prolonged wakefulness with altered glutamatergic function. Glutamatergic changes are similar after treatments that have rapid antidepressant properties such as total SD and ECT (Murck et al. 2009). SD also increases the availability of metabotropic glutamate receptors (Hefti et al. 2013) and alters glutamatergic function (Benedetti et al. 2009). Increased prior wake time is associated with increased cortical excitability (Huber et al. 2013), as with ketamine (Cornwell et al. 2012). It is not known whether the antidepressant effects of sequentially applied CTs (such as PSD or SPA) also maintain cortical excitability or whether ketamine-associated cortical excitability might also be maintained by CTs.

7.8   Summary

CTs produce rapid antidepressant effects and, when applied sequentially, maintain an enduring antidepressant response following the initial CT intervention or in association with traditional drug therapies. Rapid antidepressant effects of CTs (SD, PSD, SPA), and the novel therapeutic ketamine, are present in both MDD and BD. The effects of ketamine and CTs on SWS, BDNF, cortical excitability, and neuronal plasticity are present in MDD. Whether slow wave effects are also present in drug-free BD or healthy controls requires further investigation. The existing literature suggests that there are important differences in MDD versus BD patients in the regulation of SWS that may underlie diagnostic differences in slow wave response to ketamine. These differences further suggest that mood stabilizers may affect the expression of slow waves and moderation of mood cycles in BD. As research in this promising area moves forward, continued efforts are required to determine the contribution of clock genes, neurotrophins, and cortical excitability to the rapid-acting and durable antidepressant benefits of novel therapeutics.

Acknowledgements This work was funded by the Intramural Research Program, National Institute of Mental Health, National Institutes of Health (IRP-NIMH-NIH).

References

Aeschbach D, Cutler AJ, Ronda JM (2008) A role for non-rapid-eye-movement sleep homeostasis in perceptual learning. J Neurosci 28(11):2766–2772

Autry AE, Adachi M, Nosyreva E, Na ES, Los MF, Cheng PF, Kavalali ET, Monteggia LM (2011) NMDA receptor blockade at rest triggers rapid behavioural antidepressant responses. Nature 475(7354):91–95

Bachmann V, Klein C, Bodenmann S, Schafer N, Berger W, Brugger P, Landolt HP (2012) The BDNF Val66Met polymorphism modulates sleep intensity: EEG frequency- and state-specificity. Sleep 35(3):335–344

Barbini B, Benedetti F, Colombo C, Dotoli D, Bernasconi A, Cigala-Fulgosi M, Florita M, Smeraldi E (2005) Dark therapy for mania: a pilot study. Bipolar Disord 7(1):98–101

Beck J, Hemmeter U, Brand S, Muheim F, Hatzinger M, Holsboer-Trachsler E (2010) Modafinil reduces microsleep during partial sleep deprivation in depressed patients. J Psychiatr Res 44 (13):853–864

Benedetti F, Colombo C (2011) Sleep deprivation in mood disorders. Neuropsychobiology 64 (3):141–151

Benedetti F, Colombo C, Barbini B, Campori E, Smeraldi E (1999) Ongoing lithium treatment prevents relapse after total sleep deprivation. J Clin Psychopharmacol 19(3):240–245

Benedetti F, Barbini B, Campori E, Fulgosi MC, Pontiggia A, Colombo C (2001) Sleep phase advance and lithium to sustain the antidepressant effect of total sleep deprivation in bipolar depression: new findings supporting the internal coincidence model? J Psychiatr Res 35 (6):323–329

Benedetti F, Dallaspezia S, Fulgosi MC, Barbini B, Colombo C, Smeraldi E (2007) Phase advance is an actimetric correlate of antidepressant response to sleep deprivation and light therapy in bipolar depression. Chronobiol Int 24(5):921–937

Benedetti F, Calabrese G, Bernasconi A, Cadioli M, Colombo C, Dallaspezia S, Falini A, Radaelli D, Scotti G, Smeraldi E (2009) Spectroscopic correlates of antidepressant response to sleep deprivation and light therapy: a 3.0 Tesla study of bipolar depression. Psychiatry Res 173(3):238–242

Benedetti F, Dallaspezia S, Lorenzi C, Pirovano A, Radaelli D, Locatelli C, Poletti S, Colombo C, Smeraldi E (2012) Gene-gene interaction of glycogen synthase kinase 3-beta and serotonin transporter on human antidepressant response to sleep deprivation. J Affect Disord 136 (3):514–519

Benedetti F, Riccaboni R, Locatelli C, Poletti S, Dallaspezia S, Colombo C (2014) Rapid treatment response of suicidal symptoms to lithium, sleep deprivation, and light therapy (chronotherapeutics) in drug-resistant bipolar depression. J Clin Psychiatry 75(2):133–140

Berger M, Vollmann J, Hohagen F, Konig A, Lohner H, Voderholzer U, Riemann D (1997) Sleep deprivation combined with consecutive sleep phase advance as a fast-acting therapy in depression: an open pilot trial in medicated and unmedicated patients. Am J Psychiatry 154 (6):870–872

Borbely A (1982) Two process model of sleep regulation. Hum Neurobiol 1:195–204

Borbely AA (1987) The S-deficiency hypothesis of depression and the two-process model of sleep regulation. Pharmacopsychiatry 20(1):23–29

Borbely AA, Wirz-Justice A (1982) Sleep, sleep deprivation and depression. A hypothesis derived from a model of sleep regulation. Hum Neurobiol 1(3):205–210

Bunney BG, Bunney WE (2012) Rapid-acting antidepressant strategies: mechanisms of action. Int J Neuropsychopharmacol 15(5):695–713

Bunney BG, Bunney WE (2013) Mechanisms of rapid antidepressant effects of sleep deprivation therapy: clock genes and circadian rhythms. Biol Psychiatry 73(12):1164–1171

Campbell IG, Feinberg I (1996) NREM delta stimulation following MK-801 is a response of sleep systems. J Neurophysiol 76(6):3714–3720

Chen ZY, Patel PD, Sant G, Meng CX, Teng KK, Hempstead BL, Lee FS (2004) Variant brain-derived neurotrophic factor (BDNF) (Met66) alters the intracellular trafficking and activity-dependent secretion of wild-type BDNF in neurosecretory cells and cortical neurons. J Neurosci 24(18):4401–4411

Clow A, Law R, Evans P, Vallence AM, Hodyl NA, Goldsworthy MR, Rothwell JR, Ridding MC (2014) Day differences in the cortisol awakening response predict day differences in synaptic plasticity in the brain. Stress 17(3):219–223

Cohen DA, Freitas C, Tormos JM, Oberman L, Eldaief M, Pascual-Leone A (2010) Enhancing plasticity through repeated rTMS sessions: the benefits of a night of sleep. Clin Neurophysiol 121(12):2159–2164

Cohrs S, Tergau F, Riech S, Kastner S, Paulus W, Ziemann U, Ruther E, Hajak G (1998) High-frequency repetitive transcranial magnetic stimulation delays rapid eye movement sleep. Neuroreport 9(15):3439–3443

Cornwell BR, Salvadore G, Furey M, Marquardt CA, Brutsche NE, Grillon C, Zarate CA Jr (2012) Synaptic potentiation is critical for rapid antidepressant response to ketamine in treatment-resistant major depression. Biol Psychiatry 72(7):555–561

Dallaspezia S, Benedetti F (2011) Chronobiological therapy for mood disorders. Expert Rev Neurother 11(7):961–970

de Maertelaer V, Hoffman G, Lemaire M, Mendlewicz J (1987) Sleep spindle activity changes in patients with affective disorders. Sleep 10(5):443–451

Duman RS, Aghajanian GK (2012) Synaptic dysfunction in depression: potential therapeutic targets. Science 338(6103):68–72

Duncan WC Jr, Zarate CA Jr (2013) Ketamine, sleep, and depression: current status and new questions. Curr Psychiatry Rep 15(9):394

Duncan WC Jr, Pettigrew KD, Gillin JC (1979) REM architecture changes in bipolar and unipolar depression. Am J Psychiatry 136(11):1424–1427

Duncan WC Jr, Selter J, Brutsche N, Sarasso S, Zarate CA Jr (2013a) Baseline delta sleep ratio predicts acute ketamine mood response in major depressive disorder. J Affect Disord 145 (1):115–119

Duncan WC, Sarasso S, Ferrarelli F, Selter J, Riedner BA, Hejazi NS, Yuan P, Brutsche N, Manji HK, Tononi G, Zarate CA (2013b) Concomitant BDNF and sleep slow wave changes indicate ketamine-induced plasticity in major depressive disorder. Int J Neuropsychopharmacol 16 (2):301–311

Egan MF, Kojima M, Callicott JH, Goldberg TE, Kolachana BS, Bertolino A, Zaitsev E, Gold B, Goldman D, Dean M, Lu B, Weinberger DR (2003) The BDNF val66met polymorphism affects activity-dependent secretion of BDNF and human memory and hippocampal function. Cell 112(2):257–269

Eichhammer P, Kharraz A, Wiegand R, Langguth B, Frick U, Aigner JM, Hajak G (2002) Sleep deprivation in depression stabilizing antidepressant effects by repetitive transcranial magnetic stimulation. Life Sci 70(15):1741–1749

Eidelman P, Talbot LS, Gruber J, Hairston I, Harvey AG (2010) Sleep architecture as correlate and predictor of symptoms and impairment in inter-episode bipolar disorder: taking on the challenge of medication effects. J Sleep Res 19(4):516–524

Esser SK, Hill SL, Tononi G (2007) Sleep homeostasis and cortical synchronization: I. Modeling the effects of synaptic strength on sleep slow waves. Sleep 30(12):1617–1630

Etain B, Jamain S, Milhiet V, Lajnef M, Boudebesse C, Dumaine A, Mathieu F, Gombert A, Ledudal K, Gard S, Kahn JP, Henry C, Boland A, Zelenika D, Lechner D, Lathrop M, Leboyer M, Bellivier F (2014) Association between circadian genes, bipolar disorders and chronotypes. Chronobiol Int 31(7):807–814

Even C, Thuile J, Santos J, Bourgin P (2005) Modafinil as an adjunctive treatment to sleep deprivation in depression. J Psychiatry Neurosci 30(6):432–433

Faraguna U, Vyazovskiy VV, Nelson AB, Tononi G, Cirelli C (2008) A causal role for brain-derived neurotrophic factor in the homeostatic regulation of sleep. J Neurosci 28 (15):4088–4095

Feinberg I, Campbell IG (1993) Ketamine administration during waking increases delta EEG intensity in rat sleep. Neuropsychopharmacology 9:41–48

Fossion P, Staner L, Dramaix M, Kempenaers C, Kerkhofs M, Hubain P, Verbanck P, Mendlewicz J, Linkowski P (1998) Does sleep EEG data distinguish between UP, BPI or BPII major depressions? An age and gender controlled study. J Affect Disord 49(3):181–187

Frank E, Swartz HA, Kupfer DJ (2000) Interpersonal and social rhythm therapy: managing the chaos of bipolar disorder. Biol Psychiatry 48(6):593–604

Franken P (2013) A role for clock genes in sleep homeostasis. Curr Opin Neurobiol 23(5):864–872

Franken P, Dijk DJ (2009) Circadian clock genes and sleep homeostasis. Eur J Neurosci 29 (9):1820–1829

Frey S, Birchler-Pedross A, Hofstetter M, Brunner P, Gotz T, Munch M, Blatter K, Knoblauch V, Wirz-Justice A, Cajochen C (2012a) Challenging the sleep homeostat: sleep in depression is not premature aging. Sleep Med 13(7):933–945

Frey S, Birchler-Pedross A, Hofstetter M, Brunner P, Gotz T, Munch M, Blatter K, Knoblauch V, Wirz-Justice A, Cajochen C (2012b) Young women with major depression live on higher homeostatic sleep pressure than healthy controls. Chronobiol Int 29(3):278–294

Giese M, Beck J, Brand S, Muheim F, Hemmeter U, Hatzinger M, Holsboer-Trachsler E, Eckert A (2014) Fast BDNF serum level increase and diurnal BDNF oscillations are associated with therapeutic response after partial sleep deprivation. J Psychiatr Res 59:1–7

Giles DE, Rush AJ, Roffwarg HP (1986) Sleep parameters in bipolar I, bipolar II, and unipolar depressions. Biol Psychiatry 21(13):1340–1343

Gillin JC, Mazure C, Post RM, Jimerson D, Bunney WE Jr (1977) An EEG sleep study of a bipolar (manic-depressive) patient with a nocturnal switch process. Biol Psychiatry 12(6):711–718

Goldschmied JR, Cheng P, Armitage R, Deldin PJ (2014) Examining the effects of sleep delay on depressed males and females and healthy controls. J Sleep Res 23(6):664–672

Gomez-Bernal G (2009) Dark therapy for schizoaffective disorder. A case report. Med Hypotheses 72(1):105–106

Gorgulu Y, Caliyurt O (2009) Rapid antidepressant effects of sleep deprivation therapy correlates with serum BDNF changes in major depression. Brain Res Bull 80(3):158–162

Gould TD, Manji HK (2002) The Wnt signaling pathway in bipolar disorder. Neuroscientist 8 (5):497–511

Gould TD, Manji HK (2005) Glycogen synthase kinase-3: a putative molecular target for lithium mimetic drugs. Neuropsychopharmacology 30(7):1223–1237

Haile CN, Murrough JW, Iosifescu DV, Chang LC, Al Jurdi RK, Foulkes A, Iqbal S, Mahoney JJ III, De La Garza R II, Charney DS, Newton TF, Mathew SJ (2014) Plasma brain derived neurotrophic factor (BDNF) and response to ketamine in treatment-resistant depression. Int J Neuropsychopharmacol 17(2):331–336

Hampp G, Ripperger JA, Houben T, Schmutz I, Blex C, Perreau-Lenz S, Brunk I, Spanagel R, Ahnert-Hilger G, Meijer JH, Albrecht U (2008) Regulation of monoamine oxidase A by circadian-clock components implies clock influence on mood. Curr Biol 18(9):678–683

Harvey AG (2008) Sleep and circadian rhythms in bipolar disorder: seeking synchrony, harmony, and regulation. Am J Psychiatry 165(7):820–829

Hefti K, Holst SC, Sovago J, Bachmann V, Buck A, Ametamey SM, Scheidegger M, Berthold T, Gomez-Mancilla B, Seifritz E, Landolt HP (2013) Increased metabotropic glutamate receptor subtype 5 availability in human brain after one night without sleep. Biol Psychiatry 73 (2):161–168

Hemmeter U, Hatzinger M, Brand S, Holsboer-Trachsler E (2007) Effect of flumazenil-augmentation on microsleep and mood in depressed patients during partial sleep deprivation. J Psychiatr Res 41(10):876–884

Huber R, Ghilardi MF, Massimini M, Tononi G (2004) Local sleep and learning. Nature 430 (6995):78–81

Huber R, Ghilardi MF, Massimini M, Ferrarelli F, Riedner BA, Peterson MJ, Tononi G (2006) Arm immobilization causes cortical plastic changes and locally decreases sleep slow wave activity. Nat Neurosci 9:1169–1176

Huber R, Tononi G, Cirelli C (2007) Exploratory behavior, cortical BDNF expression, and sleep homeostasis. Sleep 30(2):129–139

Huber R, Maki H, Rosanova M, Casarotto S, Canali P, Casali AG, Tononi G, Massimini M (2013) Human cortical excitability increases with time awake. Cereb Cortex 23(2):332–338

Hudson JI, Lipinski JF, Frankenburg FR, Grochocinski VJ, Kupfer DJ (1988) Electroencephalographic sleep in mania. Arch Gen Psychiatry 45(3):267–273

Hudson JI, Lipinski JF, Keck PE Jr, Aizley HG, Lukas SE, Rothschild AJ, Waternaux CM, Kupfer DJ (1992) Polysomnographic characteristics of young manic patients. Comparison with unipolar depressed patients and normal control subjects. Arch Gen Psychiatry 49(5):378–383

Jernajczyk W (1986) Latency of eye movement and other REM sleep parameters in bipolar depression. Biol Psychiatry 21(5–6):465–472

Jovanovic UJ (1977) The sleep profile in manic-depressive patients in the depressive phase. Waking Sleeping 1:199–210

Kaplan KA, Harvey AG (2013) Behavioral treatment of insomnia in bipolar disorder. Am J Psychiatry 170(7):716–720

Kavalali ET, Monteggia LM (2012) Synaptic mechanisms underlying rapid antidepressant action of ketamine. Am J Psychiatry 169(11):1150–1156

Knowles JB, Cairns J, MacLean AW, Delva N, Prowse A, Waldron J, Letemendia FJ (1986) The sleep of remitted bipolar depressives: comparison with sex and age-matched controls. Can J Psychiatry 31(4):295–298

Kreuzer PM, Langguth B, Schecklmann M, Eichhammer P, Hajak G, Landgrebe M (2012) Can repetitive transcranial magnetic stimulation prolong the antidepressant effects of sleep deprivation. Brain Stimul 5:141–147

Kupfer DJ, Ehlers CL (1989) Two roads to rapid eye movement latency. Arch Gen Psychiatry 46 (10):945–948

Laje G, Lally N, Mathews D, Brutsche N, Chemerinski A, Akula N, Kelmendi B, Simen A, McMahon FJ, Sanacora G, Zarate C Jr (2012) Brain-derived neurotrophic factor Val66Met polymorphism and antidepressant efficacy of ketamine in depressed patients. Biol Psychiatry 72(11):e27–e28

Lauer CJ, Wiegand M, Krieg JC (1992) All-night electroencephalographic sleep and cranial computed tomography in depression. A study of unipolar and bipolar patients. Eur Arch Psychiatry Clin Neurosci 242(2–3):59–68

Lenox RH, Gould TD, Manji HK (2002) Endophenotypes in bipolar disorder. Am J Med Genet 114(4):391–406

Li N, Lee B, Liu RJ, Banasr M, Dwyer JM, Iwata M, Li XY, Aghajanian G, Duman RS (2010) mTOR-dependent synapse formation underlies the rapid antidepressant effects of NMDA antagonists. Science 329(5994):959–964

Li N, Liu RJ, Dwyer JM, Banasr M, Lee B, Son H, Li XY, Aghajanian G, Duman RS (2011) Glutamate N-methyl-D-aspartate receptor antagonists rapidly reverse behavioral and synaptic deficits caused by chronic stress exposure. Biol Psychiatry 69(8):754–761

Linkowski P, Kerkhofs M, Rielaert C, Mendlewicz J (1986) Sleep during mania in manic-depressive males. Eur Arch Psychiatry Neurol Sci 235(6):339–341

Maeng S, Zarate CA Jr (2007) The role of glutamate in mood disorders: results from the ketamine in major depression study and the presumed cellular mechanism underlying its antidepressant effects. Curr Psychiatry Rep 9(6):467–474

McCarthy MJ, Nievergelt CM, Kelsoe JR, Welsh DK (2012) A survey of genomic studies supports association of circadian clock genes with bipolar disorder spectrum illnesses and lithium response. PLoS One 7(2), e32091

McClung CA (2011) Circadian rhythms and mood regulation: insights from pre-clinical models. Eur Neuropsychopharmacol 21(Suppl 4):S683–S693

McClung CA et al (2005) Regulation of dopaminergic transmission and cocaine reward by the Clock gene. Proc Natl Acad Sci USA 102:9377–9381

Moghaddam B, Adams B, Verma A, Daly D (1997) Activation of glutamatergic neurotransmission by ketamine: a novel step in the pathway from NMDA receptor blockade to dopaminergic and cognitive disruptions associated with the prefrontal cortex. J Neurosci 17(8):2921–2927

Morin AK, Jarvis CI, Lynch AM (2007) Therapeutic options for sleep-maintenance and sleep-onset insomnia. Pharmacotherapy 27(1):89–110

Murck H, Schubert MI, Schmid D, Schussler P, Steiger A, Auer DP (2009) The glutamatergic system and its relation to the clinical effect of therapeutic-sleep deprivation in depression – an MR spectroscopy study. J Psychiatr Res 43(3):175–180

Novakova M, Prasko J, Latalova K, Sladek M, Sumova A (2015) The circadian system of patients with bipolar disorder differs in episodes of mania and depression. Bipolar Disord 17 (3):303–314

Partonen T (2012) Clock gene variants in mood and anxiety disorders. J Neural Transm 119 (10):1133–1145

Phelps J (2008) Dark therapy for bipolar disorder using amber lenses for blue light blockade. Med Hypotheses 70(2):224–229

Rao U, Dahl RE, Ryan ND, Birmaher B, Williamson DE, Rao R, Kaufman J (2002) Heterogeneity in EEG sleep findings in adolescent depression: unipolar versus bipolar clinical course. J Affect Disord 70(3):273–280

Saeki T, Nakamura M, Hirai N, Noda Y, Hayasaka S, Iwanari H, Hirayasu Y (2013) Localized potentiation of sleep slow-wave activity induced by prefrontal repetitive transcranial magnetic stimulation in patients with a major depressive episode. Brain Stimul 6(3):390–396

Sahlem GL, Kalivas B, Fox JB, Lamb K, Roper A, Williams EN, Williams NR, Korte JE, Zuschlag ZD, El Sabbagh S, Guille C, Barth KS, Uhde TW, George MS, Short EB (2014) Adjunctive triple chronotherapy (combined total sleep deprivation, sleep phase advance, and bright light therapy) rapidly improves mood and suicidality in suicidal depressed inpatients: an open label pilot study. J Psychiatr Res 59:101–107

Sitaram N, Nurnberger JI Jr, Gershon ES, Gillin JC (1982) Cholinergic regulation of mood and REM sleep: potential model and marker of vulnerability to affective disorder. Am J Psychiatry 139(5):571–576

Small JG, Milstein V, Malloy FW, Medlock CE, Klapper MH (1999) Clinical and quantitative EEG studies of mania. J Affect Disord 53(3):217–224

Soehner AM, Kaplan KA, Harvey AG (2013) Insomnia comorbid to severe psychiatric illness. Sleep Med Clin 8(3):361–371

Soreca I (2014) Circadian rhythms and sleep in bipolar disorder: implications for pathophysiology and treatment. Curr Opin Psychiatry 27(6):467–471

Spielman AJ, Saskin P, Thorpy MJ (1987) Treatment of chronic insomnia by restriction of time in bed. Sleep 10(1):45–56

Thase ME, Himmelhoch JM, Mallinger AG, Jarrett DB, Kupfer DJ (1989) Sleep EEG and DST findings in anergic bipolar depression. Am J Psychiatry 146(3):329–333

Tononi G, Cirelli C (2006) Sleep function and synaptic homeostasis. Sleep Med Rev 10(1):49–62

van der Lely S, Frey S, Garbazza C, Wirz-Justice A, Jenni OG, Steiner R, Wolf S, Cajochen C, Bromundt V, Schmidt C (2015) Blue blocker glasses as a countermeasure for alerting effects of evening light-emitting diode screen exposure in male teenagers. J Adolesc Health 56 (1):113–119

Vyazovskiy VV, Riedner BA, Cirelli C, Tononi G (2007) Sleep homeostasis and cortical synchronization: II. A local field potential study of sleep slow waves in the rat. Sleep 30 (12):1631–1642

Wehr TA, Turner EH, Shimada JM, Lowe CH, Barker C, Leibenluft E (1998) Treatment of a rapidly cycling bipolar patient by using extended bed rest and darkness to stabilize the timing and duration of sleep. Biol Psychiatry 43:822–828

Wirz-Justice A, Quinto C, Cajochen C, Werth E, Hock C (1999) A rapid-cycling bipolar patient treated with long nights, bedrest and light. Biol Psychiatry 45:1075–1077

Wu JC, Kelsoe JR, Schachat C, Bunney BG, DeModena A, Golshan S, Gillin JC, Potkin SG, Bunney WE (2009) Rapid and sustained antidepressant response with sleep deprivation and chronotherapy in bipolar disorder. Biol Psychiatry 66:298–301

Zarate CA Jr, Singh JB, Carlson PJ, Brutsche NE, Ameli R, Luckenbaugh DA, Charney DS, Manji HK (2006) A randomized trial of an N-methyl-D-aspartate antagonist in treatment-resistant major depression. Arch Gen Psychiatry 63(8):856–864

Zhao C, Eisinger BE, Driessen TM, Gammie SC (2014) Addiction and reward-related genes show altered expression in the postpartum nucleus accumbens. Front Behav Neurosci 8:388

Chapter 8
Neuroimaging Studies of Bipolar Depression: Therapeutic Implications

Jonathan Savitz, Harvey M. Morris, and Wayne C. Drevets

Abstract Bipolar disorder (BD) is characterized by pathophysiological changes to the visceromotor network, disrupting the regulation of endocrine and autonomic responses to stress and, hence, emotion and behavior. Specifically, reductions in gray matter volume and/or cortical thickness and a concomitant increase in glutamatergic neurotransmission are observed in the pregenual (pgACC) and subgenual anterior cingulate cortex (sgACC); the orbitofrontal, frontal polar, and ventrolateral prefrontal cortex (PFC); and the posterior cingulate, ventral striatum, and hippocampus. Neuroreceptor imaging data provide preliminary evidence for serotonin, serotonin transporter (5-HT), dopamine receptor, and cholinergic system dysfunction in BD. Recent PET imaging data also suggest microglial cell activation in mood disorders. Oft-reported abnormalities of the deep frontal and basal ganglia white matter, and enlargement of the third and lateral ventricles are likely associated with cerebrovascular disease. Mood stabilizers and antidepressant drugs may attenuate pathological limbic activity, increase neurotrophic processes, and decrease inflammation, restoring balance to the system.

Keywords Bipolar disorder (BD) • Neuroimaging • Amygdala • Hippocampus • Prefrontal cortex • Glutamate

J. Savitz, PhD (✉)
Laureate Institute for Brain Research, 6655 S. Yale Avenue, Tulsa, OK 74136, USA

Faculty of Community Medicine, The University of Tulsa, Tulsa, OK, USA
e-mail: jsavitz@laureateinstitute.org

H.M. Morris, PhD
Laureate Institute for Brain Research, 6655 S. Yale Avenue, Tulsa, OK 74136, USA

W.C. Drevets, MD
Janssen Pharmaceuticals of J&J, Inc., Titusville, NJ, USA

© Springer International Publishing Switzerland 2016
C.A. Zarate Jr., H.K. Manji (eds.), *Bipolar Depression: Molecular Neurobiology, Clinical Diagnosis, and Pharmacotherapy*, Milestones in Drug Therapy,
DOI 10.1007/978-3-319-31689-5_8

8.1 Introduction

Since this chapter's first edition in 2010, the field of neuroimaging studies in bipolar disorder (BD) has rapidly evolved. Some of the major changes include: a reduced emphasis on structural magnetic resonance imaging (MRI) studies, as advances in technology have led to greater interest in techniques such as magnetic resonance spectroscopy (MRS), diffusion tensor imaging (DTI), and functional MRI (fMRI) assessment of the blood oxygen level dependence (BOLD) signal during the resting state to investigate the functional connectivity across brain regions. One exception to this trend, however, has been the development of automated techniques that allow for the measurement of cortical thickness and gyrification. The past five years have also seen a general increase in the number of fMRI studies, a greater interest in correlating molecular-level phenotypes with imaging data, and a quest to identify transdiagnostic biomarkers of psychiatric illness. The evolution of the research trajectory has been influenced in part by the development of new classes of antidepressant medication underlined by the promising results of ketamine, an N-methyl-D-aspartate (NMDA) receptor antagonist, for the treatment of bipolar depression (Ionescu et al. 2015; Zarate et al. 2012). Here, we incorporate some of the results from these new areas into our review of the neuroimaging and neuropathological abnormalities in BD.

The World Health Organization ranks BD as one of the leading causes of disability (The World Health Organization (WHO) 2001); yet, our knowledge about this condition's pathogenesis remains modest. Because BD is not associated with gross brain pathology or with clear animal models for spontaneous, recurrent mood episodes, the availability of tools allowing noninvasive assessment of the human brain is critical to elucidating its neurobiology. The development of neuroimaging technologies that permit in vivo characterization of the anatomical, physiological, and neurochemical correlates of BD has thus enabled advances toward illuminating the pathophysiology of this condition. Notably, the results of neuroimaging studies and the postmortem studies that have been guided by neuroimaging results have given rise to neurocircuitry-based models in which both functional and structural brain pathologies play roles in the development of BD.

The symptomatology of the clinical syndromes that manifest in BD, namely, the major depressive and manic episodes, implicates brain systems involved in the regulation of mood and emotional expression, reward processing, attention, motivation, stress response, social cognition, and neurovegetative function (e.g., sleep, appetite, energy, libido). The symptomatology of mania thus implicates the same domains of brain function as those underlying depressive episodes but generally with opposite valence. Anxiety symptoms are also prominent during the depressed phase of BD, and this disorder commonly occurs comorbidly with anxiety disorders such as panic disorder, social phobia, posttraumatic stress disorder (PTSD), and obsessive–compulsive disorder (OCD) (Kessler et al. 2005). Consistent with the clinical phenomenology of BD, a variety of neurophysiological, neuropathological, and neurochemical abnormalities have been discovered in BD within the neural systems that modulate emotional behavior.

To date, none of these abnormalities have shown sufficient sensitivity and specificity to prove useful as a diagnostic test, and neuroimaging is not recommended within either the USA or the European practice guidelines for positively defining diagnosis of any primary psychiatric disorder (Savitz et al. 2013). The variable presence and magnitude of such abnormalities in mood disorders likely reflects the heterogeneity encompassed within the BD syndrome with respect to pathophysiology and etiology. As long as psychiatric nosology depends on syndrome-based classifications, diagnosing BD may continue to encompass patients with a range of conditions that appear clinically related but are neurobiologically distinct. This lack of precise and biologically verifiable definition of illness presumably contributes to the extant inconsistencies within the literature pertaining to neurobiological abnormalities associated with BD and to the variable responses of BD patients to psychopharmacological treatment options. Ultimately, the discovery of illness subtypes that are associated with specific genotypes is expected to improve the sensitivity and specificity of research findings, as well as of therapeutic approaches.

8.2 Neural Circuits Implicated in Bipolar Disorder

Evidence from neuroimaging, neuropathological, and lesion analysis studies implicates brain networks that normally regulate the evaluative, expressive, and experiential aspects of emotional behavior in the pathophysiology of BD (Phillips et al. 2003). These circuits include the limbic–cortical–striatal–pallidal–thalamic (LCSPT) circuits formed by the orbital and medial prefrontal cortex (OMPFC), amygdala, hippocampal subiculum, ventromedial striatum, mediodorsal thalamic nucleus, and ventral pallidum (Ongur et al. 2003). The LCSPT circuits initially were related to *emotional behavior* on the basis of their anatomical connectivity with limbic structures that mediate emotional expression, such as the hypothalamus and periaqueductal gray (PAG) (Nauta and Domesick 1984). They were also initially implicated in the *pathophysiology of depression* by the observation that degenerative basal ganglia diseases and lesions of the striatum and orbitofrontal cortex (OFC) increased the risk of developing major depressive or manic syndromes (Folstein et al. 1985).

In addition to involving LCSPT circuitry, the functional and structural brain abnormalities associated with mood disorders also affect an extended anatomical network formed by neural projections linking the LCSPT components to areas of the mid and posterior cingulate cortex, superior and medial temporal gyrus, parahippocampal cortex, medial thalamic nuclei, and habenula (Ongur et al. 2003). This extended "visceromotor" network functions to regulate autonomic, endocrine, neurotransmitter, and behavioral responses to aversive and rewarding stimuli and contexts by modulating neuronal activity within the limbic and brainstem structures that mediate and organize emotional expression (e.g., amygdala, bed nucleus of the stria terminalis (BNST), PAG, hypothalamus) (Ongur et al. 2003). Thus, impaired function within this network could disinhibit or alter emotional expression and experience, conceivably giving rise to the clinical manifestations of depression or mania. Compatible with this hypothesis, pharmacological, neurosurgical, and electrical stimulation treatments for mood disorders appear to inhibit pathological activity within visceromotor network structures such as the amygdala and subgenual anterior cingulate cortex (sgACC) (Price and Drevets 2012).

8.3 Structural Neuroimaging in Bipolar Disorder

Patients with BD show abnormalities of morphology or morphometry in multiple structures that form the extended visceromotor network (Drevets and Price 2005) (Tables 8.1 and 8.2). The extent or prevalence of these abnormalities depends partly on clinical characteristics such as age at onset of illness, risk for developing psychosis as well as mania, and evidence for familial aggregation of illness. For example, elderly BD or major depressive disorder (MDD) subjects with late-onset mood disorders show an increased prevalence of neuroimaging correlates of cerebrovascular disease relative to both age-matched healthy controls and to elderly individuals with MDD with an early age of onset (Drevets et al. 2004). Similarly, individuals with MDD or BD who manifest either psychosis (delusions and/or hallucinations) or a late-life onset of illness show nonspecific signs of atrophy, such as lateral ventricle enlargement, which are absent in early onset, nonpsychotic MDD cases.

8.3.1 Volumetric MRI Abnormalities Identified in Bipolar Disorder

Early onset, nonpsychotic BD cases also show volumetric abnormalities that are localized to some PFC, cingulate, temporal lobe, and striatal structures (Tables 8.1 and 8.2). The most prominent volumetric abnormality reported to date has been a reduction in gray matter (GM) in the *left* anterior cingulate cortex (ACC) ventral to the corpus callosum *genu* (i.e., "subgenual"), which is evident in MDD and BD with evidence of familial clustering or with psychotic features (Botteron et al. 2002; Coryell et al. 2005; Drevets et al. 1997; Hirayasu et al. 1999). This volumetric reduction exists early in the course of the illness and in young adults at high familial risk for BD or MDD (Botteron et al. 2002; Hirayasu et al. 1999). This abnormality is also evident for both BD-I and BD-II samples (Haznedar et al. 2005; Lyoo et al. 2004a). Reductions in cortical thickness of the sgACC and the rostral ACC

Table 8.1 Neuroimaging and histopathological abnormalities evident in the visceromotor network in early onset, recurrent MDD, and/or BD

Brain region	Gray matter volume	Cell counts, cell markers	Glucose metabolism, CBF	
			Dep vs. Con	Dep vs. Rem
	Dep vs. Con	Dep vs. Con		
Dorsal medial/anterolateral PFC (BA9)	Decreased	Decreased	Decreased	Increased
Frontal polar cortex (BA 10)		Decreased	Increased	Increased
Subgenual anterior cingulate cortex (sgACC)	Decreased	Decreased	Mixed findings[a]	Increased
Pregenual anterior cingulate cortex (pgACC)	Decreased	Decreased	Increased	Increased
Orbital C/Ventrolateral PFC	Decreased	Decreased	Increased	Increased
Posterior cingulate	Decreased		Increased	Increased
Parahippocampal cortex	Decreased	Decreased in BD	Increased	Increased
Amygdala	Mixed findings[b]	Decreased in MDD	Increased	Increased
Ventromedial striatum	Decreased		Increased	Increased
Hippocampus	Decreased	Decreased in BD	n.s.	n.s.
Superior temporal gyrus/ Temporopolar cortex	Decreased			Increased
Medial thalamus			Increased	Increased

[a]In the sgACC, the apparent reduction in cerebral blood flow and metabolism in PET images of subjects with MDD is thought to be accounted for by the reduction in tissue volume in the corresponding cortex. After partial volume correction for the reduction in gray matter, the metabolism appears increased relative to controls
[b]The literature disagrees with respect to amygdala volume in mood disorders (see text)
Abbreviations: BD: bipolar disorder; Dep vs. Con: unmedicated individuals with MDD versus healthy controls; Dep vs. Rem: unmedicated individuals with MDD versus themselves in either the medicated or unmedicated remitted phases; n.s.: differences generally not significant; PFC: prefrontal cortex
Empty cells indicate insufficient data. Modified from Drevets (2007)

more generally have also been reported using automated analysis software such as FreeSurfer, which is able to distinguish cortical area and cortical thickness (Elvsashagen et al. 2013; Foland-Ross et al. 2011) (Table 8.2).

Conventional antidepressant drug treatment and symptom remission do not appear to alter the reductions in GM volume in the sgACC (Drevets et al. 1997). Interestingly, chronic lithium treatment, which exerts robust neurotrophic effects in animal models (Moore et al. 2000), largely normalizes sgACC volume in treatment responders (Moore et al. 2009). These data are supported by MRS studies, which find that higher levels of N-acetylaspartate (NAA) in the sgACC, a marker of neuronal integrity, are associated with lithium treatment (Moore and Galloway 2002; Forester et al. 2008) (Fig. 8.1).

Also consistent with the structural MRI literature are the results of postmortem studies. Notably, a reduction in the number of Nissl-stained glia identified

Table 8.2 Summary of cortical thickness measurements in adult BD

Study	Regions of decreased thickness	Sample size	Mood state	Mean age
Giakoumatos et al. (2015)	LOC (B), lingual (R) in BD-no Li versus HC	186 BD, 342 HC	Psychotic	37, 36
Oertel-Knochel et al. (2015)	MOG (L), IFG (B), precuneus (B), STG (L), rACC (R)	32 BD, 35 HC	Remitted	44, 42
Janssen et al. (2014)	Frontal cortex	20 BD, 52 HC	Psychotic	16, 15
Lan et al. (2014)	IPC (B), caudal middle-frontal (R), SPC (L), PCC (R), supramarginal (R)	18 BD, 54 HC	Depressed	38, 32
Maller et al. (2014)	Parietal cortex (L), supramarginal gyrus (R), SFG (R), precuneus (R)	31 BD, 31 HC	Depressed	43, 40
Elvsashagen et al. (2013)	sgACC (L), dorsomedial PFC (B), DLPFC (B), temporal gyrus (L)	36 BD II, 42 HC	Depressed, remitted, hypomanic	33, 31
Hatton et al. (2013)	Calcarine sulcus (L), angular gyrus (L), supramarginal gyrus (R), SPC (R), precuneus (R), precentral gyrus (R), fusiform gyrus (R)	73 BD, 49 HC	Psychotic	22, 24
Foland-Ross et al. (2011)	ACC (L), pregenual ACC (L), OFC (B), frontopolar (L), dorsomedial PFC (L), temporal pole (L)	34 BD, 31 HC	Euthymic	31, 38
Rimol et al. (2010)	Frontal lobe, posterior temporal, and temporoparietal regions	139 BD, 207 HC	NS	35, 36
Fornito et al. (2009)	Male patients had increased thickness in sgACC (R)	26 BD, 26 HC	Manic (1st episode)	22, 22
Lyoo et al. (2006)	Postcentral cortex (B), dorsal ACC (L), pregenual ACC (L), PCC (L), occipital cortex (L), OFC (R), fusiform gyrus (R)	25 BD, 21 HC	Depressed	34, 32

(L) = left, (B) = bilateral, (R) = right, NS = not stated, LOC = lateral occipital cortex, MOG = medial orbital gyrus, STG = superior temporal gyrus, rACC = rostral anterior cingulate cortex, SFG = superior frontal gyrus, IFG = inferior frontal gyrus, IPC = inferior parietal cortex, SPC = superior parietal cortex, PCC = posterior cingulate cortex, sgACC = subgenual anterior cingulate cortex, DLPFC = dorsolateral prefrontal cortex, ICV = intracranial volume, Li = lithium, OFC = orbital frontal cortex, PFC = prefrontal cortex, BD = bipolar disorder, HC = healthy control

morphometrically together with an increase in neuronal density in the sgACC was found in two independent samples of patients with familial BD as well as familial MDD (Ongur et al. 1998). The reported reduction in glial cells most clearly implicated the perineuronal and myelinating oligodendroglia and may, thus, compromise the structural integrity of white matter (WM) fibers. Certainly, DTI studies have produced evidence of structural abnormalities of WM tracts connecting the sgACC with limbic nuclei (Wang et al. 2008, 2009).

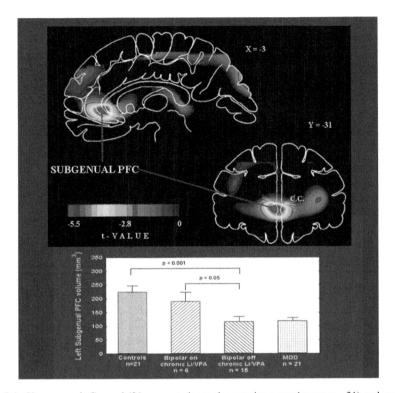

Fig. 8.1 *Upper panel:* Coronal (31 mm anterior to the anterior commissure; $y = 31$) and sagittal (3 mm left of midline; $x = -3$) sections showing negative voxel t-values where glucose metabolism is decreased in individuals with depression relative to controls. The reduction in activity in this prefrontal cortex (PFC) region located in the anterior cingulate gyrus ventral to the genu of the corpus callosum (i.e., subgenual) appeared to be accounted for by a corresponding reduction in cortex volume (Table 8.1; reproduced from Drevets et al. (1997)). Anterior or left is to left. *Lower Panel:* Although the PET data shown in the *upper panel* were obtained exclusively in unmedicated subjects, the volumetric MRI data from this study were obtained in a larger sample that included six cases who had been chronically receiving lithium or valproate prior to scanning. The bar histogram shows the mean subgenual anterior cingulate cortex (sgACC) volumes in mm³ for the healthy controls, individuals with major depressive disorder (MDD), unmedicated individuals with bipolar depression, and subjects with bipolar disorder (BD) chronically medicated with lithium or valproate

GM volume and cortical thickness is also reduced in the OFC (BA 11, 47) and ventrolateral PFC (VLPFC; BA 45, 47) in MDD (Drevets and Todd 2005) and BD (Lyoo et al. 2004a; Foland-Ross et al. 2011; Oertel-Knochel et al. 2015), in the frontal polar/dorsal anterolateral PFC (BA 9, 10) in MDD (Drevets et al. 2004), and in the posterior cingulate cortex and superior temporal gyrus in BD (Nugent et al. 2006; Oertel-Knochel et al. 2015). In BD, the peak difference in GM loss in the lateral OFC was found in the sulcal BA47 cortex (Nugent et al. 2006), a region that appears to function as part of both the visceromotor and "sensory" networks within the OMPFC (Ongur et al. 2003). Compatible with these data, the MRS study by Cecil and colleagues found reduced NAA and choline concentrations in the orbitofrontal GM in BD, suggesting decreased neuronal integrity (Cecil et al. 2002).

Decreases in the volume of the dorsal PFC have also been reported in BD and MDD (Savitz and Drevets 2009a). For example, Frangou and colleagues (2005) and Haznedar and colleagues (2005) described GM volume reductions of the dorsolateral PFC (DLPFC) (BA 8, 9, 45, 46) in medicated and remitted BD-I patients and partially medicated, "stable" bipolar-spectrum individuals, respectively. More circumscribed volume reductions of BA 9 were also reported in a medicated, euthymic pediatric BD sample (Dickstein et al. 2005). In a mixed BD-I and BD-II sample, Lochhead and colleagues reported reduced GM volume of the ACC immediately dorsal to the corpus callosum (CC) (Lochhead et al. 2004). Reductions in cortical thickness of the dorsomedial PFC and DLPFC were also reported in a lithium-free sample of BD subjects (Foland-Ross et al. 2011).

In the hippocampus, at least half of the studies reported reductions in the volume of the whole hippocampus in MDD. In BD, however, the volumetric reductions appear more specific to the anterior subiculum/ventral CA1 region (Drevets et al. 2004). In contrast, the whole hippocampal volume was reported to be smaller in BD subjects than controls in some studies, but not different from controls in most studies (Savitz and Drevets 2009a). The reasons for these apparent differences between neuromorphometric studies of MDD versus BD remain unclear. In MDD, the reduction in hippocampal volume was limited to depressed women who suffered early life trauma in some studies (Vythilingam et al. 2002) and was correlated inversely with time spent depressed in other studies (e.g., Sheline et al. (2003)), but it remains unclear whether these relationships also extend to BD. Regarding the postmortem literature, Fatemi and colleagues reported a 39 % decrease in CA4 volume together with a decreased density of reelin-expressing gamma aminobutyric acid (GABA)ergic neurons (Fatemi et al. 2000). Similarly, Pantazopoulos and colleagues detected a decrease in both the total number and the density of inhibitory, GABAergic neurons in the superficial layers of the entorhinal cortex in a BD sample (Pantazopoulos et al. 2007).

One potential reason for the apparent discrepant imaging findings between MDD and BD may be the neurotrophic/neuroprotective effects associated with mood stabilizing treatments. Animal studies demonstrate that lithium promotes hippocampal neurogenesis (Kim et al. 2004) and long-term potentiation (LTP) (Son et al. 2003). A sample of BD patients treated for four weeks with lithium showed a 3 % (24 cm^3) increase in whole brain GM volumes from baseline (Moore et al. 2000), an effect that appeared to result from the neurotrophic effect of the drug (Manji et al. 2000). Four more recent studies (Bearden et al. 2007; Beyer et al. 2004; Sassi et al. 2004; Yucel et al. 2008) comparing lithium-treated and non-lithium treated groups demonstrated similar effects in large cortical areas including the hippocampus. The phenomenon may not be restricted to lithium; comparable effects have been noted with other classes of mood stabilizers, especially valproate (Hao et al. 2004; Mark et al. 1995). In contrast, with the exception of the as yet rarely prescribed tianeptine (McEwen et al. 2002; McEwen and Olie 2005; Watanabe et al. 1992), the neurotrophic properties of antidepressants are less persuasive (although see Stewart and Reid (2000)) and Duman and Monteggia (2006)). The potential effects of lithium on hippocampal volume are supported by a recent meta-analysis that found that in aggregate, hippocampal volume is reduced in BD after controlling for the effects of lithium (Hajek et al. 2012).

Amygdala volume has been reported to be increased in some studies but decreased in others in individuals with MDD relative to controls (Drevets et al. 2004). In general, these data suggest that the amygdala volume in patients with BD shows an age-related dichotomy of findings. In adults, the predominant pattern is one of increased amygdala volume while in children and adolescents the reverse is true (Savitz and Drevets 2009a). The findings in adults seem to hold even in samples with a long history of illness (Altshuler et al. 2000; Brambilla et al. 2003; Frangou et al. 2005). As in the case of the hippocampus, reports of increased GM volume in BD subjects versus healthy controls may be an artifact of treatment with medications. Relative to healthy controls, we found larger amygdalar volumes in lithium or valproate-treated BD patients but smaller amygdalar volumes in unmedicated subjects with BD (Savitz et al. 2010). These data are supported by the postmortem literature: a 19 % reduction in oligodendrocyte density in the amygdala that reached trend level significance was found in one study (Hamidi et al. 2004), and in another study, the density of glial cells in the amygdala was significantly reduced in MDD but not in BD (Bowley et al. 2002). However, samples from two unmedicated patients with BD were indicative of reduced oligodendrocyte density (Bowley et al. 2002).

In the basal ganglia (i.e., caudate, putamen, or globus pallidus), subjects with BD generally have not shown morphometric differences relative to controls (Savitz and Drevets 2009a). These data appear consistent with the reported absence of NAA abnormalities in the basal ganglia of BD samples (Kato et al. 1996; Hamakawa et al. 1998; Ohara et al. 1998). Nevertheless, a postmortem study of a combined MDD and BD sample reported volumetric reductions of the left accumbens, bilateral pallidum, and right putamen (Baumann et al. 1999).

There has, however, been some suggestion of striatal enlargement in adult and pediatric samples with BD (Savitz and Drevets 2009a). As in the case of the hippocampus, treatment effects may conceivably confound these analyses. Enlargement of basal ganglia structures is a well-known effect of antipsychotic drugs (Jernigan et al. 1991; Swayze et al. 1992; Chakos et al. 1995; Frazier et al. 1996) and, notably, three of the studies reporting basal ganglia enlargement used samples that included subjects who were manic and treated with antipsychotic medications (Strakowski et al. 2002; DelBello et al. 2004; Wilke et al. 2004).

Enlargement of the third and lateral ventricles has commonly been observed in BD (Savitz and Drevets 2009a). Many studies reporting ventriculomegaly included subjects with early age of onset. Nevertheless, the extent to which chronic alcohol abuse (Anstey et al. 2006), incipient neurological disorders with prodromal depression, or cerebrovascular disease (Salerno et al. 1992; Goldstein et al. 2002; Knopman et al. 2005) (see ensuing section on WM abnormalities) contributes to ventricular enlargement has not been established. Notably, Strakowski and colleagues found that the lateral ventricles were significantly larger in patients with multiple-episode BD than in first-episode BD patients or healthy controls, and larger lateral ventricles were associated with a higher number of prior manic episodes (Strakowski et al. 2002). The multiple-episode patients in this study also had a smaller total cerebral volume than the healthy subjects, but not the first-episode patients. These cross-sectional data imply that a progressive loss of cerebral tissue volume occurs during repeated manic episodes, although this observation requires confirmation in longitudinal studies.

8.3.1.1 Lithium's Effects on T1 Signal Intensity of Gray Matter in MRI Image Analysis

In considering the design and interpretation of studies reporting lithium's effects on regional GM volume, the nonspecific effect of lithium on T1 signal intensity merits comment. Cousins and colleagues showed that lithium reduced T1 relaxation in cerebral GM (but not in WM), and that this effect could distort measures of GM volume obtained using analysis techniques that depend heavily or exclusively on MRI signal intensity, such as voxel-based morphometry (VBM) (Cousins et al. 2013). In this study, MRI scans were obtained in healthy controls before and after receiving lithium for 11 days. Analysis of cerebral volumes performed using VBM showed an increase in total GM volume by a mean of 1.1 % versus placebo (along with a corresponding reduction in CSF volume and no significant change in WM volume). In contrast, no significant difference between lithium and placebo was noted when the same images were analyzed using Structural Image Evaluation using Normalization of Atrophy (SIENA), a technique that additionally employed paired edge finding methods (analogous to the boundary finding capabilities of the human eye). This effect of lithium on the T1 signal must, therefore, be considered in the design and interpretation of volumetric studies of lithium's effects on GM volume.

One study that used a semi-automated segmentation approach (analogous to VBM) addressed this limitation by including a relevant control group that enabled controlling for lithium's nonspecific effects on T1 signal intensity. Moore and colleagues imaged 28 BD patients before and following a four-week course of lithium. Although in the entire sample the total brain GM volume increased significantly in the post- versus pre-treatment condition, the separate comparisons of lithium responders ($n = 11$) and nonresponders ($n = 17$) proved most instructive (Moore et al. 2009). These subgroups did not differ significantly in lithium level, age, sex, or baseline total brain volume. In the nonresponders, the treatment-associated changes in GM volume were not statistically significant but were nominally in the 0.5–1 % range in the entire PFC and subgenual PFC regions, respectively, compatible with the findings of Cousins and colleagues (2013). In the lithium responders, however, GM volume increased *significantly* in the PFC; moreover, the magnitude of the GM changes was greater in the treatment responder group, being >4 % in the PFC more broadly and 8 % in the sgPFC specifically. Thus, the nonresponder group controlled for the nonspecific effects of lithium on

MR signal intensity and enabled a compelling demonstration that, in treatment responders, lithium increased GM volume in regions where GM had been reported in previous studies to be abnormally decreased in BD.

Another useful approach was exemplified by Savitz and colleagues, who applied two useful approaches for dealing with the nonspecific effect of lithium on T1 signal in a study employing a cross-sectional design (Savitz et al. 2010). First, in addition to examining the effects of lithium on amygdala volume, this study also investigated the effects of another mood stabilizing medication, divalproex, which showed neuroprotective effects similar to those of lithium in preclinical models, but has not been associated with confounding effects on MRI signal intensity. Second, this study used a manual segmentation technique in images that were high in both spatial and tissue contrast resolution. The manual segmentation technique relies on the capabilities of the human eye to implement information about both signal intensity and boundary shape, analogous to the SIENA approach. In *unmedicated* BD subjects, the mean right amygdala volume was significantly smaller than in matched healthy controls. In contrast, the mean right amygdala volume was significantly greater (with a similar trend on the left) in BD patients treated with either lithium or divalproex, relative to the BD subjects who were unmedicated. In a post hoc analysis, the effect was at least as robust for divalproex as for lithium (the volumes were nominally larger in the divalproex-treated subsample). Since preclinical data showed that lithium and divalproex exert similar neuroplasticity effects under a variety of physiological and psychological stress models in rodents, these data appear compatible with the hypothesis that these neuroplasticity effects extend to humans with BD. Notably, the medicated BD subjects did not show any difference relative to matched healthy controls in whole brain volume, suggesting that mood stabilizer treatment may correct an abnormal reduction in amygdala volume in BD without nonspecifically changing GM volume all over the brain.

8.3.2 Neuromorphological MRI Abnormalities in BD: White Matter Pathology

Abnormalities of WM tracts in BD as indexed by reduced fractional anisotropy (FA) and increased radial diffusivity (RD) have been reported in regions such as the cingulum, genu, and splenium of the corpus callosum and inferior and superior longitudinal fasciculi (Emsell et al. 2014; Torgerson et al. 2013; Barysheva et al. 2013; Versace et al. 2014). FA is a measure of the degree to which the diffusion process is constrained by the WM fibers, such that a low FA may indicate decreases in fiber density and/or myelination in WM. Interestingly, a recent study reported that psychotic BD patients showed similar WM connectivity abnormalities to schizophrenics with decreases in FA of the callosal, posterior thalamic/optic, paralimbic, and fronto-occipital tracts (Kumar et al. 2015). These findings were partially consistent with a previous study reporting reduced FA in the anterior limb of the internal capsule, anterior thalamic radiation, and in the region of the uncinate fasciculus in patients with BD and those with schizophrenia compared with controls (Sussmann et al. 2009), thus suggesting that the presence of WM abnormalities may cut across traditional diagnostic boundaries.

In morphological MRI studies, an elevation in the incidence of WM hyperintensities (WMH), especially in the deep frontal cortex and basal ganglia, has commonly been reported in BD and in late-onset MDD samples (Krishnan et al. 1991; Figiel et al. 1991; Hickie et al. 1995; Steffens et al. 1999; Hannestad et al. 2006). Seen as high-intensity signals on T2-weighted MRI scans, WMH are caused by circumscribed increases in water content that putatively indicate a decrease in WM density due to demyelination, atrophy of the neuropil, ischemia-associated microangiopathy, or other causes (Ovbiagele and Saver 2006). This phenomenon normally is prevalent in elderly, nondepressed populations (Kertesz et al. 1988) but shows an abnormally high prevalence in MDD cases with a late age of onset and in BD samples of all ages.

The incidence of WMH may relate in part to cerebrovascular disease. BD is associated with a significantly increased prevalence of cardiovascular disease risk factors such as smoking, obesity, diabetes mellitus (DM), hypertension, and dyslipidemia (reviewed in Kilbourne et al. (2004) and Newcomer (2006)). Hypertension (Dufouil et al. 2001; Gunstad et al. 2005), obesity (Gustafson et al. 2004), smoking (Dager and Friedman 2000), and diabetes mellitus (Novak et al. 2006) have in turn been directly associated with the development of WMH. Although most published studies of BD attempt to exclude patients with such potentially confounding conditions, the whole gamut of risk conditions is rarely controlled for, raising the possibility that WMH in BD are an artifact of medical comorbidity or some obscure ischemic risk factor. Moreover, drug abuse is prevalent in BD populations and stimulant drug-induced vasoconstriction may lead to WMH (Dupont et al. 1990; Lyoo et al. 2004b). Notably, marijuana use also may interact in an additive fashion with WMH to predispose to depressive symptomatology (Medina et al. 2007). In addition, Lenze and colleagues (1999) and Nemeroff and colleagues (2000) speculated that excess depression-associated secretion of serotonin by blood platelets (Biegon et al. 1990; Musselman et al. 1996) facilitates platelet aggregation and thereby predisposes to thrombotic events and vasoconstriction. Finally, cerebrovascular reactivity, which describes the compensatory dilatory capacity of arterioles to dilatory stimuli, is reportedly reduced in acutely depressed patients without any neurological, cardiac, or vascular risk factors (de Castro et al. 2008), raising the possibility that impaired regulation of vascular tone also plays a role in the pathogenesis of WMH in BD.

Nevertheless, studies that attempted to match patients and controls for the presence of cardiovascular risk factors still find elevated rates of WMH in their depressed samples (reviewed in Savitz and Drevets (2009a)). Moreover, the hypothesis that WMH reflect cerebrovascular disease fails to account for the WM pathology noted in pediatric BD samples (Botteron et al. 1995; Lyoo et al. 2002; Pillai et al. 2002) as well as the high concentration of WMH in both BD subjects and their unaffected relatives (Ahearn et al. 1998). A significant minority of young

BD patients with a relatively typical age of onset show WM abnormalities on MRI scans (Savitz and Drevets 2009b). Thus, while a proportion of adults with BD with significant WM pathology will present with risk factors for cerebrovascular disease, WMH may also less commonly arise in pediatric or young adult BD samples due to developmental insults or via some as yet unknown pathophysiological mechanism.

Obstetric complications are well known to be associated with schizophrenia (Cannon et al. 2002), but with a few exceptions (Kinney et al. 1993, 1998), appear less salient in BD. Nevertheless, it is possible that perinatal hypoxic events precipitate BD in a vulnerable minority (Pavuluri et al. 2006).

Another possible explanation for demyelination as evidenced by WMH in BD may be changes in oligodendrocyte function. Postmortem studies have reported a downregulation of oligodendrocyte-related expression of genes impacting myelin or oligodendrocyte function and decreased oligodendrocyte density in both BD and MDD (Tkachev et al. 2003; Aston et al. 2005; Cotter et al. 2002; Hamidi et al. 2004; Uranova et al. 2004; Vostrikov et al. 2007). Conversely, oligodendrocyte density and $2',3'$-cyclic-nucleotide $3'$-phosphodiesterase (CNPase), a putative marker of oligodendrocytes, were increased in the WM underlying the DLPFC of individuals with BD prescribed mood stabilizers, suggesting a treatment effect (Hercher et al. 2014). Variants of some of these genes such as oligodendrocyte lineage transcription factor 2 (*OLIG2*) [NCBI accession number 10215], Neuregulin 1 (*NRG1*) [3084], and v-erb-a erythroblastic leukemia viral oncogene homolog 4 (*ERBB4*) [2066] have been directly associated with mood disorders and may determine how resilient these cells are to environmental stressors (see Carter (2007a, b) and Sokolov (2007) for a review). WM is decreased in the genu of the corpus callosum in both adults with BD or MDD, their high-risk children, and their adolescent offspring (particularly in females), and is also decreased in the splenium of the corpus callosum in adults with BD or MDD. Finally, the high incidence of familial WMH seen in the Ahearn and colleagues sample supports a role for genetic factors and suggests that genetic variance in genes related to oligodendrocyte function may contribute to the development of WMH in BD (Ahearn et al. 1998).

An unresolved issue is whether the relationship between WM pathology and mood disorders is one of cause or effect. Certainly, new cases of BD may be precipitated by subcortical infarcts (Starkstein and Robinson 1989). Moreover, depressive and bipolar syndromes are relatively common sequelae of the genetic disorder, cerebral autosomal dominant arteriopathy with subcortical infarcts and leukoencephalopathy (CADASIL) (Chabriat et al. 1995; Desmond et al. 1999). The deep frontal WM pathology commonly seen in mood disorders may conceivably disrupt the pathways linking subcortical regions such as the striatum to functionally homologous regions of the PFC, giving rise to dysregulation of emotional behavior in BD (Adler et al. 2004; Geschwind 1965a, b).

8.4   Neurophysiological Imaging in Bipolar Depression

Many regions where structural abnormalities are apparent in mood disorders also contain abnormalities of cerebral blood flow (CBF) and glucose metabolism (Table 8.1; Fig. 8.1). In most of these structures, and particularly those that form the extended visceromotor network, the basal activity appears abnormally increased during the depressed phase of BD. In MDD, this pattern of differences has also been demonstrated in cross-sectional studies of depressed MDD subjects relative to controls, longitudinal studies of patients imaged before versus after treatment (Drevets et al. 2002a), and challenge studies of remitted patients scanned before versus during depressive relapse (Neumeister et al. 2004; Hasler et al. 2008).

Nevertheless, the reduction in GM volume in some structures is sufficiently prominent to produce partial volume effects in functional brain images due to their relatively low spatial resolution, yielding complex relationships between physiological measures and depression severity. For example, relative to controls, depressed BD and MDD subjects show metabolic activity that appears *reduced* in the sgACC (Drevets et al. 1997; Kegeles et al. 2003). However, this abnormal reduction in flow and metabolism may be partly attributable to the partial volume averaging effect associated with the reduction in the corresponding GM (Drevets and Price 2005). This effect may contribute to the complex relationship observed between metabolic activity and clinical state, as activity is decreased further in the remitted versus the depressed phase of mood disorders in the sgACC, as assessed following effective treatment (Drevets et al. 2002a; Holthoff et al. 2004; Nobler et al. 2001; Mayberg et al. 2005). Conversely, metabolic activity is increased in the sgACC in remitted MDD cases during depressive relapse induced by tryptophan depletion or catecholamine depletion (Neumeister et al. 2004; Hasler et al. 2008). The volumetric reductions in the OFC and VLPFC may also contribute to the complexity of relationships observed between metabolism and illness severity, as metabolism appears elevated in depressed samples of mild-to-moderate severity, but reduced in more severe, treatment-refractory cases (Ketter and Drevets 2002).

Although the pattern of activity in the extended visceromotor network generally is one in which metabolism is elevated during the depressed versus the remitted phases, the relationship between activity and symptom severity differs in valence across some structures, compatible with preclinical evidence that distinct structures are involved in opponent processes with respect to emotion modulation (Vidal-Gonzalez et al. 2006). Regions where metabolism correlates positively with depression severity include the amygdala, sgACC, and ventromedial frontal polar cortex (Hasler et al. 2008; Drevets et al. 2002a). Metabolism and flow decrease in these regions during effective treatment (Mayberg et al. 2005; Drevets et al. 2002a). Conversely, in recovered MDD cases who experience depressive relapse under serotonin or catecholamine depletion, metabolic activity generally increases in these regions as depressive symptoms return (Neumeister et al. 2004, 2006; Hasler et al. 2008).

In the amygdala, abnormal elevations of resting metabolism can be seen in depressed samples categorized as having BD, familial pure depressive disease (FPDD), MDD-melancholic type, or MDD that responds to a night of total sleep deprivation (Drevets 2001). In such cases, amygdala metabolism decreases toward normative levels during effective antidepressant treatment (Drevets et al. 2002a). In BD, these findings of increased baseline amygdalar activity have largely been limited to adults (Ketter et al. 2001; Sheline et al. 2001; Drevets et al. 2002b; Bauer et al. 2005; Mah et al. 2007), in whom resting activity has correlated positively with severity of depression (Ketter et al. 2001). Furthermore, increased hemodynamic responses of the amygdala to negatively valenced faces have been reported in BD subjects relative to healthy controls (Yurgelun-Todd et al. 2000; Lawrence et al. 2004; Rich et al. 2006; Pavuluri et al. 2007). Notably, both adolescents with BD and unaffected adolescents with a family history of BD showed an elevated hemodynamic response in the amygdala when presented with faces expressing negatively valenced emotion, implying that this abnormality either arises very early in the illness or reflects a heritable trait-like biomarker (Olsavsky et al. 2012; Manelis et al. 2015). This finding extended to the offspring of parents with non-bipolar mood disorder as well; however, in the functional connectivity analysis from the same study, an increased amygdala–VLPFC and reduced amygdala–ACC functional connectivity pattern previously shown in individuals with BD also differentiated the offspring of BD parents from those of non-bipolar mood disorders (Manelis et al. 2015). If these findings prove reproducible, they would imply that these abnormalities may constitute risk markers for the development of BD.

In the accumbens, medial thalamus, and posterior cingulate cortex, resting metabolism and perfusion appear abnormally elevated in the depressed phase of MDD and BD (Drevets et al. 2002b, 2004; Mah et al. 2007). In the OFC, Blumberg and colleagues showed that manic patients have reduced rCBF (Blumberg et al. 1999), while induction of a sad mood through psychological means resulted in decreased rCBF to the medial OFC in euthymic but not depressed BD subjects versus controls (Kruger et al. 2003). Finally, reductions in metabolism have been reported in the dorsolateral PFC, and abnormalities in the hemodynamic responses to various cognitive–behavioral tasks have been reported in BD in both the OMPFC and the DLPFC (Phillips et al. 2003).

The patterns of hemodynamic response to a variety of positively or negatively valenced stimuli have consistently differed between mood disordered and healthy control samples and, in some cases, between depressed subjects with MDD versus BD; these consistently highlight altered function within the limbic–cortical circuits that involve the OMPFC (Phillips and Swartz 2014). For example, in an fMRI study that compared patterns of hemodynamic change during a reward anticipation task between BD-I, MDD, and healthy control groups, the reward expectancy-related activation in the ACC observed in healthy individuals was significantly reduced in depressed patients with either BD-I or MDD, despite showing no significant difference in the ventral striatum across the three groups (Chase et al. 2013). Notably, the anticipation-related increase in hemodynamic activity in the left

VLPFC was significantly exaggerated in the BD-I depressed group compared to the other two groups. While medication effects have been a confounding factor in most fMRI studies of BD, some have begun to take such factors into account in the study design. For example, Hafeman and colleagues studied the impact of medication in youth with BD and reported an area in the right VLPFC in which unmedicated BD youth showed decreased activation relative to both healthy and psychiatric controls during the processing of negative face stimuli; a separate sample of medicated BD subjects also showed decreased activation in this cluster relative to healthy controls and non-BD youth, but the magnitude of this differences was diminished in the medicated BD group compared to the unmedicated group with respect to the control groups (Hafeman et al. 2014).

8.4.1 Patterning Hemodynamic Responses to Classify Individual Subjects with BD or Those at High Risk for BD

The pattern of hemodynamic abnormalities within and outside the medial prefrontal network detected in BD using fMRI has been explored for its potential as a biomarker signature capable of classifying individual subjects. For example, in one study, the hemodynamic response of the default mode and temporal lobe networks during an auditory oddball paradigm was applied a priori to a sample of 14 medicated patients with BD-I, 21 medicated patients with schizophrenia, and 26 healthy controls (Calhoun et al. 2008). The authors were able to distinguish BD patients from patients with schizophrenia and healthy controls with 83 % sensitivity and 100 % specificity. The accuracy of the BD versus healthy control classification was not provided, however. Similarly, Hahn and colleagues used three independent fMRI paradigms in an attempt to maximize classification accuracy: the passive viewing of emotionally valenced faces and two different versions of the monetary incentive delay task emphasizing potential winnings and potential losses, respectively (Hahn et al. 2011). A decision tree algorithm derived from the combination of the imaging task classifiers produced a diagnostic sensitivity of 80 % and a specificity of 87 % in a sample of 30 patients with depression (both MDD and BD) and 30 healthy controls. The algorithm's ability to distinguish subjects with MDD from BD was not reported.

Another study applied a Gaussian Process Classifiers (GPCs) machine-based learning approach to distinguish healthy adolescents with and without a parent with BD from each other with 75 % sensitivity and 75 % specificity (Mourao-Miranda et al. 2012). A discriminating pattern of BOLD activation was found in the superior temporal sulcus and ventromedial PFC when subjects were presented with neutral faces in the context of happy faces. Six out of 13 of the high-risk adolescents who were followed clinically subsequently met DSM-IV criteria for MDD or an anxiety disorder. These six individuals had higher GPC risk scores than the seven high-risk subjects who did not become ill. Conversely, three out of the four high-risk subjects that the GPC algorithm incorrectly classified as low-risk remained healthy at follow-up. While these sample sizes are small, the study highlights the potential utility of such approaches for developing predictive biomarkers in samples at high familial risk for BD.

8.4.2 Neuropathological Correlations in Mood Disorders

Most regions where MRI studies demonstrated volumetric abnormalities in BD have also been shown to contain histopathological changes or GM volumetric reductions in postmortem studies of MDD and BD. For example, reductions of GM volume, thickness, or wet weight have been reported in the sgACC, postero-lateral orbital cortex, and ventral striatum in MDD and/or BD subjects relative to controls (Baumann et al. 1999; Bowen et al. 1989; Ongur et al. 1998; Rajkowska et al. 1999). The histopathological correlates of these abnormalities included reductions in synapses or synaptic proteins, reductions in glial cells, elevations in neuronal density in some regions, and reductions in neuronal size in MDD and/or BD samples (Ongur et al. 1998; Eastwood and Harrison 2000; Uranova et al. 2004; Rajkowska and Miguel-Hidalgo 2007). Reductions in glial cell counts and density and/or glia-to-neuron ratios were also found in MDD subjects versus controls in the pgACC [BA24] (Cotter et al. 2001a), the dorsal anterolateral PFC (BA9) (Cotter et al. 2002; Uranova et al. 2004), and the amygdala (Bowley et al. 2002; Hamidi et al. 2004). Finally, the density of non-pyramidal neurons was decreased in the ACC and hippocampus in BD (Benes et al. 2001; Todtenkopf et al. 2005), and in the dorsal anterolateral PFC (BA9) of individuals with MDD (Rajkowska and Miguel-Hidalgo 2007). Reductions in synapses and synaptic proteins were evident in BD subjects in the hippocampal subiculum/ventral CA1 region (Eastwood and Harrison 2000; Czeh and Lucassen 2007).

The glial type that specifically differed between mood disordered and control samples in many of these studies was the oligodendrocyte (Uranova et al. 2004; Hamidi et al. 2004). Oligodendroglia are best characterized for their role in myelination, and the reduction in oligodendrocytes may conceivably arise secondary to an effect on myelin, either through demyelination, abnormal development, or atrophy in the number of myelinated axons. Notably, myelin basic protein concentration was found to be decreased in the frontal polar cortex (BA 10) (Honer et al. 1999), and the expression of genes related to oligodendrocyte function (i.e., genes that encoded structural components of myelin, enzymes involved in the synthesis of myelin constituents or in the regulation of myelin formation, transcription factors regulating other myelination-related genes, or factors involved in oligodendrocyte differentiation) was decreased in the middle temporal gyrus in MDD subjects relative to controls (Aston et al. 2005). Similarly, a quantitative PCR analysis of BA 9 demonstrated a significant reduction in mRNA expression of protein markers of myelination and oligodendrocyte function (Tkachev et al. 2003).

Expression of proteolipid protein 1 (PLP1), myelin associated glycoprotein (MAG), oligodendrocyte specific protein (CLDN11), myelin oligodendrocyte glycoprotein (MOG), and transferrin (TF) were reduced by approximately two- to four-fold in BD patients relative to psychiatrically healthy controls (Tkachev et al. 2003). Further, expression of the OLIG2 and SOX10 genes, which code for transcription factors involved in oligodendrocyte differentiation and maturation, was downregulated by two- to three-fold in BD. Similarly, MacDonald and colleagues reported that mRNA transcripts of oligodendrocyte-specific proteins such as gelsolin, MAG, and ERBB3 were downregulated in BD (MacDonald et al. 2006).

Compatible with these data, myelin staining was decreased in the deep WM of the DLPFC in MDD and BD subjects (Regenold et al. 2007), and the WM volume of the genual and splenial portions of the corpus callosum was abnormally reduced in MDD and BD (Brambilla et al. 2004). These regions of the corpus callosum were also smaller in child and adolescent offspring of women with MDD who had not yet developed a mood disorder relative to age-matched controls, suggesting that the reduction in WM in MDD reflects a developmental defect that exists prior to illness onset (Martinez et al. 2002).

Finally, satellite oligodendrocytes were also implicated in the pathophysiology of mood disorders by an electron microscopic study of the PFC in BD, which revealed decreased nuclear size, clumping of chromatin, and other types of damage to satellite oligodendrocytes, including indications of both apoptotic and necrotic degeneration (Uranova et al. 2001; Vostrikov et al. 2007). Satellite oligodendrocytes are immunohistochemically reactive for glutamine synthetase, suggesting that they function like astrocytes to take up synaptically released glutamate for conversion to glutamine and cycling back into neurons (Janus et al. 2000).

In other brain regions, reductions in astroglia have been reported by postmortem studies of mood disorders. In the frontal cortex, one study found that four forms of the astrocytic product glial fibrillary acidic protein (GFAP) were decreased in subjects with mood disorders relative to controls, although it was not determined whether this decrement reflected a reduction in astrocyte density or GFAP expression (Johnston-Wilson et al. 2000). However, another study that used immunohistochemical staining for GFAP found no significant differences in cortical astrocytes between controls and MDD or BD cases (Webster et al. 2001). Other studies also found no differences in GFAP between mood disorder cases and controls (Cotter et al. 2001b).

Factors that may conceivably contribute to a loss of oligodendroglia in mood disorders include elevated glucocorticoid secretion, microglial activation, and glutamatergic transmission evident during depression and mania. Glucocorticoids affect both glia and neurons (Cheng and de Vellis 2000), and elevated glucocorticoid concentrations and repeated stress decrease the proliferation of oligodendrocyte precursors (Alonso 2000; Banasr et al. 2004). Recent PET studies have found evidence for microglial activation in the PFC and ACC of patients with MDD and the hippocampus in patients with BD as indexed by an increase in the distribution of volume of the ligand for the translocator protein ligand, TPSO (Setiawan et al. 2015; Haarman et al. 2014). Activated microglia, in turn, releases the neurotoxin and NMDA receptor agonist quinolinic acid (QA) (Dantzer et al. 2008). In line with these data, a postmortem immunohistochemistry study showed that relative to controls, a mixed sample of MDD and BD subjects had increased QA-positive cell densities in the anterior mid-cingulate cortex and sgACC, suggesting microglial cell activation (Steiner et al. 2011). Activation of the kynurenine pathway may affect the structure of regions such as the hippocampus and medial PFC. We previously reported that the ratio of QA to kynurenic acid (KynA) in the serum is inversely correlated with hippocampal and amygdalar volume in both unmedicated and medicated patients with BD and have as yet unpublished data showing an inverse correlation between QA and thickness of the BA32 (Savitz et al. 2014a). Moreover, oligodendrocytes express α-amino-3-hydroxy-5-methyl-4-isoxazolepropionic acid (AMPA) and kainate type glutamate receptors and are sensitive to excitotoxic damage from excess glutamate (Hamidi et al. 2004). The targeted nature of the reductions in GM volume and glial cells to specific areas of the limbic–cortical circuits that show increased glucose metabolism during depressive episodes is noteworthy given the evidence reviewed below that the glucose metabolic signal is dominated by glutamatergic transmission.

8.4.3   Correlations with Rodent Models of Chronic and Repeated Stress

In regions that appear homologous to the areas where GM reductions are evident in humans with BD (i.e., medial PFC, hippocampus), repeated stress results in dendritic atrophy and reductions in glial cell counts or proliferation in rodents (Banasr et al. 2004; Czeh et al. 2005; McEwen and Magarinos 2001; Wellman 2001; Radley et al. 2008). In contrast, in the basolateral amygdala (BLA), chronic, unpredictable stress also produced dendritic atrophy, but chronic immobilization stress instead *increased* dendritic branching (Vyas et al. 2002, 2003).

Dendritic atrophy would be reflected by a decrease in the volume of the neuropil, which occupies most of the GM volume. The similarities between the histopathological changes that accompany stress-induced dendritic atrophy in rats and those found in humans suffering from depression thus led to the hypotheses that homologous processes underlie the reductions in GM volume in hippocampal and PFC structures in MDD and BD (McEwen and Magarinos 2001). In rats, the stress-induced dendritic atrophy in the medial PFC was associated with impaired modulation (i.e., extinction) of behavioral responses to fear-conditioned stimuli (Izquierdo et al. 2006; Miracle et al. 2006). Notably, healthy humans with thinner ventromedial PFC tissue also showed a greater galvanic skin response to conditioned stimuli during extinction learning (Milad et al. 2005). Finally, when rats were subjected to repeated stress beyond four weeks, the dendritic atrophy could be reversed by lithium (McEwen and Magarinos 2001), resembling the effects on sgACC volume in depressed humans.

In rodent stress models, these dendritic reshaping processes depend on interactions between increased NMDA receptor stimulation and glucocorticoid secretion associated with repeated stress (Wellman 2001; McEwen and Magarinos 2001). Elevations of glutamate transmission and cortisol secretion in mood disorders also may contribute to reductions in GM volume and synaptic markers by inducing dendritic atrophy in some brain structures, given that the depressive subtypes (e.g., BD, FPDD) who show regional reductions in GM volume also show evidence of cortisol hypersecretion under stressed conditions (reviewed in Drevets et al. (2002a)) and increased glutamate transmission. Subjects with familial BD also show elevations of glucose metabolism, which largely reflect glutamate transmission (see above) in the medial and orbital PFC, amygdala, and cingulate cortex regions that show reductions in GM volume and cellular elements. The findings that GM reductions appear to occur specifically in regions that show hypermetabolism during BD thus raise the possibility that excitatory amino acid transmission plays a role in the neuropathology of BD.

8.4.4   Implications for Treatment Mechanisms that Target Plasticity Around the Glutamatergic Synapse

Reverberatory glutamatergic transmission is thought to underlie the pathophysiological activation of limbic-thalamo-cortical circuits of the medial prefrontal (visceromotor) network in BD and MDD (Savitz et al. 2014a, b). The anatomical projections between the OMPFC, striatum, and amygdala implicated in mood disorders are formed by predominantly excitatory projections (Ongur et al. 2003). Because cerebral glucose metabolism largely reflects the energy requirements associated with glutamatergic transmission (Shulman et al. 2004), elevated metabolism in limbic-thalamo-cortical circuits in depression would imply that glutamatergic transmission is increased in these circuits (Drevets et al. 1992). Compatible with this hypothesis, postmortem studies of the NMDA receptor complex and other elements of the glutamatergic synapse in suicide victims show changes in synaptic gene expression and sensitivity that putatively reflect compensatory responses to abnormally elevated excitatory signaling in depression; these data implicate disturbances in glutamate metabolism, NMDA, and mGluR1,5 receptors in depression and suicide and suggest that glutamatergic transmission is increased in the PFC antemortem (Paul and Skolnick 2003; Duric et al. 2013). Furthermore, increased levels of glutamate have been found in postmortem tissue of the frontal cortex of individuals with MDD or BD (Hashimoto et al. 2007).

Additional postmortem data suggest that the impairment in glutamate transport results in excessive or dysregulated glutamate receptor signaling. Glutamate reuptake is critical for regulating glutamate concentrations in the synaptic cleft and maintaining normal synaptic activity. Evidence for impaired glutamate reuptake in mood disorders has been obtained in postmortem studies of mood disorders by direct measures of reduced glutamate transport by astroglia (Chandley et al. 2013; Rajkowska and Stockmeier 2013), reduced expression of glial excitatory amino acid transporters 1 and 2 (EAAT1 and EAAT2), and reductions in cell counts, density, and gene expression of the astrocyte and perineuronal oligodendroglia that express EAATs and glutamine synthetase (Miguel-Hidalgo et al. 2010; Kim and Webster 2010; Pitt et al. 2003). Microarray analysis of anterior cingulate and dorsolateral PFC (BA9, 46) tissue obtained from MDD patients postmortem demonstrated concomitant downregulation of EAAT1 (SLC1A3) and EAAT2 (SLC1A2) along with decreased expression of glutamine synthetase (L-glutamate–ammonia ligase), the enzyme that converts glutamate to glutamine (Choudary et al. 2005). The mechanisms that may potentially impair EAAT function in patients with mood disorders include glial cell dysfunction and loss, which some evidence suggests may arise secondary to the elevated release of glucocorticoid hormones and proinflammatory cytokines extant in subgroups of patients with mood disorders (Pitt et al. 2003; Boehmer et al. 2006; Boycott et al. 2008).

Notably, the astrocyte-based glutamate transporter GLT-1 (aka. EAAT-2) is thought to be responsible for nearly 90 % of glutamate uptake in the brain, and the importance of this and other glial-based glutamate transporters in the development of depressive behaviors is supported by preclinical models. In rodents, the local infusion of the astrocyte-specific toxin L-alpha-aminoadipic acid into the medial PFC induced depression-like behaviors (in contrast, infusion of ibotenic acid, a neuronal toxin, had no effect in this model) and pharmacological blockade of astrocytic glutamate uptake by dihydrokainic acid, an EAAT2 (GLT-1) inhibitor, induced anhedonia-like behavior (Banasr and Duman 2008). Moreover, Banasr and colleagues showed that chronic stress impairs cortical glial function in rodents in association with depression-like behaviors and impaired glutamate metabolism and that the cellular, metabolic, and behavioral alterations induced by chronic unpredictable stress were reversed and/or prevented by chronic administration of riluzole, which increases the expression of the astrocyte-based glutamate transporter GLT-1 (aka. EAAT-2) (Banasr et al. 2010). Riluzole also reversed depressive-like symptoms induced by chronic stress and in the olfactory bulbectomy model (reviewed in Pilc et al. (2013)).

Reverberatory glutamatergic transmission may also be a downstream effect of circuits being released from inhibition. For example, mRNA expression of parvalbumin (PV), a putative marker for a subset of GABA neurons that powerfully inhibits pyramidal cells via innervation at the cell soma and axon initial segment (Markram et al. 2004; Lewis et al. 2012), was reduced in postmortem samples of DLPFC from individuals with BD (Sibille et al. 2011). This study suggests that GABA neurotransmission in at least a subset of local circuit neurons is attenuated; however, the functional implications are not well understood. In addition to disinhibition of pyramidal neurons, other studies found that PV-containing interneurons are critical for generating gamma oscillations (Gonzalez-Burgos et al. 2015; Gonzalez-Burgos and Lewis 2008), which are implicated in normal cortical function (e.g., working memory (Yamamoto et al. 2014)). Interestingly, expression of mRNA for neuronal activity-regulated pentraxin (NARP), a protein that is secreted at presynaptic glutamate synapses that terminate on PV-containing interneurons, is reduced in the DLPFC of subjects with BD (Kimoto et al. 2015), suggesting that excitatory drive onto this interneuron subclass is disrupted; this, in turn, could lead to a disruption of gamma oscillations and associated cortical function.

Such a pathophysiological process may contribute to reverberatory excitatory transmission, which in the presence of the diminished excitatory amino acid transporter function and expression found in MDD and BD postmortem, may conceivably result in glutamate-induced excitotoxicity. Such a mechanism could account for the GM loss and accompanying histopathological changes in mood disorders, such as losses of neuropil (Stockmeier et al. 2004) and glial cells (Ongur et al. 1998). Chronic activation of AMPA receptors is neurotoxic and could be related to the reduced volume of the hippocampus and medial PFC that appears early in the course of depression but then becomes more prominent over the course of chronic or recurrent depressive episodes (Price and Drevets 2012; Stockmeier et al. 2004). Clinical correlations with these MRI data indicate that GM reductions are associated with disease recurrence and chronicity. The findings of reduced GM volume in association with impaired glutamate transport, dendritic atrophy, synapse loss, and cellular loss (especially of glia and interneurons) have led to hypotheses that dysregulation of glutamatergic signaling results in reverberatory activity in limbic-thalamo-cortical circuits implicated in mood disorders (Price and Drevets 2012).

The in vivo MRS literature highlights intriguing differences between MDD and BD within the glutamatergic system. The results of proton MRS studies have largely shown a decrease in Glx (which is constituted predominantly by the intracellular components of glutamate and glutamine), particularly within the medial PFC and DLPFC in depressed patients with MDD, but an increase in the Glx signal in BD (Yuksel and Ongur 2010; Taylor et al. 2009; Hasler et al. 2007). In addition, the MRS data differentiate depression from mania based on the ratio of glutamine to glutamate, which is abnormally reduced in studies of depression (both MDD and BD) but elevated in mania. These patterns suggest that the glutamate-related metabolite pool (only part of which is relevant to neurotransmission) is constricted in MDD and expanded in BD, but that depressive and manic episodes may be characterized by modulation of the glutamine/glutamate ratio in opposite directions, possibly suggesting reduced versus elevated glutamate conversion to glutamine by glial cells, respectively (Yuksel and Ongur 2010). Finally, studies showing that MRS measures of GABA are abnormally decreased in MDD suggest a decrease in GABAergic signaling (Sanacora et al. 1999; Hasler et al. 2007).

8.4.4.1 Antidepressant and Mood Stabilizing Drug Effects on Glutamatergic Transmission

In mood disorders, several experimental and conventional treatments may exert their clinical effects via mechanisms that depend on altering glutamatergic transmission. As described above, riluzole, which increases the expression of the astrocyte-based glutamate transporter GLT-1 (EAAT-2), appeared effective as adjunctive and monotherapy for treatment-resistant depression and as adjunctive therapy for bipolar depression, mainly in open-label studies (Zarate et al. 2004; Yuksel and Ongur 2010). In addition, lamotrigine is an anticonvulsant that reduces glutamate release via sodium, calcium, and potassium channel modulation and has shown efficacy as a mood stabilizer in BD, especially for prevention of depressive episodes and as adjunctive therapy for both BD and MDD (Yuksel and Ongur 2010). Notably, a single, subanesthetic dose infusion of ketamine has been shown to have rapid and potent antidepressant effects in treatment-resistant MDD and BD patients (Lee et al. 2015; Ionescu et al. 2015; McGirr et al. 2015; Zarate et al. 2012). Further, antidepressant and mood stabilizing drugs that have diverse primary pharmacological actions are hypothesized to have a final common pathway of reducing NMDA receptor sensitivity and/or transmission, and many of these agents also increase GABA levels or transmission (Krystal et al. 2002; Paul and Skolnick 2003). For example, chronic treatment with some conventional antidepressant drugs reduces both glutamate release in the rat brain and NMDA receptor subunit expression in depressed humans (Bonanno et al. 2005; Golembiowska and Dziubina 2000; Paul and Skolnick 2003), suggesting that a persistent effect that modulates dysregulation of glutamate release can maintain antidepressant effects. Compatible with these data, during effective antidepressant drug or electroconvulsive therapy, glucose metabolic activity decreases in the regions of the extended visceromotor network (Table 8.1; (Drevets et al. 2002b, 2004)), which would be expected if treatment-induced NMDA receptor desensitization modulated glutamatergic transmission (Paul and Skolnick 2003). As described in the ensuing sections, elevated glutamatergic transmission within discrete anatomical circuits may partly explain the targeted nature of GM changes within mood disorders (e.g., affecting left more than right sgACC) (McEwen and Magarinos 2001; Drevets and Price 2005), so one important mechanism of effective treatment in bipolar depression may involve modulation of excessive excitatory transmission (McEwen and Magarinos 2001).

In contrast, preclinical data suggest that the mechanism underlying the antidepressant effect of acute administration of ketamine and some other NMDA receptor antagonists, which persist for at least several days beyond the clearance of the drug from the plasma in some BD and MDD patients, instead may depend on a transient disinhibition of glutamate release that stimulates a persistent effect on synaptic plasticity. Duman and colleagues found that a single administration of ketamine can reverse the loss of synapses and synaptic function induced by repeated stress by inducing synaptogenesis in the medial PFC (Abdallah et al. 2015). Conceivably, ketamine's antidepressant effect in MDD and BD may reflect such an effect on synaptic plasticity that addresses the histopathological changes in mood disorders, but that may be initiated and potentially maintained by acute administration followed by pulsed, as opposed to chronic, treatment.

8.5 Neuroreceptor Imaging in Bipolar Depression

Of the neurochemical systems that modulate neural transmission within the visceromotor network, mood disorders have been associated with abnormalities of serotonergic, dopaminergic, noradrenergic, cholinergic, glutamatergic, GABAergic, glucocorticoid, and peptidergic (e.g., corticotrophin releasing factor, CRF) functions. Some receptors of the monoaminergic neurotransmitter systems have been imaged in BD using PET or SPECT.

8.5.1 Serotonergic System

The central serotonin (5-HT) system has received particular interest in depression research because selective serotonin reuptake inhibitors (SSRIs) exert antidepressant effects and because some other antidepressant drug classes also increase postsynaptic 5-HT$_{1A}$ receptor transmission (Drevets et al. 2007). This effect of antidepressant drugs may augment endogenous serotonin release during the stress of depression, analogous to the enhanced serotonergic transmission that occurs in some brain regions during stress in rodents (Cannon et al. 2007; Barton et al. 2008). Enhancement of serotonin transmission in MDD also may compensate for abnormalities in density and sensitivity of some serotonin receptor subtypes evidenced by postmortem, neuroimaging, and pharmacological challenge studies of depression (Stockmeier 2003; Drevets et al. 2007). For example, postsynaptic 5-HT$_{1A}$ receptor binding or mRNA expression is decreased in the insula, hippocampus, cingulate, parieto-occipital, and orbital/ventrolateral prefrontal cortices in some neuroimaging studies of MDD and BD (Drevets et al. 2007; Bhagwagar et al. 2004; Sargent et al. 2000; Hirvonen et al. 2008). Of note is our group's recent replication of our previous findings using a slightly different methodology. Specifically, we found that mean 5-HT$_{1A}$ receptor binding potential (measured by BP$_P$ as well as BP$_{ND}$) was significantly lower in BD subjects compared to controls in cortical regions where 5-HT$_{1A}$ receptors are expressed postsynaptically, most prominently in the mesiotemporal cortex (Nugent et al. 2013). Further, BP$_P$ in the mesiotemporal cortex was inversely correlated with trough plasma cortisol levels, consistent with preclinical literature indicating that hippocampal 5-HT$_{1A}$ receptor expression is inhibited by glucocorticoid receptor stimulation (Nugent et al. 2013).

These data receive support from a PET study of previously healthy subjects exposed to a severe recent stressor. Compared with nonstressed subjects, stressed subjects displayed reduced binding in the ACC, insula, and hippocampus, with a trend toward significance in the amygdala, DLPFC, parietal cortex, temporal cortex, and raphe (Jovanovic et al. 2011). In addition, a reduction in 5-HT$_{1A}$ receptor distribution volume (V_T) was found in the raphe nuclei, amygdala, hippocampus, and ACC in a PET study of subordinate cynomolgus monkeys who showed behavioral signs of depression after exposure to social defeat (Shively et al. 2006).

In postmortem studies, 5-HT$_{1A}$ receptor binding was increased in the rostral, ventrolateral, and dorsal subnuclei of the raphe but decreased in the caudal subnucleus of the raphe (Boldrini et al. 2008; Stockmeier et al. 1998).

Although the interruption of 5-HT$_{1A}$ receptor function during neurodevelopment has been shown to persistently alter the function of emotion-modulating systems in genetically engineered mice (Gross et al. 2002), the reduction in postsynaptic 5-HT$_{1A}$ receptor binding and mRNA expression in mood disorders may arise secondary to cortisol hypersecretion (Lopez et al. 1998). 5-HT$_{1A}$ receptor mRNA expression and density are tonically inhibited by glucocorticoid receptor stimulation. In experimental animals, elevated CORT secretion during chronic or repeated stress resulted in reduced 5-HT$_{1A}$ receptor density and mRNA expression (Lopez et al. 1998; Flugge 1995). Thus, the mood disordered subgroups with reduced postsynaptic 5-HT$_{1A}$ receptor binding may be limited to those with a diathesis to hypersecrete cortisol (e.g., Lopez et al. (1998), Drevets et al. (1999, 2007)).

Altered serotonin transporter (5-HTT) function is also thought to play a role in the pathophysiology of mood disorders (Cannon et al. 2006b; Stockmeier 2003). For example, depressed BD subjects showed elevated 5-HTT binding in the striatum, thalamus, and insula, as well as reduced binding in the vicinity of the pontine raphe (Cannon et al. 2006b, 2007) (Fig. 8.2). PET studies performed using 5-HTT radioligands with high selectivity for 5-HTT sites, such as [^{11}C]DASB, similarly

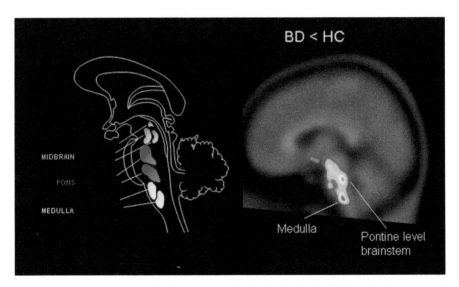

Fig. 8.2 Serotonin transporter binding is reduced in bipolar depression. This section from a voxelwise analysis of [^{11}C]DASB parametric binding potential images shows regions where individuals with bipolar disorder (BD) have reduced serotonin transporter (5-HTT) binding relative to controls at $p < 0.05$ (*right panel*), together with a schematic illustration showing approximate locations of the raphe nuclei within the brainstem (*left panel*; after Carpenter and Sutin (1983)). Reproduced with permission from Cannon et al. (2006b)

reported abnormally increased 5-HTT binding in the striatum, thalamus, insula, and ACC of individuals with early onset MDD and/or those MDD patients with negativistic attitudes; however, the reduction in 5-HTT binding in the pontine raphe found in individuals with bipolar depression did not extend to MDD cases (Cannon et al. 2007). Nevertheless, although another study found no difference in 5-HTT binding between MDD patients and healthy controls, scores on the Dysfunctional Attitude Scale were correlated positively with 5-HTT binding in the PFC, ACC, putamen, and thalamus (Meyer et al. 2004). The possible increase in 5-HTT binding in BD is also consistent with increases in 5-HTT binding in the DLPFC, amygdala, hypothalamus, raphe, and posterior cingulate cortex in clinically depressed patients with Parkinson's disease (PD) (Politis et al. 2010; Boileau et al. 2008).

8.5.2 Dopamine Receptor Imaging

Neuroimaging studies have discovered abnormalities involving multiple aspects of the central dopaminergic system in depression, which converge with other types of evidence to implicate this system in the pathophysiology of mood disorders. However, few of these studies specifically assessed bipolar depression. With respect to dopamine D1 receptors, Suhara and colleagues reported that binding of [^{11}C]SCH-23990 was decreased in the frontal cortex of BD subjects studied in various illness phases, a finding that awaits replication using more selective D_1 receptor ligands (Suhara et al. 1992). [^{11}C]SCH-23390 displays poor specificity, which potentially compromises the reliability of the D_1 receptor measurement in extrastriatal areas such as the frontal cortex, where the density of these receptors is significantly lower than in the striatum. In a postmortem study, the percentage of D1-expressing neurons together with D1 mRNA expression was reported to be increased by 25 % in the CA3 region of the hippocampus (Pantazopoulos et al. 2004) in BD subjects versus controls, but did not differ from controls in the amygdala of MDD subjects (Xiang et al. 2008).

Pearlson and colleagues showed that *psychotic* individuals with BD had increased striatal uptake of the dopamine D2/D3 receptor ligand, [^{11}C]-*N*-methylspiperone, relative to healthy controls and nonpsychotic individuals with BD but that the nonpsychotic BD cases did not differ from healthy controls (Pearlson et al. 1995). In contrast, no difference in the binding of the more selective ligand, [^{11}C]raclopride, was found between manic patients with BD and healthy controls (Yatham et al. 2002). Similarly, a SPECT-[^{123}I]IBZM study found no difference in striatal dopamine D2/D3 receptor binding at baseline, and no difference in change in [^{123}I]IBZM binding under amphetamine challenge between medicated, euthymic BD patients and healthy controls (Anand et al. 2000). It is unclear if D2 receptor binding differs in subjects with MDD compared to BD. Unmedicated MDD patients with motor retardation displayed increased binding in the caudate and striatum compared with healthy controls (Meyer et al. 2006).

Regarding the dopamine transporter, DAT, decreased binding was reported in the caudate, but not the putamen of BD patients relative to healthy controls (Anand et al. 2011), raising the possibility that abnormalities of dopamine reuptake constitute a risk factor for the development of BD.

8.5.3 Cholinergic System

The cholinergic system is also implicated in the pathophysiology of mood disorders, with evidence indicating that the muscarinic cholinergic system is overactive or hyperresponsive in depression. Janowsky and colleagues (1994) reported that increasing cholinergic activity using the acetylcholinesterase inhibitor, physostigmine, resulted in the rapid induction of depressive symptoms in currently manic BD subjects and in a worsening of symptoms in individuals with MDD. The administration of the M_2R antagonist, procaine, elicited emotional responses in humans ranging from sadness, fear, and severe anxiety to euphoria and increased the physiological activity of the cingulate cortex (Benson et al. 2004; Ketter et al. 1996), a region densely innervated by cholinergic projections. In individuals with bipolar depression, decreased M_2R binding has been reported in the cingulate cortex (Cannon et al. 2006a) (Fig. 8.3). This finding appeared attributable to an interaction involving genetic variation in the cholinergic-muscarinic type 2 (M2) receptor gene (CHRM2). A differential impact of the M2-receptor polymorphism at rs324650 on the V_T of the M2 receptor was evident in individuals with BD versus healthy controls, such that the BD subjects homozygous for the T-allele showed markedly lower V_T (by 27–37 % across regions) than healthy controls of the same genotype (Cannon et al. 2011). Post hoc analyses suggested that within the BD sample, T homozygosity was associated with a more severe illness course. Multiple M_2R gene polymorphisms have been associated with increased risk for developing major depressive episodes (Cannon et al. 2011), but thus far, these single nucleotide polymorphisms (SNPs) have not been associated with BD. Finally, the muscarinic cholinergic receptor antagonist, scopolamine, exerts rapid and robust antidepressant effects in depressed MDD and BD patients, although the ~24 hour delay in onset of these effects raises the possibility that a secondary mechanism of action underlies the antidepressant response (Furey and Drevets 2006). Preclinical evidence suggests that this antidepressant effect depends on the induction of synaptic plasticity changes analogous to those induced by ketamine administration (Duman 2014).

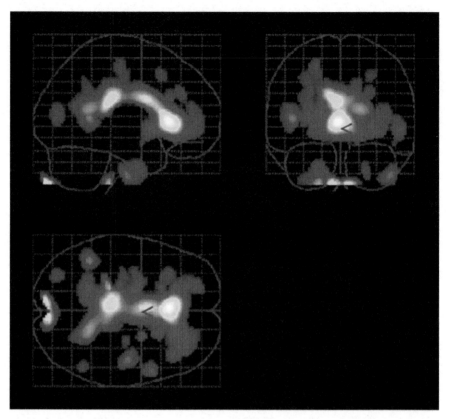

Fig. 8.3 Reduced muscarinic type 2 (M2) receptor binding in the cingulate cortex in individuals with bipolar depression relative to healthy controls. The statistical parametric map shows voxel t-values corresponding to areas where the uptake of $[^{18}F]FP$-TZTP, a PET radioligand that selectively binds M2 receptors, was significantly reduced (at $p < 0.005$) in individuals with bipolar depression relative to healthy controls. The areas of maximal difference between groups were located in the anterior cingulate cortex. Reproduced from Cannon et al. (2006a)

8.6 Implications for Neurocircuitry Models of Depression

Taken together, the neuropathological, neurochemical, and neurophysiological abnormalities extant within the extended visceromotor network may impair this network's modulation of autonomic, endocrine, immune, neurotransmitter, emotional, and cognitive responses to aversive and reward-related stimuli or contexts (Ongur et al. 2003), potentially accounting for the disturbances within these domains seen in BD (Fig. 8.4). The neuroimaging abnormalities in the VLPFC, OFC, sgACC, pgACC, amygdala, ventral striatum, and medial thalamus evident in BD implicate a limbic-thalamo-cortical circuit involving the amygdala, the mediodorsal nucleus of the thalamus (MD), and the OMPFC, and a limbic–striatal–pallidal–thalamic circuit involving related parts of the striatum and ventral

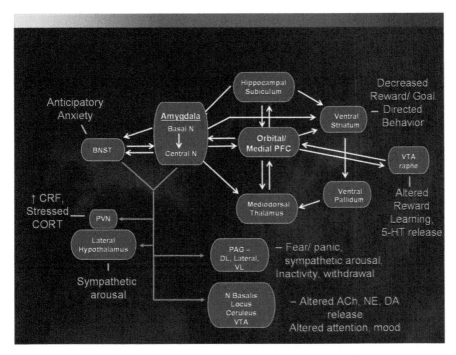

Fig. 8.4 Anatomical circuits involving the orbitomedial PFC (OMPFC) and amygdala reviewed within the context of a model in which OMPFC dysfunction results in disinhibition of limbic transmission through the amygdala, yielding the emotional, cognitive, endocrine, autonomic, and neurochemical manifestations of depression. The basolateral amygdala sends efferent projections to the central nucleus of the amygdala (ACe) and the bed nucleus of the stria terminalis (BNST). The efferent projections from these structures to the hypothalamus, periaqueductal gray (PAG), nucleus basalis, locus coeruleus, raphe, and other diencephalic and brainstem nuclei then organize the neuroendocrine, neurotransmitter, autonomic, and behavioral responses to stressors and emotional stimuli (Davis and Shi 1999; LeDoux 2003). The OMPFC shares reciprocal projections with all of these structures (although only the connections with the amygdala are illustrated), which function to modulate each component of emotional expression (Ongur et al. 2003). Impaired OMPFC function may thus disinhibit or dysregulate limbic outflow through the ACe and BNST. *Solid white lines* indicate some of the major anatomical connections between structures, with *closed arrowheads* indicating the direction of projecting axons. *Solid yellow lines* show efferent pathways of the ACe and BNST, which are generally monosynaptic, but in some cases are bisynaptic connections (e.g., Herman and Cullinan (1997)). Other abbreviations: 5-HT—serotonin; ACh—acetylcholine; DA—dopamine; DL—dorsolateral column of PAG; N—nucleus, NE—norepinephrine; NTS—nucleus tractus solitarius; PVN—paraventricular N of the hypothalamus; VL—ventrolateral column of PAG; VTA—ventral tegmental area. Reproduced with permission from Drevets (2007)

pallidum along with the components of the other circuit (Drevets et al. 1992). The first of these circuits can be conceptualized as an excitatory triangular circuit (Fig. 8.4), whereby the BLA and the OMPFC are interconnected by excitatory (especially glutamatergic) projections with each other and with the MD (Drevets and Price 2005), so increased glucose metabolism in these structures would presumably reflect increased synaptic transmission through the limbic-thalamo-cortical circuit.

The basolateral nuclei of the amygdala send anatomical projections to the central nucleus of the amygdala (ACe) and the BNST, and projections from these structures to the hypothalamus, PAG, locus coeruleus, raphe, nucleus basalis, and other diencephalic and brainstem nuclei play major roles in organizing neuroendocrine, neurotransmitter, autonomic, and behavioral responses to stressors and emotional stimuli (Davis and Shi 1999; LeDoux 2003). The OMPFC sends overlapping projections to each of these structures and to the amygdala that function to modulate each component of emotional expression (Price and Drevets 2010). The neuropathological changes evident in the OMPFC in BD and mood disorders arising secondary to neurological disorders may thus impair the modulatory role of the OMPFC over emotional expression, disinhibiting or dysregulating limbic responses to stressors and to social and emotional stimuli (Drevets and Price 2005; Price and Drevets 2012). These data suggest that during depressive episodes, the increased activity seen within some OMPFC areas reflects a compensatory response that modulates depressive symptoms and that impaired function of these emotion regulatory regions (possibly due to the neuropathological changes in BD) may result in more severe and treatment-refractory illness (Phillips et al. 2008).

8.7 Summary

Convergent results from studies conducted using neuroimaging, lesion analysis, and postmortem techniques support models in which the clinical phenomenology of BD emanates from dysfunction affecting the extended medial prefrontal network that interferes with this system's modulation of visceromotor and emotional behavior (Savitz et al. 2014b). At a molecular level, these abnormalities may be partly driven by a dysregulation of glutamatergic neurotransmission that in turn may ultimately be related to inflammatory processes. Mood stabilizing and antidepressant therapies may compensate for this dysfunction by attenuating pathological excitatory transmission in cortico-limbic circuits (Drevets et al. 2002a) and increasing expression of neurotrophic/neuroprotective factors that preserve the structure and function of the OMPFC (Manji et al. 2001).

References

Abdallah CG, Sanacora G, Duman RS, Krystal JH (2015) Ketamine and rapid-acting antidepressants: a window into a new neurobiology for mood disorder therapeutics. Annu Rev Med 66:509–523

Adler CM, Holland SK, Schmithorst V, Wilke M, Weiss KL, Pan H, Strakowski SM (2004) Abnormal frontal white matter tracts in bipolar disorder: a diffusion tensor imaging study. Bipolar Disord 6(3):197–203

Ahearn EP, Steffens DC, Cassidy F, Van Meter SA, Provenzale JM, Seldin MF, Weisler RH, Krishnan KR (1998) Familial leukoencephalopathy in bipolar disorder. Am J Psychiatry 155(11):1605–1607

Alonso G (2000) Prolonged corticosterone treatment of adult rats inhibits the proliferation of oligodendrocyte progenitors present throughout white and gray matter regions of the brain. Glia 31(3):219–231

Altshuler LL, Bartzokis G, Grieder T, Curran J, Jimenez T, Leight K, Wilkins J, Gerner R, Mintz J (2000) An MRI study of temporal lobe structures in men with bipolar disorder or schizophrenia. Biol Psychiatry 48(2):147–162

Anand A, Verhoeff P, Seneca N, Zoghbi SS, Seibyl JP, Charney DS, Innis RB (2000) Brain SPECT imaging of amphetamine-induced dopamine release in euthymic bipolar disorder patients. Am J Psychiatry 157(7):1108–1114

Anand A, Barkay G, Dzemidzic M, Albrecht D, Karne H, Zheng QH, Hutchins GD, Normandin MD, Yoder KK (2011) Striatal dopamine transporter availability in unmedicated bipolar disorder. Bipolar Disord 13(4):406–413

Anstey KJ, Jorm AF, Reglade-Meslin C, Maller J, Kumar R, von Sanden C, Windsor TD, Rodgers B, Wen W, Sachdev P (2006) Weekly alcohol consumption, brain atrophy, and white matter hyperintensities in a community-based sample aged 60 to 64 years. Psychosom Med 68(5):778–785

Aston C, Jiang L, Sokolov BP (2005) Transcriptional profiling reveals evidence for signaling and oligodendroglial abnormalities in the temporal cortex from patients with major depressive disorder. Mol Psychiatry 10(3):309–322

Banasr M, Duman RS (2008) Glial loss in the prefrontal cortex is sufficient to induce depressive-like behaviors. Biol Psychiatry 64(10):863–870

Banasr M, Hery M, Printemps R, Daszuta A (2004) Serotonin-induced increases in adult cell proliferation and neurogenesis are mediated through different and common 5-HT receptor subtypes in the dentate gyrus and the subventricular zone. Neuropsychopharmacology 29(3):450–460

Banasr M, Chowdhury GM, Terwilliger R, Newton SS, Duman RS, Behar KL, Sanacora G (2010) Glial pathology in an animal model of depression: reversal of stress-induced cellular, metabolic and behavioral deficits by the glutamate-modulating drug riluzole. Mol Psychiatry 15(5):501–511

Barton DA, Esler MD, Dawood T, Lambert EA, Haikerwal D, Brenchley C, Socratous F, Hastings J, Guo L, Wiesner G, Kaye DM, Bayles R, Schlaich MP, Lambert GW (2008) Elevated brain serotonin turnover in patients with depression: effect of genotype and therapy. Arch Gen Psychiatry 65(1):38–46

Barysheva M, Jahanshad N, Foland-Ross L, Altshuler LL, Thompson PM (2013) White matter microstructural abnormalities in bipolar disorder: A whole brain diffusion tensor imaging study. Neuroimage Clin 2:558–568

Bauer M, London ED, Rasgon N, Berman SM, Frye MA, Altshuler LL, Mandelkern MA, Bramen J, Voytek B, Woods R, Mazziotta JC, Whybrow PC (2005) Supraphysiological doses of levothyroxine alter regional cerebral metabolism and improve mood in bipolar depression. Mol Psychiatry 10(5):456–469

Baumann B, Danos P, Krell D, Diekmann S, Leschinger A, Stauch R, Wurthmann C, Bernstein HG, Bogerts B (1999) Reduced volume of limbic system-affiliated basal ganglia in mood disorders: preliminary data from a postmortem study. J Neuropsychiatry Clin Neurosci 11(1):71–78

Bearden CE, Thompson PM, Dalwani M, Hayashi KM, Lee AD, Nicoletti M, Trakhtenbroit M, Glahn DC, Brambilla P, Sassi RB, Mallinger AG, Frank E, Kupfer DJ, Soares JC (2007) Greater cortical gray matter density in lithium-treated patients with bipolar disorder. Biol Psychiatry 62:7–16

Benes FM, Vincent SL, Todtenkopf M (2001) The density of pyramidal and nonpyramidal neurons in anterior cingulate cortex of schizophrenic and bipolar subjects. Biol Psychiatry 50 (6):395–406

Benson BE, Carson RE, Kiesewetter DO, Herscovitch P, Eckelman WC, Post RM, Ketter TA (2004) A potential cholinergic mechanism of procaine's limbic activation. Neuropsychopharmacology 29:1239–1250

Beyer JL, Kuchibhatla M, Payne ME, Moo-Young M, Cassidy F, Macfall J, Krishnan KR (2004) Hippocampal volume measurement in older adults with bipolar disorder. Am J Geriatr Psychiatry 12(6):613–620

Bhagwagar Z, Rabiner EA, Sargent PA, Grasby PM, Cowen PJ (2004) Persistent reduction in brain serotonin1A receptor binding in recovered depressed men measured by positron emission tomography with [11C]WAY-100635. Mol Psychiatry 9(4):386–392

Biegon A, Essar N, Israeli M, Elizur A, Bruch S, Bar-Nathan AA (1990) Serotonin 5-HT2 receptor binding on blood platelets as a state dependent marker in major affective disorder. Psychopharmacology (Berl) 102(1):73–75

Blumberg HP, Stern E, Ricketts S, Martinez D, de Asis J, White T, Epstein J, Isenberg N, McBride PA, Kemperman I, Emmerich S, Dhawan V, Eidelberg D, Kocsis JH, Silbersweig DA (1999) Rostral and orbital prefrontal cortex dysfunction in the manic state of bipolar disorder. Am J Psychiatry 156(12):1986–1988

Boehmer C, Palmada M, Rajamanickam J, Schniepp R, Amara S, Lang F (2006) Post-translational regulation of EAAT2 function by co-expressed ubiquitin ligase Nedd4-2 is impacted by SGK kinases. J Neurochem 97(4):911–921

Boileau I, Warsh JJ, Guttman M, Saint-Cyr JA, McCluskey T, Rusjan P, Houle S, Wilson AA, Meyer JH, Kish SJ (2008) Elevated serotonin transporter binding in depressed patients with Parkinson's disease: a preliminary PET study with [11C]DASB. Mov Disord 23 (12):1776–1780

Boldrini M, Underwood MD, Mann JJ, Arango V (2008) Serotonin-1A autoreceptor binding in the dorsal raphe nucleus of depressed suicides. J Psychiatr Res 42(6):433–442

Bonanno G, Giambelli R, Raiteri L, Tiraboschi E, Zappettini S, Musazzi L, Raiteri M, Racagni G, Popoli M (2005) Chronic antidepressants reduce depolarization-evoked glutamate release and protein interactions favoring formation of SNARE complex in hippocampus. J Neurosci 25(13):3270–3279

Botteron KN, Vannier MW, Geller B, Todd RD, Lee BC (1995) Preliminary study of magnetic resonance imaging characteristics in 8- to 16-year-olds with mania. J Am Acad Child Adolesc Psychiatry 34(6):742–749

Botteron KN, Raichle ME, Drevets WC, Heath AC, Todd RD (2002) Volumetric reduction in left subgenual prefrontal cortex in early onset depression. Biol Psychiatry 51(4):342–344

Bowen DM, Najlerahim A, Procter AW, Francis PT, Murphy E (1989) Circumscribed changes of the cerebral cortex in neuropsychiatric disorders of later life. Proc Natl Acad Sci USA 86(23):9504–9508

Bowley MP, Drevets WC, Ongur D, Price JL (2002) Low glial numbers in the amygdala in major depressive disorder. Biol Psychiatry 52(5):404–412

Boycott HE, Wilkinson JA, Boyle JP, Pearson HA, Peers C (2008) Differential involvement of TNF alpha in hypoxic suppression of astrocyte glutamate transporters. Glia 56(9):998–1004

Brambilla P, Harenski K, Nicoletti M, Sassi RB, Mallinger AG, Frank E, Kupfer DJ, Keshavan MS, Soares JC (2003) MRI investigation of temporal lobe structures in bipolar patients. J Psychiatr Res 37(4):287–295

Brambilla P, Nicoletti M, Sassi RB, Mallinger AG, Frank E, Keshavan MS, Soares JC (2004) Corpus callosum signal intensity in patients with bipolar and unipolar disorder. J Neurol Neurosurg Psychiatry 75(2):221–225

Calhoun VD, Maciejewski PK, Pearlson GD, Kiehl KA (2008) Temporal lobe and "default" hemodynamic brain modes discriminate between schizophrenia and bipolar disorder. Hum Brain Mapp 29(11):1265–1275

Cannon M, Jones PB, Murray RM (2002) Obstetric complications and schizophrenia: historical and meta-analytic review. Am J Psychiatry 159(7):1080–1092

Cannon DM, Carson RE, Nugent AC, Eckelman WC, Kiesewetter DO, Williams J, Rollis D, Drevets M, Gandhi S, Solorio G, Drevets WC (2006a) Reduced muscarinic type 2 receptor binding in subjects with bipolar disorder. Arch Gen Psychiatry 63(7):741–747

Cannon DM, Ichise M, Fromm SJ, Nugent AC, Rollis D, Gandhi SK, Klaver JM, Charney DS, Manji HK, Drevets WC (2006b) Serotonin transporter binding in bipolar disorder assessed using [11C]DASB and positron emission tomography. Biol Psychiatry 60(3):207–217

Cannon DM, Ichise M, Rollis D, Klaver JM, Gandhi SK, Charney DS, Manji HK, Drevets WC (2007) Elevated serotonin transporter binding in major depressive disorder assessed using positron emission tomography and [11C]DASB; comparison with bipolar disorder. Biol Psychiatry 62(8):870–877

Cannon DM, Klaver JK, Gandhi SK, Solorio G, Peck SA, Erickson K, Akula N, Savitz J, Eckelman WC, Furey ML, Sahakian BJ, McMahon FJ, Drevets WC (2011) Genetic variation in cholinergic muscarinic-2 receptor gene modulates M2 receptor binding in vivo and accounts for reduced binding in bipolar disorder. Mol Psychiatry 16(4):407–418

Carpenter M, Sutin J (1983) Human neuroanatomy, 8th edn. Williams and Wilkins, Baltimore, MD

Carter CJ (2007a) EIF2B and oligodendrocyte survival: where nature and nurture meet in bipolar disorder and schizophrenia? Schizophr Bull 33:1343–1353

Carter CJ (2007b) Multiple genes and factors associated with bipolar disorder converge on growth factor and stress activated kinase pathways controlling translation initiation: implications for oligodendrocyte viability. Neurochem Int 50(3):461–490

Cecil KM, DelBello MP, Morey R, Strakowski SM (2002) Frontal lobe differences in bipolar disorder as determined by proton MR spectroscopy. Bipolar Disord 4(6):357–365

Chabriat H, Vahedi K, Iba-Zizen MT, Joutel A, Nibbio A, Nagy TG, Krebs MO, Julien J, Dubois B, Ducrocq X et al (1995) Clinical spectrum of CADASIL: a study of 7 families. Cerebral autosomal dominant arteriopathy with subcortical infarcts and leukoencephalopathy. Lancet 346(8980):934–939

Chakos MH, Lieberman JA, Alvir J, Bilder R, Ashtari M (1995) Caudate nuclei volumes in schizophrenic patients treated with typical antipsychotics or clozapine. Lancet 345 (8947):456–457

Chandley MJ, Szebeni K, Szebeni A, Crawford J, Stockmeier CA, Turecki G, Miguel-Hidalgo JJ, Ordway GA (2013) Gene expression deficits in pontine locus coeruleus astrocytes in men with major depressive disorder. J Psychiatry Neurosci 38(4):276–284

Chase HW, Nusslock R, Almeida JR, Forbes EE, LaBarbara EJ, Phillips ML (2013) Dissociable patterns of abnormal frontal cortical activation during anticipation of an uncertain reward or loss in bipolar versus major depression. Bipolar Disord 15(8):839–854

Cheng JD, de Vellis J (2000) Oligodendrocytes as glucocorticoids target cells: functional analysis of the glycerol phosphate dehydrogenase gene. J Neurosci Res 59(3):436–445

Choudary PV, Molnar M, Evans SJ, Tomita H, Li JZ, Vawter MP, Myers RM, Bunney WE Jr, Akil H, Watson SJ, Jones EG (2005) Altered cortical glutamatergic and GABAergic signal transmission with glial involvement in depression. Proc Natl Acad Sci USA 102 (43):15653–15658

Coryell W, Nopoulos P, Drevets W, Wilson T, Andreasen NC (2005) Subgenual prefrontal cortex volumes in major depressive disorder and schizophrenia: diagnostic specificity and prognostic implications. Am J Psychiatry 162(9):1706–1712

Cotter D, Mackay D, Landau S, Kerwin R, Everall I (2001a) Reduced glial cell density and neuronal size in the anterior cingulate cortex in major depressive disorder. Arch Gen Psychiatry 58(6):545–553

Cotter DR, Pariante CM, Everall IP (2001b) Glial cell abnormalities in major psychiatric disorders: the evidence and implications. Brain Res Bull 55(5):585–595

Cotter D, Mackay D, Chana G, Beasley C, Landau S, Everall IP (2002) Reduced neuronal size and glial cell density in area 9 of the dorsolateral prefrontal cortex in subjects with major depressive disorder. Cereb Cortex 12(4):386–394

Cousins DA, Aribisala B, Nicol Ferrier I, Blamire AM (2013) Lithium, gray matter, and magnetic resonance imaging signal. Biol Psychiatry 73(7):652–657

Czeh B, Lucassen PJ (2007) What causes the hippocampal volume decrease in depression? Are neurogenesis, glial changes and apoptosis implicated? Eur Arch Psychiatry Clin Neurosci 257:250–260

Czeh B, Pudovkina O, van der Hart MG, Simon M, Heilbronner U, Michaelis T, Watanabe T, Frahm J, Fuchs E (2005) Examining SLV-323, a novel NK1 receptor antagonist, in a chronic psychosocial stress model for depression. Psychopharmacology (Berl) 180(3):548–557

Dager SR, Friedman SD (2000) Brain imaging and the effects of caffeine and nicotine. Ann Med 32(9):592–599

Dantzer R, O'Connor JC, Freund GG, Johnson RW, Kelley KW (2008) From inflammation to sickness and depression: when the immune system subjugates the brain. Nat Rev Neurosci 9(1):46–56

Davis M, Shi C (1999) The extended amygdala: are the central nucleus of the amygdala and the bed nucleus of the stria terminalis differentially involved in fear versus anxiety? Ann N Y Acad Sci 877:281–291

de Castro AG, Bajbouj M, Schlattmann P, Lemke H, Heuser I, Neu P (2008) Cerebrovascular reactivity in depressed patients without vascular risk factors. J Psychiatr Res 42:78–82

DelBello MP, Zimmerman ME, Mills NP, Getz GE, Strakowski SM (2004) Magnetic resonance imaging analysis of amygdala and other subcortical brain regions in adolescents with bipolar disorder. Bipolar Disord 6(1):43–52

Desmond DW, Moroney JT, Lynch T, Chan S, Chin SS, Mohr JP (1999) The natural history of CADASIL: a pooled analysis of previously published cases. Stroke 30(6):1230–1233

Dickstein DP, Milham MP, Nugent AC, Drevets WC, Charney DS, Pine DS, Leibenluft E (2005) Frontotemporal alterations in pediatric bipolar disorder: results of a voxel-based morphometry study. Arch Gen Psychiatry 62(7):734–741

Drevets WC (2001) Neuroimaging and neuropathological studies of depression: implications for the cognitive-emotional features of mood disorders. Curr Opin Neurobiol 11(2):240–249

Drevets WC (2007) Orbitofrontal cortex function and structure in depression. Ann N Y Acad Sci 1121:499–527

Drevets WC, Price JL (2005) Neuroimaging and neuropathological studies of mood disorders. In: Licinio JWM (ed) Biology of depression: from novel insights to therapeutic strategies, vol 1. Wiley-VCH Verlag GmbH & Co., Weinheim, pp 427–466

Drevets WC, Todd RD (2005) Depression, mania and related disorders. In: Rubin E, Zorumski C (eds) Adult psychiatry, 2nd edn. Blackwell, Oxford, pp 91–129

Drevets WC, Videen TO, Price JL, Preskorn SH, Carmichael ST, Raichle ME (1992) A functional anatomical study of unipolar depression. J Neurosci 12(9):3628–3641

Drevets WC, Price JL, Simpson JR Jr, Todd RD, Reich T, Vannier M, Raichle ME (1997) Subgenual prefrontal cortex abnormalities in mood disorders. Nature 386(6627):824–827

Drevets WC, Frank E, Price JC, Kupfer DJ, Holt D, Greer PJ, Huang Y, Gautier C, Mathis C (1999) PET imaging of serotonin 1A receptor binding in depression. Biol Psychiatry 46 (10):1375–1387

Drevets WC, Bogers W, Raichle ME (2002a) Functional anatomical correlates of antidepressant drug treatment assessed using PET measures of regional glucose metabolism. Eur Neuro-psychopharmacol 12(6):527–544

Drevets WC, Price JL, Bardgett ME, Reich T, Todd RD, Raichle ME (2002b) Glucose metabolism in the amygdala in depression: relationship to diagnostic subtype and plasma cortisol levels. Pharmacol Biochem Behav 71(3):431–447

Drevets WC, Gadde K, Krishnan KRR (2004) Neuroimaging studies of depression. In: Charney DS, Nestler EJ, Bunney BJ (eds) The neurobiological foundation of mental illness, 2nd edn. Oxford University Press, New York, pp 461–490

Drevets WC, Thase ME, Moses-Kolko EL, Price J, Frank E, Kupfer DJ, Mathis C (2007) Serotonin-1A receptor imaging in recurrent depression: replication and literature review. Nucl Med Biol 34(7):865–877

Dufouil C, de Kersaint-Gilly A, Besancon V, Levy C, Auffray E, Brunnereau L, Alperovitch A, Tzourio C (2001) Longitudinal study of blood pressure and white matter hyperintensities: the EVA MRI Cohort. Neurology 56(7):921–926

Duman RS (2014) Neurobiology of stress, depression, and rapid acting antidepressants: remodeling synaptic connections. Depress Anxiety 31(4):291–296

Duman RS, Monteggia LM (2006) A neurotrophic model for stress-related mood disorders. Biol Psychiatry 59(12):1116–1127

Dupont RM, Jernigan TL, Butters N, Delis D, Hesselink JR, Heindel W, Gillin JC (1990) Subcortical abnormalities detected in bipolar affective disorder using magnetic resonance imaging. Clinical and neuropsychological significance. Arch Gen Psychiatry 47(1):55–59

Duric V, Banasr M, Stockmeier CA, Simen AA, Newton SS, Overholser JC, Jurjus GJ, Dieter L, Duman RS (2013) Altered expression of synapse and glutamate related genes in post-mortem hippocampus of depressed subjects. Int J Neuropsychopharmacol 16(1):69–82

Eastwood SL, Harrison PJ (2000) Hippocampal synaptic pathology in schizophrenia, bipolar disorder and major depression: a study of complexin mRNAs. Mol Psychiatry 5(4):425–432

Elvsashagen T, Westlye LT, Boen E, Hol PK, Andreassen OA, Boye B, Malt UF (2013) Bipolar II disorder is associated with thinning of prefrontal and temporal cortices involved in affect regulation. Bipolar Disord 15(8):855–864

Emsell L, Chaddock C, Forde N, Van Hecke W, Barker GJ, Leemans A, Sunaert S, Walshe M, Bramon E, Cannon D, Murray R, McDonald C (2014) White matter microstructural abnormalities in families multiply affected with bipolar I disorder: a diffusion tensor tractography study. Psychol Med 44(10):2139–2150

Fatemi SH, Earle JA, McMenomy T (2000) Reduction in Reelin immunoreactivity in hippocampus of subjects with schizophrenia, bipolar disorder and major depression. Mol Psychiatry 5(6):654–663, 571

Figiel GS, Krishnan KR, Rao VP, Doraiswamy M, Ellinwood EH Jr, Nemeroff CB, Evans D, Boyko O (1991) Subcortical hyperintensities on brain magnetic resonance imaging: a comparison of normal and bipolar subjects. J Neuropsychiatry Clin Neurosci 3(1):18–22

Flugge G (1995) Dynamics of central nervous 5-HT1A-receptors under psychosocial stress. J Neurosci 15(11):7132–7140

Foland-Ross LC, Thompson PM, Sugar CA, Madsen SK, Shen JK, Penfold C, Ahlf K, Rasser PE, Fischer J, Yang Y, Townsend J, Bookheimer SY, Altshuler LL (2011) Investigation of cortical thickness abnormalities in lithium-free adults with bipolar I disorder using cortical pattern matching. Am J Psychiatry 168(5):530–539

Folstein MF, Robinson R, Folstein S, McHugh PR (1985) Depression and neurological disorders. New treatment opportunities for elderly depressed patients. J Affect Disord Suppl 1:S11–S14

Forester BP, Finn CT, Berlow YA, Wardrop M, Renshaw PF, Moore CM (2008) Brain lithium, N-acetyl aspartate and myo-inositol levels in older adults with bipolar disorder treated with lithium: a lithium-7 and proton magnetic resonance spectroscopy study. Bipolar Disord 10(6):691–700

Fornito A, Yucel M, Wood SJ, Bechdolf A, Carter S, Adamson C, Velakoulis D, Saling MM, McGorry PD, Pantelis C (2009) Anterior cingulate cortex abnormalities associated with a first psychotic episode in bipolar disorder. Br J Psychiatry 194(5):426–433

Frangou S, Donaldson S, Hadjulis M, Landau S, Goldstein LH (2005) The Maudsley Bipolar Disorder Project: executive dysfunction in bipolar disorder I and its clinical correlates. Biol Psychiatry 58(11):859–864

Frazier JA, Giedd JN, Kaysen D, Albus K, Hamburger S, Alaghband-Rad J, Lenane MC, McKenna K, Breier A, Rapoport JL (1996) Childhood-onset schizophrenia: brain MRI rescan after 2 years of clozapine maintenance treatment. Am J Psychiatry 153(4):564–566

Furey ML, Drevets WC (2006) Antidepressant efficacy of the antimuscarinic drug scopolamine: a randomized, placebo-controlled clinical trial. Arch Gen Psychiatry 63(10):1121–1129

Geschwind N (1965a) Disconnexion syndromes in animals and man. I. Brain 88:237–294

Geschwind N (1965b) Disconnexion syndromes in animals and man. II. Brain 88:585–644

Giakoumatos CI, Nanda P, Mathew IT, Tandon N, Shah J, Bishop JR, Clementz BA, Pearlson GD, Sweeney JA, Tamminga CA, Keshavan MS (2015) Effects of lithium on cortical thickness and hippocampal subfield volumes in psychotic bipolar disorder. J Psychiatr Res 61:180–187

Goldstein IB, Bartzokis G, Guthrie D, Shapiro D (2002) Ambulatory blood pressure and brain atrophy in the healthy elderly. Neurology 59(5):713–719

Golembiowska K, Dziubina A (2000) Effect of acute and chronic administration of citalopram on glutamate and aspartate release in the rat prefrontal cortex. Pol J Pharmacol 52(6):441–448

Gonzalez-Burgos G, Lewis DA (2008) GABA neurons and the mechanisms of network oscillations: implications for understanding cortical dysfunction in schizophrenia. Schizophr Bull 34(5):944–961

Gonzalez-Burgos G, Cho RY, Lewis DA (2015) Alterations in cortical network oscillations and parvalbumin neurons in schizophrenia. Biol Psychiatry 77(12):1031–1040

Gross C, Zhuang X, Stark K, Ramboz S, Oosting R, Kirby L, Santarelli L, Beck S, Hen R (2002) Serotonin1A receptor acts during development to establish normal anxiety-like behaviour in the adult. Nature 416(6879):396–400

Gunstad J, Cohen RA, Tate DF, Paul RH, Poppas A, Hoth K, Macgregor KL, Jefferson AL (2005) Blood pressure variability and white matter hyperintensities in older adults with cardiovascular disease. Blood Press 14(6):353–358

Gustafson DR, Steen B, Skoog I (2004) Body mass index and white matter lesions in elderly women. An 18-year longitudinal study. Int Psychogeriatr 16(3):327–336

Haarman BC, Riemersma-Van der Lek RF, de Groot JC, Ruhe HG, Klein HC, Zandstra TE, Burger H, Schoevers RA, de Vries EF, Drexhage HA, Nolen WA, Doorduin J (2014) Neuroinflammation in bipolar disorder – A [(11)C]-(R)-PK11195 positron emission tomography study. Brain Behav Immun 40:219–225

Hafeman DM, Bebko G, Bertocci MA, Fournier JC, Bonar L, Perlman SB, Travis M, Gill MK, Diwadkar VA, Sunshine JL, Holland SK, Kowatch RA, Birmaher B, Axelson D, Horwitz SM, Arnold LE, Fristad MA, Frazier TW, Youngstrom EA, Findling RL, Drevets W, Phillips ML (2014) Abnormal deactivation of the inferior frontal gyrus during implicit emotion processing in youth with bipolar disorder: attenuated by medication. J Psychiatr Res 58:129–136

Hahn T, Marquand AF, Ehlis AC, Dresler T, Kittel-Schneider S, Jarczok TA, Lesch KP, Jakob PM, Mourao-Miranda J, Brammer MJ, Fallgatter AJ (2011) Integrating neurobiological markers of depression. Arch Gen Psychiatry 68(4):361–368

Hajek T, Kopecek M, Hoschl C, Alda M (2012) Smaller hippocampal volumes in patients with bipolar disorder are masked by exposure to lithium: a meta-analysis. J Psychiatry Neurosci 37(5):333–343

Hamakawa H, Kato T, Murashita J, Kato N (1998) Quantitative proton magnetic resonance spectroscopy of the basal ganglia in patients with affective disorders. Eur Arch Psychiatry Clin Neurosci 248(1):53–58

Hamidi M, Drevets WC, Price JL (2004) Glial reduction in amygdala in major depressive disorder is due to oligodendrocytes. Biol Psychiatry 55(6):563–569

Hannestad J, Taylor WD, McQuoid DR, Payne ME, Krishnan KR, Steffens DC, Macfall JR (2006) White matter lesion volumes and caudate volumes in late-life depression. Int J Geriatr Psychiatry 21(12):1193–1198

Hao Y, Creson T, Zhang L, Li P, Du F, Yuan P, Gould TD, Manji HK, Chen G (2004) Mood stabilizer valproate promotes ERK pathway-dependent cortical neuronal growth and neurogenesis. J Neurosci 24(29):6590–6599

Hashimoto K, Sawa A, Iyo M (2007) Increased levels of glutamate in brains from patients with mood disorders. Biol Psychiatry 62(11):1310–1316

Hasler G, van der Veen JW, Tumonis T, Meyers N, Shen J, Drevets WC (2007) Reduced prefrontal glutamate/glutamine and gamma-aminobutyric acid levels in major depression determined using proton magnetic resonance spectroscopy. Arch Gen Psychiatry 64(2):193–200

Hasler G, Fromm S, Carlson PJ, Luckenbaugh DA, Waldeck T, Geraci M, Roiser JP, Neumeister A, Meyers N, Charney DS, Drevets WC (2008) Neural response to catecholamine depletion in unmedicated subjects with major depressive disorder in remission and healthy subjects. Arch Gen Psychiatry 65(5):521–531

Hatton SN, Lagopoulos J, Hermens DF, Scott E, Hickie IB, Bennett MR (2013) Cortical thinning in young psychosis and bipolar patients correlate with common neurocognitive deficits. Int J Bipolar Disord 1:3

Haznedar MM, Roversi F, Pallanti S, Baldini-Rossi N, Schnur DB, Licalzi EM, Tang C, Hof PR, Hollander E, Buchsbaum MS (2005) Fronto-thalamo-striatal gray and white matter volumes and anisotropy of their connections in bipolar spectrum illnesses. Biol Psychiatry 57 (7):733–742

Hercher C, Chopra V, Beasley CL (2014) Evidence for morphological alterations in prefrontal white matter glia in schizophrenia and bipolar disorder. J Psychiatry Neurosci 39(6):376–385

Herman JP, Cullinan WE (1997) Neurocircuitry of stress: central control of the hypothalamo-pituitary- adrenocortical axis. Trends Neurosci 20(2):78–84

Hickie I, Scott E, Mitchell P, Wilhelm K, Austin MP, Bennett B (1995) Subcortical hyperintensities on magnetic resonance imaging: clinical correlates and prognostic significance in patients with severe depression. Biol Psychiatry 37(3):151–160

Hirayasu Y, Shenton ME, Salisbury DF, Kwon JS, Wible CG, Fischer IA, Yurgelun-Todd D, Zarate C, Kikinis R, Jolesz FA, McCarley RW (1999) Subgenual cingulate cortex volume in first-episode psychosis. Am J Psychiatry 156(7):1091–1093

Hirvonen J, Karlsson H, Kajander J, Lepola A, Markkula J, Rasi-Hakala H, Nagren K, Salminen JK, Hietala J (2008) Decreased brain serotonin 5-HT1A receptor availability in medication-naive patients with major depressive disorder: an in-vivo imaging study using PET and [carbonyl-11C]WAY-100635. Int J Neuropsychopharmacol 11(4):465–476

Holthoff VA, Beuthien-Baumann B, Zundorf G, Triemer A, Ludecke S, Winiecki P, Koch R, Fuchtner F, Herholz K (2004) Changes in brain metabolism associated with remission in unipolar major depression. Acta Psychiatr Scand 110(3):184–194

Honer WG, Falkai P, Chen C, Arango V, Mann JJ, Dwork AJ (1999) Synaptic and plasticity-associated proteins in anterior frontal cortex in severe mental illness. Neuroscience 91 (4):1247–1255

Ionescu DF, Luckenbaugh DA, Niciu MJ, Richards EM, Zarate CA Jr (2015) A single infusion of ketamine improves depression scores in patients with anxious bipolar depression. Bipolar Disord 17(4):438–443

Izquierdo A, Wellman CL, Holmes A (2006) Brief uncontrollable stress causes dendritic retraction in infralimbic cortex and resistance to fear extinction in mice. J Neurosci 26(21):5733–5738

Janowsky DS, Overstreet DH, Nurnberger JIJ (1994) Is cholinergic sensitivity a genetic marker for the affective disorders? Am J Med Genet 54:335–344

Janssen J, Aleman-Gomez Y, Schnack H, Balaban E, Pina-Camacho L, Alfaro-Almagro F, Castro-Fornieles J, Otero S, Baeza I, Moreno D, Bargallo N, Parellada M, Arango C, Desco M (2014) Cortical morphology of adolescents with bipolar disorder and with schizophrenia. Schizophr Res 158(1–3):91–99

Janus C, D'Amelio S, Amitay O, Chishti MA, Strome R, Fraser P, Carlson GA, Roder JC, St George-Hyslop P, Westaway D (2000) Spatial learning in transgenic mice expressing human presenilin 1 (PS1) transgenes. Neurobiol Aging 21(4):541–549

Jernigan TL, Zisook S, Heaton RK, Moranville JT, Hesselink JR, Braff DL (1991) Magnetic resonance imaging abnormalities in lenticular nuclei and cerebral cortex in schizophrenia. Arch Gen Psychiatry 48(10):881–890

Johnston-Wilson NL, Sims CD, Hofmann JP, Anderson L, Shore AD, Torrey EF, Yolken RH (2000) Disease-specific alterations in frontal cortex brain proteins in schizophrenia, bipolar disorder, and major depressive disorder. The Stanley Neuropathology Consortium. Mol Psychiatry 5(2):142–149

Jovanovic H, Perski A, Berglund H, Savic I (2011) Chronic stress is linked to 5-HT(1A) receptor changes and functional disintegration of the limbic networks. Neuroimage 55(3):1178–1188

Kato T, Hamakawa H, Shioiri T, Murashita J, Takahashi Y, Takahashi S, Inubushi T (1996) Choline-containing compounds detected by proton magnetic resonance spectroscopy in the basal ganglia in bipolar disorder. J Psychiatry Neurosci 21(4):248–254

Kegeles LS, Malone KM, Slifstein M, Ellis SP, Xanthopoulos E, Keilp JG, Campbell C, Oquendo M, Van Heertum RL, Mann JJ (2003) Response of cortical metabolic deficits to serotonergic challenge in familial mood disorders. Am J Psychiatry 160(1):76–82

Kertesz A, Black SE, Tokar G, Benke T, Carr T, Nicholson L (1988) Periventricular and subcortical hyperintensities on magnetic resonance imaging. 'Rims, caps, and unidentified bright objects'. Arch Neurol 45(4):404–408

Kessler RC, Chiu WT, Demler O, Walters EE (2005) Prevalence, severity, and comorbidity of 12-month DSM-IV disorders in the national comorbidity survey replication. Arch Gen Psychiatry 62(6):617–627

Ketter TA, Drevets WC (2002) Neuroimaging studies of bipolar depression: functional neuropathology, treatment effects, and predictors of clinical response. Clin Neurosci Res 2:182–192

Ketter TA, Andreason PJ, George MS, Lee C, Gill DS, Parekh PI, Willis MW, Herscovitch P, Post RM (1996) Anterior paralimbic mediation of procaine-induced emotional and psychosensory experiences. Arch Gen Psychiatry 53:59–69

Ketter TA, Kimbrell TA, George MS, Dunn RT, Speer AM, Benson BE, Willis MW, Danielson A, Frye MA, Herscovitch P, Post RM (2001) Effects of mood and subtype on cerebral glucose metabolism in treatment-resistant bipolar disorder. Biol Psychiatry 49(2):97–109

Kilbourne AM, Cornelius JR, Han X, Pincus HA, Shad M, Salloum I, Conigliaro J, Haas GL (2004) Burden of general medical conditions among individuals with bipolar disorder. Bipolar Disord 6(5):368–373

Kim S, Webster MJ (2010) Correlation analysis between genome-wide expression profiles and cytoarchitectural abnormalities in the prefrontal cortex of psychiatric disorders. Mol Psychiatry 15(3):326–336

Kim JS, Chang MY, Yu IT, Kim JH, Lee SH, Lee YS, Son H (2004) Lithium selectively increases neuronal differentiation of hippocampal neural progenitor cells both in vitro and in vivo. J Neurochem 89(2):324–336

Kimoto S, Zaki MM, Bazmi HH, Lewis DA (2015) Altered markers of cortical gamma-aminobutyric acid neuronal activity in schizophrenia: role of the NARP gene. JAMA Psychiatry 72(8):747–756

Kinney DK, Yurgelun-Todd DA, Levy DL, Medoff D, Lajonchere CM, Radford-Paregol M (1993) Obstetrical complications in patients with bipolar disorder and their siblings. Psychiatry Res 48(1):47–56

Kinney DK, Yurgelun-Todd DA, Tohen M, Tramer S (1998) Pre- and perinatal complications and risk for bipolar disorder: a retrospective study. J Affect Disord 50(2–3):117–124

Knopman DS, Mosley TH, Catellier DJ, Sharrett AR (2005) Cardiovascular risk factors and cerebral atrophy in a middle-aged cohort. Neurology 65(6):876–881

Krishnan KR, Doraiswamy PM, Figiel GS, Husain MM, Shah SA, Na C, Boyko OB, McDonald WM, Nemeroff CB, Ellinwood EH Jr (1991) Hippocampal abnormalities in depression. J Neuropsychiatry Clin Neurosci 3(4):387–391

Kruger S, Seminowicz D, Goldapple K, Kennedy SH, Mayberg HS (2003) State and trait influences on mood regulation in bipolar disorder: blood flow differences with an acute mood challenge. Biol Psychiatry 54(11):1274–1283

Krystal JH, Sanacora G, Blumberg H, Anand A, Charney DS, Marek G, Epperson CN, Goddard A, Mason GF (2002) Glutamate and GABA systems as targets for novel antidepressant and mood-stabilizing treatments. Mol Psychiatry 7(Suppl 1):S71–S80

Kumar J, Iwabuchi S, Oowise S, Balain V, Palaniyappan L, Liddle PF (2015) Shared white-matter dysconnectivity in schizophrenia and bipolar disorder with psychosis. Psychol Med 45 (4):759–770

Lan MJ, Chhetry BT, Oquendo MA, Sublette ME, Sullivan G, Mann JJ, Parsey RV (2014) Cortical thickness differences between bipolar depression and major depressive disorder. Bipolar Disord 16(4):378–388

Lawrence NS, Williams AM, Surguladze S, Giampietro V, Brammer MJ, Andrew C, Frangou S, Ecker C, Phillips ML (2004) Subcortical and ventral prefrontal cortical neural responses to facial expressions distinguish patients with bipolar disorder and major depression. Biol Psychiatry 55(6):578–587

LeDoux J (2003) The emotional brain, fear, and the amygdala. Cell Mol Neurobiol 23 (4–5):727–738

Lee EE, Della Selva MP, Liu A, Himelhoch S (2015) Ketamine as a novel treatment for major depressive disorder and bipolar depression: a systematic review and quantitative meta-analysis. Gen Hosp Psychiatry 37(2):178–184

Lenze E, Cross D, McKeel D, Neuman RJ, Sheline YI (1999) White matter hyperintensities and gray matter lesions in physically healthy depressed subjects. Am J Psychiatry 156 (10):1602–1607

Lewis DA, Curley AA, Glausier JR, Volk DW (2012) Cortical parvalbumin interneurons and cognitive dysfunction in schizophrenia. Trends Neurosci 35(1):57–67

Lochhead RA, Parsey RV, Oquendo MA, Mann JJ (2004) Regional brain gray matter volume differences in patients with bipolar disorder as assessed by optimized voxel-based morphometry. Biol Psychiatry 55(12):1154–1162

Lopez JF, Chalmers DT, Little KY, Watson SJ (1998) A.E. Bennett Research Award. Regulation of serotonin1A, glucocorticoid, and mineralocorticoid receptor in rat and human hippocampus: implications for the neurobiology of depression. Biol Psychiatry 43(8):547–573

Lyoo IK, Lee HK, Jung JH, Noam GG, Renshaw PF (2002) White matter hyperintensities on magnetic resonance imaging of the brain in children with psychiatric disorders. Compr Psychiatry 43(5):361–368

Lyoo IK, Kim MJ, Stoll AL, Demopulos CM, Parow AM, Dager SR, Friedman SD, Dunner DL, Renshaw PF (2004a) Frontal lobe gray matter density decreases in bipolar I disorder. Biol Psychiatry 55(6):648–651

Lyoo IK, Streeter CC, Ahn KH, Lee HK, Pollack MH, Silveri MM, Nassar L, Levin JM, Sarid-Segal O, Ciraulo DA, Renshaw PF, Kaufman MJ (2004b) White matter hyperintensities in subjects with cocaine and opiate dependence and healthy comparison subjects. Psychiatry Res 131(2):135–145

Lyoo IK, Sung YH, Dager SR, Friedman SD, Lee JY, Kim SJ, Kim N, Dunner DL, Renshaw PF (2006) Regional cerebral cortical thinning in bipolar disorder. Bipolar Disord 8(1):65–74

MacDonald ML, Naydenov A, Chu M, Matzilevich D, Konradi C (2006) Decrease in creatine kinase messenger RNA expression in the hippocampus and dorsolateral prefrontal cortex in bipolar disorder. Bipolar Disord 8(3):255–264

Mah L, Zarate CA Jr, Singh J, Duan YF, Luckenbaugh DA, Manji HK, Drevets WC (2007) Regional cerebral glucose metabolic abnormalities in bipolar II depression. Biol Psychiatry 61(6):765–775

Maller JJ, Thaveenthiran P, Thomson RH, McQueen S, Fitzgerald PB (2014) Volumetric, cortical thickness and white matter integrity alterations in bipolar disorder type I and II. J Affect Disord 169:118–127

Manelis A, Ladouceur CD, Graur S, Monk K, Bonar LK, Hickey MB, Dwojak AC, Axelson D, Goldstein BI, Goldstein TR, Bebko G, Bertocci MA, Hafeman DM, Gill MK, Birmaher B, Phillips ML (2015) Altered amygdala-prefrontal response to facial emotion in offspring of parents with bipolar disorder. Brain 138(Pt 9):2777–2790

Manji HK, Moore GJ, Chen G (2000) Clinical and preclinical evidence for the neurotrophic effects of mood stabilizers: implications for the pathophysiology and treatment of manic-depressive illness. Biol Psychiatry 48(8):740–754

Manji HK, Drevets WC, Charney DS (2001) The cellular neurobiology of depression. Nat Med 7(5):541–547

Mark RJ, Ashford JW, Goodman Y, Mattson MP (1995) Anticonvulsants attenuate amyloid beta-peptide neurotoxicity, Ca2+ deregulation, and cytoskeletal pathology. Neurobiol Aging 16(2):187–198

Markram H, Toledo-Rodriguez M, Wang Y, Gupta A, Silberberg G, Wu C (2004) Interneurons of the neocortical inhibitory system. Nat Rev Neurosci 5(10):793–807

Martinez P, Ronsaville D, Gold PW, Hauser P, Drevets WC (2002) Morphometric abnormalities in adolescent offspring of depressed mothers. Soc Neurosci Abstr 32

Mayberg HS, Lozano AM, Voon V, McNeely HE, Seminowicz D, Hamani C, Schwalb JM, Kennedy SH (2005) Deep brain stimulation for treatment-resistant depression. Neuron 45 (5):651–660

McEwen BS, Magarinos AM (2001) Stress and hippocampal plasticity: implications for the pathophysiology of affective disorders. Hum Psychopharmacol 16(S1):S7–S19

McEwen BS, Olie JP (2005) Neurobiology of mood, anxiety, and emotions as revealed by studies of a unique antidepressant: tianeptine. Mol Psychiatry 10(6):525–537

McEwen BS, Magarinos AM, Reagan LP (2002) Structural plasticity and tianeptine: cellular and molecular targets. Eur Psychiatry 17(Suppl 3):318–330

McGirr A, Berlim MT, Bond DJ, Fleck MP, Yatham LN, Lam RW (2015) A systematic review and meta-analysis of randomized, double-blind, placebo-controlled trials of ketamine in the rapid treatment of major depressive episodes. Psychol Med 45(4):693–704

Medina KL, Nagel BJ, Park A, McQueeny T, Tapert SF (2007) Depressive symptoms in adolescents: associations with white matter volume and marijuana use. J Child Psychol Psychiatry 48 (6):592–600

Meyer JH, Houle S, Sagrati S, Carella A, Hussey DF, Ginovart N, Goulding V, Kennedy J, Wilson AA (2004) Brain serotonin transporter binding potential measured with carbon 11-labeled DASB positron emission tomography effects of major depressive episodes and severity of dysfunctional attitudes. Arch Gen Psychiatry 61:1271–1279

Meyer JH, McNeely HE, Sagrati S, Boovariwala A, Martin K, Verhoeff NP, Wilson AA, Houle S (2006) Elevated putamen D(2) receptor binding potential in major depression with motor retardation: an [11C]raclopride positron emission tomography study. Am J Psychiatry 163(9):1594–1602

Miguel-Hidalgo JJ, Waltzer R, Whittom AA, Austin MC, Rajkowska G, Stockmeier CA (2010) Glial and glutamatergic markers in depression, alcoholism, and their comorbidity. J Affect Disord 127(1–3):230–240

Milad MR, Quinn BT, Pitman RK, Orr SP, Fischl B, Rauch SL (2005) Thickness of ventromedial prefrontal cortex in humans is correlated with extinction memory. Proc Natl Acad Sci USA 102(30):10706–10711

Miracle AD, Brace MF, Huyck KD, Singler SA, Wellman CL (2006) Chronic stress impairs recall of extinction of conditioned fear. Neurobiol Learn Mem 85(3):213–218

Moore GJ, Galloway MP (2002) Magnetic resonance spectroscopy: neurochemistry and treatment effects in affective disorders. Psychopharmacol Bull 36(2):5–23

Moore GJ, Bebchuk JM, Wilds IB, Chen G, Manji HK (2000) Lithium-induced increase in human brain grey matter. Lancet 356(9237):1241–1242

Moore GJ, Cortese BM, Glitz DA, Zajac-Benitez C, Quiroz JA, Uhde TW, Drevets WC, Manji HK (2009) A longitudinal study of the effects of lithium treatment on prefrontal and subgenual prefrontal gray matter volume in treatment-responsive bipolar disorder patients. J Clin Psychiatry 70(5):699–705

Mourao-Miranda J, Oliveira L, Ladouceur CD, Marquand A, Brammer M, Birmaher B, Axelson D, Phillips ML (2012) Pattern recognition and functional neuroimaging help to discriminate healthy adolescents at risk for mood disorders from low risk adolescents. PLoS One 7(2):e29482

Musselman DL, Tomer A, Manatunga AK, Knight BT, Porter MR, Kasey S, Marzec U, Harker LA, Nemeroff CB (1996) Exaggerated platelet reactivity in major depression. Am J Psychiatry 153(10):1313–1317

Nauta WJH, Domesick V (1984) Afferent and efferent relationships of the basal ganglia. In: Evered D, O'Conner M (eds) Function of the basal ganglia. Pitman Press, London

Nemeroff CB, Musselman DL (2000) Are platelets the link between depression and ischemic heart disease? Am Heart J 140(4 Suppl):57–62

Neumeister A, Nugent AC, Waldeck T, Geraci M, Schwarz M, Bonne O, Bain EE, Luckenbaugh DA, Herscovitch P, Charney DS, Drevets WC (2004) Neural and behavioral responses to tryptophan depletion in unmedicated patients with remitted major depressive disorder and controls. Arch Gen Psychiatry 61(8):765–773

Neumeister A, Hu XZ, Luckenbaugh DA, Schwarz M, Nugent AC, Bonne O, Herscovitch P, Goldman D, Drevets WC, Charney DS (2006) Differential effects of 5-HTTLPR genotypes on the behavioral and neural responses to tryptophan depletion in patients with major depression and controls. Arch Gen Psychiatry 63(9):978–986

Newcomer JW (2006) Medical risk in patients with bipolar disorder and schizophrenia. J Clin Psychiatry 67(Suppl 9):25–30, discussion 36–42

Nobler MS, Oquendo MA, Kegeles LS, Malone KM, Campbell CC, Sackeim HA, Mann JJ (2001) Decreased regional brain metabolism after ect. Am J Psychiatry 158(2):305–308

Novak V, Last D, Alsop DC, Abduljalil AM, Hu K, Lepicovsky L, Cavallerano J, Lipsitz LA (2006) Cerebral blood flow velocity and periventricular white matter hyperintensities in type 2 diabetes. Diabetes Care 29(7):1529–1534

Nugent AC, Milham MP, Bain EE, Mah L, Cannon DM, Marrett S, Zarate CA, Pine DS, Price JL, Drevets WC (2006) Cortical abnormalities in bipolar disorder investigated with MRI and voxel-based morphometry. Neuroimage 30(2):485–497

Nugent AC, Bain EE, Carlson PJ, Neumeister A, Bonne O, Carson RE, Eckelman W, Herscovitch P, Zarate CA Jr, Charney DS, Drevets WC (2013) Reduced post-synaptic serotonin type 1A receptor binding in bipolar depression. Eur Neuropsychopharmacol 23(8):822–829

Oertel-Knochel V, Reuter J, Reinke B, Marbach K, Feddern R, Alves G, Prvulovic D, Linden DE, Knochel C (2015) Association between age of disease-onset, cognitive performance and cortical thickness in bipolar disorders. J Affect Disord 174:627–635

Ohara K, Isoda H, Suzuki Y, Takehara Y, Ochiai M, Takeda H, Igarashi Y, Ohara K (1998) Proton magnetic resonance spectroscopy of the lenticular nuclei in bipolar I affective disorder. Psychiatry Res 84(2-3):55–60

Olsavsky AK, Brotman MA, Rutenberg JG, Muhrer EJ, Deveney CM, Fromm SJ, Towbin K, Pine DS, Leibenluft E (2012) Amygdala hyperactivation during face emotion processing in unaffected youth at risk for bipolar disorder. J Am Acad Child Adolesc Psychiatry 51(3):294–303

Ongur D, Drevets WC, Price JL (1998) Glial reduction in the subgenual prefrontal cortex in mood disorders. Proc Natl Acad Sci USA 95(22):13290–13295

Ongur D, Ferry AT, Price JL (2003) Architectonic subdivision of the human orbital and medial prefrontal cortex. J Comp Neurol 460(3):425–449

Ovbiagele B, Saver JL (2006) Cerebral white matter hyperintensities on MRI: current concepts and therapeutic implications. Cerebrovasc Dis 22(2–3):83–90

Pantazopoulos H, Stone D, Walsh J, Benes FM (2004) Differences in the cellular distribution of D1 receptor mRNA in the hippocampus of bipolars and schizophrenics. Synapse 54 (3):147–155

Pantazopoulos H, Lange N, Baldessarini RJ, Berretta S (2007) Parvalbumin neurons in the entorhinal cortex of subjects diagnosed with bipolar disorder or schizophrenia. Biol Psychiatry 61(5):640–652

Paul IA, Skolnick P (2003) Glutamate and depression: clinical and preclinical studies. Ann N Y Acad Sci 1003:250–272

Pavuluri MN, Henry DB, Nadimpalli SS, O'Connor MM, Sweeney JA (2006) Biological risk factors in pediatric bipolar disorder. Biol Psychiatry 60(9):936–941

Pavuluri MN, O'Connor MM, Harral E, Sweeney JA (2007) Affective neural circuitry during facial emotion processing in pediatric bipolar disorder. Biol Psychiatry 62(2):158–167

Pearlson GD, Wong DF, Tune LE, Ross CA, Chase GA, Links JM, Dannals RF, Wilson AA, Ravert HT, Wagner HN Jr et al (1995) In vivo D2 dopamine receptor density in psychotic and nonpsychotic patients with bipolar disorder. Arch Gen Psychiatry 52(6):471–477

Phillips ML, Swartz HA (2014) A critical appraisal of neuroimaging studies of bipolar disorder: toward a new conceptualization of underlying neural circuitry and a road map for future research. Am J Psychiatry 171(8):829–843

Phillips ML, Drevets WC, Rauch SL, Lane R (2003) Neurobiology of emotion perception II: implications for major psychiatric disorders. Biol Psychiatry 54(5):515–528

Phillips ML, Ladouceur CD, Drevets WC (2008) A neural model of voluntary and automatic emotion regulation: implications for understanding the pathophysiology and neuro-development of bipolar disorder. Mol Psychiatry 13(9):829, 833–857

Pilc A, Wieronska JM, Skolnick P (2013) Glutamate-based antidepressants: preclinical psycho-pharmacology. Biol Psychiatry 73(12):1125–1132

Pillai JJ, Friedman L, Stuve TA, Trinidad S, Jesberger JA, Lewin JS, Findling RL, Swales TP, Schulz SC (2002) Increased presence of white matter hyperintensities in adolescent patients with bipolar disorder. Psychiatry Res 114(1):51–56

Pitt D, Nagelmeier IE, Wilson HC, Raine CS (2003) Glutamate uptake by oligodendrocytes: Implications for excitotoxicity in multiple sclerosis. Neurology 61(8):1113–1120

Politis M, Wu K, Loane C, Turkheimer FE, Molloy S, Brooks DJ, Piccini P (2010) Depressive symptoms in PD correlate with higher 5-HTT binding in raphe and limbic structures. Neurology 75(21):1920–1927

Price JL, Drevets WC (2010) Neurocircuitry of mood disorders. Neuropsychopharmacology 35(1):192–216

Price JL, Drevets WC (2012) Neural circuits underlying the pathophysiology of mood disorders. Trends Cogn Sci 16(1):61–71

Radley JJ, Rocher AB, Rodriguez A, Ehlenberger DB, Dammann M, McEwen BS, Morrison JH, Wearne SL, Hof PR (2008) Repeated stress alters dendritic spine morphology in the rat medial prefrontal cortex. J Comp Neurol 507(1):1141–1150

Rajkowska G, Miguel-Hidalgo JJ (2007) Gliogenesis and glial pathology in depression. CNS Neurol Disord Drug Targets 6(3):219–233

Rajkowska G, Stockmeier CA (2013) Astrocyte pathology in major depressive disorder: insights from human postmortem brain tissue. Curr Drug Targets 14(11):1225–1236

Rajkowska G, Miguel-Hidalgo JJ, Wei J, Dilley G, Pittman SD, Meltzer HY, Overholser JC, Roth BL, Stockmeier CA (1999) Morphometric evidence for neuronal and glial prefrontal cell pathology in major depression. Biol Psychiatry 45(9):1085–1098

Regenold WT, Phatak P, Marano CM, Gearhart L, Viens CH, Hisley KC (2007) Myelin staining of deep white matter in the dorsolateral prefrontal cortex in schizophrenia, bipolar disorder, and unipolar major depression. Psychiatry Res 151(3):179–188

Rich BA, Vinton DT, Roberson-Nay R, Hommer RE, Berghorst LH, McClure EB, Fromm SJ, Pine DS, Leibenluft E (2006) Limbic hyperactivation during processing of neutral facial expressions in children with bipolar disorder. Proc Natl Acad Sci USA 103(23):8900–8905

Rimol LM, Hartberg CB, Nesvag R, Fennema-Notestine C, Hagler DJ Jr, Pung CJ, Jennings RG, Haukvik UK, Lange E, Nakstad PH, Melle I, Andreassen OA, Dale AM, Agartz I (2010) Cortical thickness and subcortical volumes in schizophrenia and bipolar disorder. Biol Psychiatry 68(1):41–50

Salerno JA, Murphy DG, Horwitz B, DeCarli C, Haxby JV, Rapoport SI, Schapiro MB (1992) Brain atrophy in hypertension. A volumetric magnetic resonance imaging study. Hypertension 20(3):340–348

Sanacora G, Mason GF, Rothman DL, Behar KL, Hyder F, Petroff OA, Berman RM, Charney DS, Krystal JH (1999) Reduced cortical gamma-aminobutyric acid levels in depressed patients determined by proton magnetic resonance spectroscopy. Arch Gen Psychiatry 56 (11):1043–1047

Sargent PA, Kjaer KH, Bench CJ, Rabiner EA, Messa C, Meyer J, Gunn RN, Grasby PM, Cowen PJ (2000) Brain serotonin1A receptor binding measured by positron emission tomography with [11C]WAY-100635: effects of depression and antidepressant treatment. Arch Gen Psychiatry 57(2):174–180

Sassi RB, Brambilla P, Hatch JP, Nicoletti MA, Mallinger AG, Frank E, Kupfer DJ, Keshavan MS, Soares JC (2004) Reduced left anterior cingulate volumes in untreated bipolar patients. Biol Psychiatry 56(7):467–475

Savitz J, Drevets WC (2009a) Bipolar and major depressive disorder: neuroimaging the developmental-degenerative divide. Neurosci Biobehav Rev 33(5):699–771

Savitz JB, Drevets WC (2009b) Imaging phenotypes of major depressive disorder: genetic correlates. Neuroscience 164(1):300–330

Savitz J, Nugent AC, Bogers W, Liu A, Sills R, Luckenbaugh DA, Bain EE, Price JL, Zarate C, Manji HK, Cannon DM, Marrett S, Charney DS, Drevets WC (2010) Amygdala volume in depressed patients with bipolar disorder assessed using high resolution 3T MRI: the impact of medication. Neuroimage 49(4):2966–2976

Savitz JB, Rauch SL, Drevets WC (2013) Clinical application of brain imaging for the diagnosis of mood disorders: the current state of play. Mol Psychiatry 18(5):528–539

Savitz J, Dantzer R, Wurfel BE, Victor TA, Ford BN, Bodurka J, Bellgowan PS, Teague TK, Drevets WC (2014a) Neuroprotective kynurenine metabolite indices are abnormally reduced and positively associated with hippocampal and amygdalar volume in bipolar disorder. Psychoneuroendocrinology 52C:200–211

Savitz JB, Price JL, Drevets WC (2014b) Neuropathological and neuromorphometric abnormalities in bipolar disorder: view from the medial prefrontal cortical network. Neurosci Biobehav Rev 42:132–147

Setiawan E, Wilson AA, Mizrahi R, Rusjan PM, Miler L, Rajkowska G, Suridjan I, Kennedy JL, Rekkas PV, Houle S, Meyer JH (2015) Role of translocator protein density, a marker of neuroinflammation, in the brain during major depressive episodes. JAMA Psychiatry 72(3):268–275

Sheline YI, Barch DM, Donnelly JM, Ollinger JM, Snyder AZ, Mintun MA (2001) Increased amygdala response to masked emotional faces in depressed subjects resolves with antidepressant treatment: an fMRI study. Biol Psychiatry 50(9):651–658

Sheline YI, Gado MH, Kraemer HC (2003) Untreated depression and hippocampal volume loss. Am J Psychiatry 160(8):1516–1518

Shively CA, Friedman DP, Gage HD, Bounds MC, Brown-Proctor C, Blair JB, Henderson JA, Smith MA, Buchheimer N (2006) Behavioral depression and positron emission tomography-determined serotonin 1A receptor binding potential in cynomolgus monkeys. Arch Gen Psychiatry 63(4):396–403

Shulman RG, Rothman DL, Behar KL, Hyder F (2004) Energetic basis of brain activity: implications for neuroimaging. Trends Neurosci 27(8):489–495

Sibille E, Morris HM, Kota RS, Lewis DA (2011) GABA-related transcripts in the dorsolateral prefrontal cortex in mood disorders. Int J Neuropsychopharmacol 14(6):721–734

Sokolov BP (2007) Oligodendroglial abnormalities in schizophrenia, mood disorders and substance abuse. Comorbidity, shared traits, or molecular phenocopies? Int J Neuropsychopharmacol 10(4):547–555

Son H, Yu IT, Hwang SJ, Kim JS, Lee SH, Lee YS, Kaang BK (2003) Lithium enhances long-term potentiation independently of hippocampal neurogenesis in the rat dentate gyrus. J Neurochem 85(4):872–881

Starkstein SE, Robinson RG (1989) Affective disorders and cerebral vascular disease. Br J Psychiatry 154:170–182

Steffens DC, Helms MJ, Krishnan KR, Burke GL (1999) Cerebrovascular disease and depression symptoms in the cardiovascular health study. Stroke 30(10):2159–2166

Steiner J, Walter M, Gos T, Guillemin GJ, Bernstein HG, Sarnyai Z, Mawrin C, Brisch R, Bielau H, Meyer zu Schwabedissen L, Bogerts B, Myint AM (2011) Severe depression is associated with increased microglial quinolinic acid in subregions of the anterior cingulate gyrus: evidence for an immune-modulated glutamatergic neurotransmission? J Neuroinflammation 8:94

Stewart CA, Reid IC (2000) Repeated ECS and fluoxetine administration have equivalent effects on hippocampal synaptic plasticity. Psychopharmacology (Berl) 148(3):217–223

Stockmeier CA (2003) Involvement of serotonin in depression: evidence from postmortem and imaging studies of serotonin receptors and the serotonin transporter. J Psychiatr Res 37(5):357–373

Stockmeier CA, Shapiro LA, Dilley GE, Kolli TN, Friedman L, Rajkowska G (1998) Increase in serotonin-1A autoreceptors in the midbrain of suicide victims with major depression-postmortem evidence for decreased serotonin activity. J Neurosci 18(18):7394–7401

Stockmeier CA, Mahajan GJ, Konick LC, Overholser JC, Jurjus GJ, Meltzer HY, Uylings HB, Friedman L, Rajkowska G (2004) Cellular changes in the postmortem hippocampus in major depression. Biol Psychiatry 56(9):640–650

Strakowski SM, DelBello MP, Zimmerman ME, Getz GE, Mills NP, Ret J, Shear P, Adler CM (2002) Ventricular and periventricular structural volumes in first- versus multiple-episode bipolar disorder. Am J Psychiatry 159(11):1841–1847

Suhara T, Nakayama K, Inoue O, Fukuda H, Shimizu M, Mori A, Tateno Y (1992) D1 dopamine receptor binding in mood disorders measured by positron emission tomography. Psychopharmacology (Berl) 106(1):14–18

Sussmann JE, Lymer GK, McKirdy J, Moorhead TW, Munoz Maniega S, Job D, Hall J, Bastin ME, Johnstone EC, Lawrie SM, McIntosh AM (2009) White matter abnormalities in bipolar disorder and schizophrenia detected using diffusion tensor magnetic resonance imaging. Bipolar Disord 11(1):11–18

Swayze VW, Andreasen NC, Alliger RJ, Yuh WT, Ehrhardt JC (1992) Subcortical and temporal lobe structures in affective disorder and schizophrenia: A magnetic resonance imaging study. Biol Psychiatry 31:221–240

Taylor MJ, Selvaraj S, Norbury R, Jezzard P, Cowen PJ (2009) Normal glutamate but elevated myo-inositol in anterior cingulate cortex in recovered depressed patients. J Affect Disord 119(1–3):186–189

The World Health Organization (WHO) (2001) The World health report 2001. wwwwhoint, Switzerland, Geneva

Tkachev D, Mimmack ML, Ryan MM, Wayland M, Freeman T, Jones PB, Starkey M, Webster MJ, Yolken RH, Bahn S (2003) Oligodendrocyte dysfunction in schizophrenia and bipolar disorder. Lancet 362(9386):798–805

Todtenkopf MS, Vincent SL, Benes FM (2005) A cross-study meta-analysis and three-dimensional comparison of cell counting in the anterior cingulate cortex of schizophrenic and bipolar brain. Schizophr Res 73(1):79–89

Torgerson CM, Irimia A, Leow AD, Bartzokis G, Moody TD, Jennings RG, Alger JR, Van Horn JD, Altshuler LL (2013) DTI tractography and white matter fiber tract characteristics in euthymic bipolar I patients and healthy control subjects. Brain Imaging Behav 7(2):129–139

Uranova N, Orlovskaya D, Vikhreva O, Zimina I, Kolomeets N, Vostrikov V, Rachmanova V (2001) Electron microscopy of oligodendroglia in severe mental illness. Brain Res Bull 55(5):597–610

Uranova NA, Vostrikov VM, Orlovskaya DD, Rachmanova VI (2004) Oligodendroglial density in the prefrontal cortex in schizophrenia and mood disorders: a study from the Stanley Neuropathology Consortium. Schizophr Res 67(2–3):269–275

Versace A, Andreazza AC, Young LT, Fournier JC, Almeida JR, Stiffler RS, Lockovich JC, Aslam HA, Pollock MH, Park H, Nimgaonkar VL, Kupfer DJ, Phillips ML (2014) Elevated serum measures of lipid peroxidation and abnormal prefrontal white matter in euthymic bipolar adults: toward peripheral biomarkers of bipolar disorder. Mol Psychiatry 19(2):200–208

Vidal-Gonzalez I, Vidal-Gonzalez B, Rauch SL, Quirk GJ (2006) Microstimulation reveals opposing influences of prelimbic and infralimbic cortex on the expression of conditioned fear. Learn Mem 13(6):728–733

Vostrikov VM, Uranova NA, Orlovskaya DD (2007) Deficit of perineuronal oligodendrocytes in the prefrontal cortex in schizophrenia and mood disorders. Schizophr Res 94(1–3):273–280

Vyas A, Mitra R, Shankaranarayana Rao BS, Chattarji S (2002) Chronic stress induces contrasting patterns of dendritic remodeling in hippocampal and amygdaloid neurons. J Neurosci 22(15):6810–6818

Vyas A, Bernal S, Chattarji S (2003) Effects of chronic stress on dendritic arborization in the central and extended amygdala. Brain Res 965(1–2):290–294

Vythilingam M, Heim C, Newport J, Miller AH, Anderson E, Bronen R, Brummer M, Staib L, Vermetten E, Charney DS, Nemeroff CB, Bremner JD (2002) Childhood trauma associated with smaller hippocampal volume in women with major depression. Am J Psychiatry 159(12):2072–2080

Wang F, Jackowski M, Kalmar JH, Chepenik LG, Tie K, Qiu M, Gong G, Pittman BP, Jones MM, Shah MP, Spencer L, Papademetris X, Constable RT, Blumberg HP (2008) Abnormal anterior cingulum integrity in bipolar disorder determined through diffusion tensor imaging. Br J Psychiatry 193(2):126–129

Wang F, Kalmar JH, He Y, Jackowski M, Chepenik LG, Edmiston EE, Tie K, Gong G, Shah MP, Jones M, Uderman J, Constable RT, Blumberg HP (2009) Functional and structural connectivity between the perigenual anterior cingulate and amygdala in bipolar disorder. Biol Psychiatry 66(5):516–521

Watanabe Y, Gould E, Daniels DC, Cameron H, McEwen BS (1992) Tianeptine attenuates stress-induced morphological changes in the hippocampus. Eur J Pharmacol 222(1):157–162

Webster MJ, Knable MB, Johnston-Wilson N, Nagata K, Inagaki M, Yolken RH (2001) Immunohistochemical localization of phosphorylated glial fibrillary acidic protein in the prefrontal cortex and hippocampus from patients with schizophrenia, bipolar disorder, and depression. Brain Behav Immun 15(4):388–400

Wellman CL (2001) Dendritic reorganization in pyramidal neurons in medial prefrontal cortex after chronic corticosterone administration. J Neurobiol 49(3):245–253

Wilke M, Kowatch RA, DelBello MP, Mills NP, Holland SK (2004) Voxel-based morphometry in adolescents with bipolar disorder: first results. Psychiatry Res 131(1):57–69

Xiang L, Szebeni K, Szebeni A, Klimek V, Stockmeier CA, Karolewicz B, Kalbfleisch J, Ordway GA (2008) Dopamine receptor gene expression in human amygdaloid nuclei: elevated D4 receptor mRNA in major depression. Brain Res 1207:214–224

Yamamoto J, Suh J, Takeuchi D, Tonegawa S (2014) Successful execution of working memory linked to synchronized high-frequency gamma oscillations. Cell 157(4):845–857

Yatham LN, Liddle PF, Lam RW, Shiah IS, Lane C, Stoessl AJ, Sossi V, Ruth TJ (2002) PET study of the effects of valproate on dopamine D(2) receptors in neuroleptic- and mood-stabilizer-naive patients with nonpsychotic mania. Am J Psychiatry 159(10):1718–1723

Yucel K, Taylor VH, McKinnon MC, Macdonald K, Alda M, Young LT, Macqueen GM (2008) Bilateral hippocampal volume increase in patients with bipolar disorder and short-term lithium treatment. Neuropsychopharmacology 33:361–367

Yuksel C, Ongur D (2010) Magnetic resonance spectroscopy studies of glutamate-related abnormalities in mood disorders. Biol Psychiatry 68(9):785–794

Yurgelun-Todd DA, Gruber SA, Kanayama G, Killgore WD, Baird AA, Young AD (2000) fMRI during affect discrimination in bipolar affective disorder. Bipolar Disord 2(3 Pt 2):237–248

Zarate CA Jr, Payne JL, Quiroz J, Sporn J, Denicoff KK, Luckenbaugh D, Charney DS, Manji HK (2004) An open-label trial of riluzole in patients with treatment-resistant major depression. Am J Psychiatry 161(1):171–174

Zarate CA Jr, Brutsche NE, Ibrahim L, Franco-Chaves J, Diazgranados N, Cravchik A, Selter J, Marquardt CA, Liberty V, Luckenbaugh DA (2012) Replication of ketamine's antidepressant efficacy in bipolar depression: a randomized controlled add-on trial. Biol Psychiatry 71(11):939–946

Part III
Acute and Long-term Treatment of Bipolar Depression

Chapter 9
Pharmacological Treatment of Acute Bipolar Depression

Gary S. Sachs

Abstract Controversy regarding management of the acute phase of bipolar depression largely reflects the natural divergence of opinions arising from large gaps in knowledge and the fact that no medication is particularly effective in this population. While there remains a great need for more and better evidence from high-quality studies to close the gaps, the past decade has seen the publication of a substantial number of fully powered placebo-controlled trials for bipolar depression, including several new agents. This chapter addresses the challenges of treating bipolar depression by describing a general schema consisting of principles for iterative personalized care that can be applied to an evolving knowledge base. This chapter also reviews the state of the evidence supporting various pharmacological treatments for bipolar depression, including quetiapine, lurasidone, lamotrigine, olanzapine, olanzapine plus fluoxetine (OFC), modafinil, armodafinil, lithium, valproate, carbamazepine, ketamine, *N*-acetyl cysteine (NAC), inositol, pramipexole, riluzole, and standard antidepressants, among others. Currently, five medications are supported by Category A criteria. In order of the strength of support evidence, these are quetiapine, lurasidone, olanzapine, OFC, and lamotrigine. Regardless of whether patients accept these treatments or begin with alternatives, a measurement-based approach to treatment provides a systematic means of working toward an optimized individual treatment plan.

Keywords Bipolar disorder • Bipolar depression • Treatment • Mood stabilizer • Antidepressant

G.S. Sachs, MD (✉)
Bipolar Clinic and Research Program, Massachusetts General Hospital, 50 Staniford Street, Suite 580, Boston, MA 02114, USA
e-mail: sachsg@aol.com

© Springer International Publishing Switzerland 2016   185
C.A. Zarate Jr., H.K. Manji (eds.), *Bipolar Depression: Molecular Neurobiology,
Clinical Diagnosis, and Pharmacotherapy*, Milestones in Drug Therapy,
DOI 10.1007/978-3-319-31689-5_9

9.1   Introduction

The preference for prescribing proven treatments over unproven treatments is a core principle of evidence-based clinical practice. Practitioners recognize that patients seeking treatment for any condition have the right to be informed with regard to interventions with proven efficacy. This right, articulated by Klerman in testimony and publications (Klerman 1990), was the product of professional debate arising from litigation brought by a patient suffering from manic depression and gave modern psychiatric care a firmer scientific foundation. Treatment recommendations can no longer be based solely on the practitioner's "school of thought," but instead grant primacy to the best available clinical trial data. In short, though we acknowledge that an unproven treatment is not necessarily ineffective, maintaining knowledge of proven treatments is a duty owed to our patients.

Bipolar depression remains a clinical challenge. Debate over its treatment continues into the twenty-first century (Pacchiarotti et al. 2013). Although abnormal mood elevation is the cardinal diagnostic feature of bipolar disorder (BD), depression is more than three times as common as episodes of mood elevation and represents the doorway through which BD patients most often enter treatment (Judd et al. 2003).

Controversy regarding management of the acute phase of bipolar depression largely reflects the natural divergence of opinions arising from large gaps in knowledge. While there remains a great need for more and better evidence from high-quality studies to close the gaps, the past decade has seen the publication of a substantial number of fully powered placebo-controlled trials for bipolar depression, including several new agents. While additional clinical trials are certainly welcome, alone they are insufficient to meet clinical needs because applying results from populations of clinical trial subjects to individual patients is fraught with problems. This chapter addresses the challenges of treating bipolar depression by describing a general schema consisting of principles for iterative personalized care that can be applied to an evolving knowledge base. This chapter also reviews the state of the evidence supporting various pharmacological treatments for bipolar depression.

9.2   Approach to Clinical Management

9.2.1   Recognition that Bipolar Depression Is Inherently Difficult to Treat

By definition, patients with bipolar depression meet criteria for a current major depressive episode and a lifetime history of BD. To diagnose bipolar depression, a clinician must determine that the patient has experienced clinically significant abnormal mood elevation and energy levels in the past. Confident diagnosis of bipolar depression requires identifying at least one specific past episode meeting criteria for mania or hypomania. Thus, the process leading to appropriate treatment requires recognition of a clinical state that can only be assessed retrospectively at the time the patient presents for treatment.

Overreliance on retrospective assessment limits the reliability of lifetime mood disorder diagnosis and likely accounts for much of the uncertainty inherent to diagnosing BD. In a rigorous epidemiological study, Takayanagi and colleagues found that diagnosis based on recall for past episodes of psychiatric illness was associated with under-recognition by a factor of two to 12 times compared to rates based on cross-sectional reporting (Takayanagi et al. 2014). In contrast, the reliability of current mood states is encouraging. Data from field trials for DSM-IV and DSM-5 showed high rates of agreement for acute mania (Keller et al. 1996), but these studies did not assess the reliability of eliciting a past history of mania or hypomania from a currently depressed patient.

Guidelines and quality standards are beginning to recognize the benefit of repeated formal longitudinal assessment and the importance of inquiring about a past personal or family history of mania when assessing every patient with acute depression (Johnson and Brickman 2006). BD cannot be ruled out based on a single self-report or in the absence of input from collateral sources, particularly in the face of severe acute depression (Bowden 2001). Care of bipolar depression can be greatly facilitated by clear documentation in the medical record of an index episode of hypomania or mania.

9.2.2 Principles for Combining Evidence from Randomized Controlled Trials and Clinical Empiricism in Treating Bipolar Depression: An Alternative to Guidelines and Algorithms

The sparse use of algorithms and guidelines is understandable. Clinicians recognize that static official consensus guideline documents are often out of date by the time of their publication, seldom acknowledge important gaps in knowledge, and mostly offer general recommendations of questionable relevance to individual patients. The alternative model presented here (see Fig. 9.1) is adapted from the general treatment schema used in the Systematic Treatment-Enhancement Program for Bipolar Disorder (STEP-BD) study (Sachs 2004b, 2007). This model is based on principles of collaborative care and encourages clinicians to offer an initial set of choices for their patients based on the clinician's knowledge of "proven treatments" and pertinent individual factors. Clinical empiricism then drives iterative changes to the treatment plan based on actual prospective outcomes that reveal the patient's pattern of response, adverse effect tolerance, and personal preferences.

As seen in Fig. 9.1, the iterative process starts with the critical decision point at the start of an acute episode of bipolar depression. The collaborative treatment model assumes the patient meets formal diagnostic criteria for BD, meets criteria for a current major depressive episode, and that one of the patient's objectives is to receive treatment for the depressive episode.

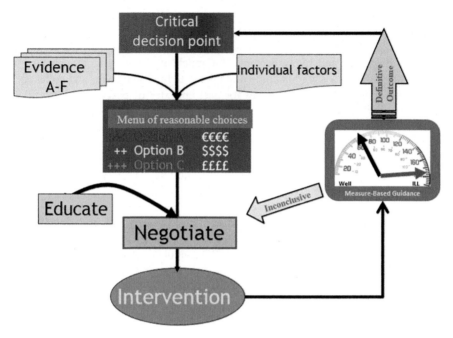

Fig. 9.1 Schema for treatment of bipolar depression: combining evidence and clinical empiricism. Permission granted—Sachs G. 2015

9.2.3 Formulate a Menu of Reasonable Choices

After making the diagnosis, the clinician's next task is to formulate a personalized menu of reasonable choices. The menu of reasonable choices consists of two to four options selected by the clinician after considering both the best available evidence pertaining to the treatment of bipolar depression and individual factors (i.e., the clinician's knowledge of the patient). Evidence-based practice recognizes an implicit duty to at least offer proven treatments first (Klerman 1990). Clinicians can meet this duty by maintaining a working knowledge of proven treatments defined in Table 9.2 as "Category A" treatments and being aware of the key individual characteristics of their patients relevant to treatment choice. At a minimum, this will include the patient's history of prior treatment response, adverse effect tolerance, pertinent general medical conditions, and personal preferences.

Essential to collaborative care is the concept of having a plan with shared decision making, communication with other professionals, and communication with any supports the patient designates as care partners. Including the patient as an active agent in his or her own care requires an engaged, well-informed patient, some time for education, and negotiation skills. Given the opportunity, patients and their care providers are often motivated to make a well-informed selection from the menu of reasonable choices and participate in a variety of self-management strategies. Revicki and colleagues showed that BD patients are generally able to weigh the risk benefit tradeoffs presented by clinicians, which is requisite for collaborative patient participation in formulating a treatment plan (Revicki et al. 2005). Thus, patients who do not respond to a treatment can be offered each of the reasonable options in a sequence based on the patient's preference.

9.2.4 Integrate Routine Outcome Measurement into the Management Plan: Guidance Based on Clinical Empiricism

After patients have agreed to use a medication, it is useful to agree on metrics for measuring progress (Rush et al. 2003, 2006). By integrating measurement into the management plan, the clinician can accumulate prospective outcomes that provide management guidance personalized for each patient. As definitive outcomes are recorded and augment the individual factors known to the clinician, the menu of reasonable choices is then revised with guidance from the process of clinical empiricism.

While formal depression severity scales such as the Hamilton Depression Rating Scale or self-report scales like the Inventory of Depressive Symptoms have the virtue of generating numbers that may relate to published research findings, even raw symptom counts help navigate the complex and changing clinical course. The clinical monitoring form and waiting room self-report form used at the Massachusetts General Hospital Bipolar Clinic are available at www.manicdepressive.org and www.ccihub.com. In addition, many mood tracking apps are now available for smart phones. Assessing the benefit of an intervention may employ many other formal scales or may consist of simply documenting serial judgments made in reference to a patient's personal goals.

9.2.5 Aim to Carry Out Each Intervention to a Definitive Endpoint

The process of clinical empiricism moves forward when interventions are carried out to one of three definitive endpoints: treatment is effective, ineffective, or intolerable. Inconclusive outcomes, however, may result when tolerable interventions are curtailed without adequate dose or duration or are simply rejected as unacceptable. As the outcome of each intervention is evaluated, beneficial treatments are retained and ineffective or intolerable treatments are withdrawn.

Several lines of evidence support the rationale for continuing well tolerated, efficacious treatments and replacing treatments that are ineffective and/or poorly tolerated (Altshuler et al. 2009; Amsterdam and Shults 2010; Tohen et al. 2006).

9.2.6   Harness the Predictive Value of Early Outcomes

Importantly, several studies have indicated that a patient's record of response to treatment has impressive predictive value. For subjects ($N = 3369$) enrolled in 10 placebo-controlled pivotal trials for bipolar depression, Calabrese and colleagues examined the value of "early response" (defined as improvement in the Depression scale score of at least 20 % from baseline after two weeks of treatment) for predicting the probability of response and remission at the end of each study (seven to 10 weeks of treatment) (Calabrese et al. 2010a). The most compelling finding in this analysis was the high negative predictive value associated with not meeting criteria for early improvement. Across all 10 active treatment groups as well as the placebo groups, subjects with less than 20 % improvement after two weeks of treatment had only a 10–20 % chance of meeting remission criteria at the end of the study (Calabrese et al. 2010a; Kemp et al. 2011b). As Fig. 9.2 makes clear, this finding has been highly replicated (Calabrese et al. 2010a; Kemp et al. 2011a; Szegedi et al. 2009; Nierenberg et al. 2000). The consistency of this pattern observed across large placebo-controlled studies for bipolar depression suggests that a determination of the need for dose adjustment or declaring treatment as ineffective could be made with acceptable confidence as rapidly as every two weeks. No other currently available biomarker or group of biomarkers offers a better means of guiding treatment decisions.

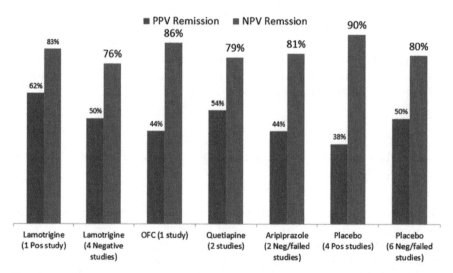

Fig. 9.2 Using prospective outcomes: "Measure-based" guidance for clinical empiricism. The absence of early improvement (≥ 20 %) by day 14 during the acute treatment of bipolar depression predicts no remission at weeks 6 through 8. Adapted from Calabrese and colleagues (2010a). *PPV* positive predictive value, *NPV* negative predictive value, *OFC* olanzapine and fluoxetine

Table 9.1 Categories of clinical evidence

A:	Double-blind, placebo-controlled trials with adequate sample size
B:	Double-blind, controlled trials without placebo or without adequate samples; controlled studies without randomization.
C:	Naturalistic or open-label trials/nonexperimental descriptive studies or case–control studies.
D:	Uncontrolled observations, case series, and single case reports.
E:	Absence of positive published studies. However, Category A evidence supports a class effect.
F:	Absence of positive published studies and Category A evidence does not support a class effect.

The pitfalls of repetitious indecisive trials can be avoided by carrying out each intervention with sufficient dose and duration to declare a definitive outcome (effective, ineffective, or intolerable).

9.3   State of the Evidence

Clinicians seeking guidance from the published literature encounter many forms of evidence that should be weighed in accordance with their scientific quality. Table 9.1 offers a simple grading system intended to help the reader distinguish between levels of evidence. The best guidance for clinical decision-making comes from double-blind, randomized, placebo-controlled trials with adequate samples. Studies meeting these criteria are referred to here as Category A studies, and drugs with positive data can be considered proven.

It is natural to ask which of the proven medications is best. Clinicians looking for guidance in choosing medications for their patients should remain mindful of important caveats bearing on evidence-based practice. Placing the results in a table (such as Table 9.2) invites casual comparison; however, comparisons across studies are not scientifically valid. As the available clinical trial knowledge base for a drug grows, the range of outcomes for that drug tends to broaden. Hence, it becomes increasingly clear that determining which drug works best is not a simple matter of comparing outcomes across studies by calculating effect sizes.

Confident head-to-head comparison of medications requires clinical trials in which subjects are randomly assigned to the agents being compared and also requires sophisticated interpretation. The clinical trial results for any active agent, as well as the results for placebo, are influenced by many aspects of study design and execution, including dosing schedule, choice of comparators, subject selection, demographics, rater expectation, and the availability of alternative treatments for the indication under study. The importance of these factors is evident as the range of results for the best-studied agents can be wider than the difference between medications. As a general rule of thumb, more confidence can be given to findings from studies with larger sample sizes and findings that have been replicated in multiple studies are more reliable than results from a single study. Following this principle, Table 9.2 presents the drugs in order of the number of positive clinical

Table 9.2 Proven treatments: at least one positive Category A study

Medication	Positive outcomes/total arms compared to placebo	Range of effect sizes (Cohen's d)	NNT	Comment positive results/total arms
Quetiapine	9/10			
300 mg (Thase et al. 2006; Young et al. 2010; McElroy et al. 2010; Suppes et al. 2010)	5/5	0.61–xx	5	Monotherapy for eight weeks
600 mg (Thase et al. 2006; Young et al. 2010; McElroy et al. 2010)	4/4	0.54–1.06	5	Monotherapy for eight weeks
150–300 mg (Gao et al. 2014)	0/1	0.19	100	Monotherapy or adjunct for BD-I and BD-II with comorbid GAD. No significant benefit for depression or anxiety
Lurasidone	4/4			
20–60 mg (Loebel et al. 2014a, 2015)	2/2	0.51–0.80		Monotherapy for six weeks, BD-I depression; monotherapy for six weeks, MDD/BD NEC depression with mixed features
80–120 mg (Loebel et al. 2014a)	1/1	0.51		Monotherapy therapy for six weeks, BD-I depression
20–120 mg (Loebel et al. 2014b)	1/1	0.34		Adjunctive therapy for six weeks, BD-I depression
Olanzapine	2/2			
5–20 mg (Tohen et al. 2003, 2012)		0.22–0.32	12	Monotherapy therapy for eight weeks, BD-I depression
			11	Monotherapy therapy for six weeks, BD-I depression
Olanzapine + Fluoxetine	1/1			
6 + 25 mg – 12 + 50 mg (Tohen et al. 2003)	1/1	0.68	5	Combination therapy for eight weeks, BD-I depression
Lamotrigine (Geddes et al. 2009)	Meta-analysis			Of five company-sponsored studies taken individually, one was positive on MADRS (secondary measure) and four were failed studies. These five studies yielded a positive result in meta-analysis.
50 mg	0/1			

(continued)

Table 9.2 (continued)

Medication	Positive outcomes/ total arms compared to placebo	Range of effect sizes (Cohen's d)	NNT	Comment positive results/ total arms
200 mg (Geddes et al. 2009)	1/4			
200 mg (van der Loos et al. 2009)	1/1		5	Adjunctive therapy
100–400 mg (Geddes et al. 2009)				
Modafinil				
100–200 mg (Frye et al. 2007)	1/1	0.47	5	Adjunctive therapy for six weeks. Small sample with robust separation at four time points. Larger effect on energy fatigue subscale.
Armodafinil	2/4			All studies compared armodafinil to placebo as adjunctive therapy. Two unpublished negative/failed studies
150 mg (Calabrese et al. 2010b, 2014)	2/4		−100	Adjunctive therapy for eight weeks. Modest effect with statistically significant result only at two time points. No significant benefit on any secondary outcome measure (MADRS, CGI-BP, QIDS-SR, HAM-A, Q-LES-Q)
		0.28	9	Adjunctive therapy × eight weeks Small effect. Separation from placebo only occurred after seven weeks of treatment.

Effect size: Cohen's d calculated in each study based on the difference between active and placebo groups on change from baseline score divided by the pooled standard deviation for the primary outcome variable

NNT: Number Needed to Treat is calculated as the inverse of the difference between response rate for the active group minus the response rate for placebo group

Abbreviations: GAD: generalized anxiety disorder; BD: bipolar disorder; MADRS: Montgomery–Asberg Depression Rating Scale; NEC: not elsewhere classified; HAM-A: Hamilton Rating Scale for Anxiety; CGI-BP: Clinical Global Impression for Bipolar Disorder; QIDS-SR: Quick Inventory of Depressive Symptomatology (Self-Report); Q-LES-Q: Quality of Life Enjoyment and Satisfaction Questionnaire

trial results in the supporting database. Taking this approach, an adequately powered study with two active arms and a placebo arm could yield no positive results, one positive result, or two positive results compared to placebo.

9.4 Proven Treatment Options

Here, I discuss pharmacological treatments associated with at least one positive Category A study (i.e., an adequately powered, double-blind, placebo-controlled clinical trial).

9.4.1 Quetiapine

Five adequately powered, double-blind, placebo-controlled trials have demonstrated the robust efficacy of quetiapine for bipolar depression (Calabrese et al. 2005; McElroy et al. 2010; Suppes et al. 2010; Thase et al. 2006; Young et al. 2013), and one marginally powered study found no evidence of efficacy (Gao et al. 2014). The latter was a study of patients with bipolar depression and comorbid Generalized Anxiety Disorder. The five positive studies included nine active quetiapine arms, two active comparator arms, and five placebo arms. In these studies, all nine quetiapine versus placebo comparisons found that quetiapine had significant advantages over placebo.

The initial study, BOLDER I (Calabrese et al. 2005), was a double-blind trial that randomized 542 patients with bipolar depression to placebo, quetiapine 300 mg, or quetiapine 600 mg. Both dosages of quetiapine resulted in significantly higher response rates (58 %) compared to placebo (36 %) at eight weeks, as well as a significant advantage over placebo on mean change from baseline in Montgomery–Asberg Depression Rating Scale (MADRS) score and significant improvement in Hamilton Anxiety Rating Scale (HAM-A) total score. No differences between the groups were found in treatment emergent affective switch (TEAS) rates, but the group receiving 600 mg experienced more adverse effects, and only 54 % completed the eight-week study compared to 67 % of subjects in the 300 mg group.

Two subsequent studies used variations on the same design with the addition of an active control; one study used lithium (Young et al. 2010) and the other used paroxetine (McElroy et al. 2010). Results for the quetiapine and placebo groups were similar to the prior study outcomes, but neither of the active control groups differed from placebo. Since the positive results for quetiapine demonstrate assay sensitivity, these results must be counted as negative studies for lithium and paroxetine rather than failed studies.

Importantly, the overall study results included outcomes for patients with bipolar I (BD-I) and bipolar II (BD-II) disorder. Since most other studies excluded patients with BD-II, it is worth noting that quetiapine is the only agent that has shown a statistically significant benefit for depression in BD-II patients and had the largest effect size (0.91 and 1.09 for 300 and 600 mg/day of quetiapine in the BOLDER I study, respectively) observed in any trial for BD-I depression. Smaller effect sizes were observed in subsequent studies, likely reflecting the impact of expectancy following the BOLDER I study and the tendency of trials with more arms to have a higher placebo response.

As noted above, Gao and colleagues conducted an investigator-initiated study of treatment for BD-I and BD-II depressed patients with co-occurring generalized anxiety disorder (GAD) (Gao et al. 2014). This study randomized 100 subjects to add flexibly dosed quetiapine XR or placebo to ongoing treatment and found that quetiapine had no benefit in treating either depression or anxiety.

9.4.2 Lurasidone

Three adequately powered, double-blind, placebo-controlled trials demonstrated the robust efficacy of lurasidone for bipolar depression, including four active arms and three placebo arms. In 2014, Loebel and colleagues published results from a monotherapy study (Loebel et al. 2014a) and an adjunct study (Loebel et al. 2014b) in which lurasidone was superior to placebo on the primary outcome (change from baseline MADRS) and all key secondary outcomes (percent meeting response criteria, Quality of Life Enjoyment and Satisfaction Questionnaire (Q-LES-Q), Sheehan Disability Scale (SDS)).

Recently presented findings show impressive results for lurasidone compared to placebo in depressed subjects not meeting criteria for BD-I or BD-II (Loebel et al. 2015). These subjects all had a major depressive episode with two to three symptoms unique to mania/hypomania (e.g., mood elevation/grandiosity, talkativeness, increased libido, racing thoughts). Although it is unclear whether such patients represent patients with a disorder on the bipolar spectrum or major depressive disorder (MDD) patients at higher risk for conversion to BD, the effect size of 0.8 suggests the possibility that patients meeting the entry criteria may be more responsive to treatment with lurasidone than those who entered trials with eligibility restricted to BD-I subjects.

9.4.3 Lamotrigine

Calabrese and colleagues first demonstrated the efficacy of lamotrigine for treating bipolar depression (Calabrese et al. 1999). In a seven-week, multicenter, double-blind, fixed-dose study, 195 patients were treated with lamotrigine (50 or 200 mg/day) or placebo. The trend favoring lamotrigine on the primary outcome measure (mean change in HAM-D score) fell just short of statistical significance; however, significant improvements over placebo were found for key secondary endpoints, such as change in MADRS and Clinical Global Impression for Bipolar Disorder (CGI-BP) mean scores. Lamotrigine was not associated with increased risk of TEAS compared to placebo. Although subsequent Glaxo-sponsored studies of lamotrigine for bipolar depression resulted in failed trials, trends in these results have consistently favored lamotrigine, and a meta-analysis across all available studies found significant antidepressant effects for lamotrigine (Geddes et al. 2009).

In a Category A study not sponsored by Glaxo, Van der Loos and colleagues showed that lamotrigine had statistically significant benefits over placebo as adjunct to lithium for bipolar depression (van der Loos et al. 2009).

Nierenberg and colleagues randomized BD-I and BD-II patients with depression who had not responded to at least two trials of standard antidepressants combined with mood stabilizers to receive open treatment with lamotrigine (150–250 mg), risperidone (up to 1–6 mg), or inositol (10–25 g) (Nierenberg et al. 2006). This STEP-BD, randomized, open-label study reported positive results for lamotrigine (24 % "recovered") compared with risperidone (5 % "recovered"). Outcomes for inositol (17 % recovered) did not significantly differ from either risperidone or lamotrigine.

Brown and colleagues randomized patients with bipolar depression to receive either lamotrigine or combined treatment with olanzapine and fluoxetine (OFC) (Brown et al. 2006). In this study, response rates for bipolar depression were generally comparable between the groups; however, lamotrigine was associated with significantly less weight gain, sedation, dry mouth, and tremor than OFC.

9.4.4 Olanzapine and OFC

The efficacy of olanzapine has been evaluated in two well-powered studies that included two placebo arms, two olanzapine arms, and one OFC arm. The placebo-controlled trial examining olanzapine and OFC is particularly valuable because it permits a valid comparison between two active arms and a placebo control. Tohen and colleagues randomized BD-I depressed subjects to receive placebo ($n = 355$), olanzapine ($n = 352$), or OFC ($n = 82$) and reported a statistically significant advantage for olanzapine over placebo (Tohen et al. 2003). OFC was found to have superior efficacy to olanzapine monotherapy as well as placebo. The groups did not differ in TEAS rates.

A replication study by Tohen and colleagues found statistically significant antidepressant efficacy for olanzapine ($n = 343$) in a six-week placebo-controlled trial ($n = 171$), but with an effect size (0.22) somewhat smaller than the original study (0.32) (Tohen et al. 2012).

9.5 Noteworthy Evidence from Non-category A Studies

9.5.1 Modafinil and Armodafinil

Modafinil and its active R-enantiomer, armodafinil, have been studied in five fully powered studies. The initial report by Frye and colleagues was positive, but its relatively small sample size gave it marginal statistical power and increased vulnerability to produce outlier results. The study randomized 85 depressed BD patients with an inadequate response to treatment with a mood stabilizer and an antidepressant to add either placebo or modafinil (100–200 mg/day (Frye et al. 2007). The modafinil treatment group had greater improvement in their depressive symptoms as well as better response and remission rates than the placebo group.

Subsequent studies with armodafinil have produced inconsistent results. In the two published studies (Calabrese et al. 2010b, 2014), positive results were obtained on the primary outcome measure (Inventory of Depressive Symptomatology-Clinician-Rated (IDS-C30)) at a few time points, but not on the MADRS or other secondary outcome measures. Two unpublished studies showed no advantage for armodafinil over placebo. It remains unclear if the discrepancy between the studies represents shortcomings of study execution or that the initial study result should be viewed as a false positive (Ostacher 2014).

9.6 Other Commonly Used Medications that Lack Positive Category A Evidence

9.6.1 Standard Antidepressant Drugs

In the past, management of bipolar depression was extrapolated from accumulated experience and research in treating MDD simply because there was no evidence specific to BD patients. More than two dozen agents have been approved by the US Food and Drug Administration (FDA) or other regulatory authorities for the treatment of MDD, also known as unipolar depression. However, the appropriateness of these agents for patients with bipolar depression cannot be determined based on studies of MDD because the mechanisms by which these drugs act remain poorly understood and because trials establishing the efficacy of these agents typically excluded BD patients.

In the twenty-first century, five Category A studies have been reported involving imipramine (one study), paroxetine (three studies), and bupropion (one study) with BD patients (see Table 9.2). The first Category A study for bipolar depression was reported by Nemeroff and colleagues (2001). This study randomized depressed BD-I subjects treated with lithium (0.5–1.2 mmol per liter) to double-blind adjunctive treatment with placebo, paroxetine, or imipramine. Overall, this study found that neither paroxetine nor imipramine had any benefit over placebo on any of the efficacy measures. Among the subgroup with lithium levels between 0.5 and 0.8 mmol/L, however, there was a significant advantage for patients receiving paroxetine. No differences were found between the groups in TEAS rate. Unfortunately, confidence in this finding is limited because the study had no formal rating scale to assess mood elevation and was likely insensitive to TEAS.

The largest placebo-controlled study of standard antidepressants was carried out by the STEP-BD study. This National Institute of Mental Health (NIMH)-sponsored, double-blind study randomized 366 patients to receive treatment with a mood stabilizer and placebo or a mood stabilizer and an antidepressant (bupropion or paroxetine). There were no group differences with regard to subjects' likelihood of achieving a durable recovery (eight consecutive weeks euthymic) (27 % in placebo group and 24 % in antidepressant groups), their TEAS rates (10–11 %, both groups), or other outcome measures. Thus, there was neither evidence of benefit nor harm in adjunctive use of bupropion or paroxetine.

In summary, no Category A study has found a statistically significant benefit to adding a standard antidepressant to lithium, carbamazepine, or valproic acid. Brown and colleagues conducted a double-blind, seven-week, controlled trial that randomized BD-I depressed patients ($n = 205$) to OFC or lamotrigine (Brown et al. 2006). Although response rates did not differ significantly between treatment groups, time to 50 % reduction in their MADRS score was significantly shorter with OFC. There was no significant difference in rates of TEAS between groups, although OFC was associated with significantly more weight gain, somnolence, dry mouth, and tremor. Thus, the combination of OFC was superior to treatment with placebo and superior to treatment with olanzapine monotherapy. This single study represents the only available high-quality evidence supporting the practice of administering a standard antidepressant medication for depressed patients with BD. The degree to which this finding generalizes to combining fluoxetine with other agents, or the possibility that olanzapine might potentiate or be potentiated by other standard antidepressants both remain unclear. No available data from randomized controlled trials address the long-term implications of this combination (Table 9.3).

9.6.2   Studies Without Placebo Control

The Stanley Foundation Bipolar Network (SFBN) conducted a randomized, double-blind comparison of add-on bupropion, sertraline, or venlafaxine to ongoing "mood stabilizer" treatment in 159 patients (Leverich et al. 2006; Post et al. 2006). Overall, TEAS into hypomania and mania occurred in 11.4 % and 7.9 %, respectively, of acute treatment trials (10 weeks) and in 21.8 % and 14.9 % of the continuation trials (one year duration). The rate of TEAS was higher in patients with BD-I (30.8 %)

Table 9.3 Other psychotropic medications commonly used to treat bipolar depression

Category of evidence	
B	Lithium
	Lithium + Paroxetine
	Lithium + Imipramine
	Moclobemide
	Imipramine + Mood stabilizer
	Venlafaxine
	Paroxetine + Mood stabilizer
	Fluoxetine
	Sertraline
	Bupropion
	Imipramine
	Tranylcypromine
	Pramipexole
	Divalproex
	Citalopram
C	Carbamazepine
F	Standard antidepressants (as a class)
	Omega-3s

than in those with BD-II (18.6 %). The risk of switch into hypomania or mania was significantly increased in subjects treated with venlafaxine (15 %) compared to bupropion (4 %) or sertraline (7 %).

In a six-week, randomized, single-blind trial, Vieta and colleagues compared the efficacy of mood stabilizers with adjunctive paroxetine ($n = 30$) to mood stabilizers with adjunctive venlafaxine ($n = 30$) (Vieta et al. 2002). No significant differences in terms of treatment response rates were found between the groups (paroxetine 43 %, venlafaxine 47 %), but lower rates of TEAS were observed in the paroxetine group (3 %) than in the venlafaxine group (13 %). Silverstone and colleagues compared the antidepressant efficacy of moclobemide ($n = 81$) and imipramine ($n = 75$) in patients with BD, 64 % of whom were prescribed concomitant mood stabilizers (Silverstone 2001). No statistically significant differences between the two groups were found on any of the efficacy measures. The trend for higher rates of study withdrawal due to TEAS did not reach statistical significance (moclobemide = 3.7 %, imipramine = 11 %).

Amsterdam and colleagues evaluated the efficacy of fluoxetine monotherapy in patients with BD-II experiencing a major depressive episode in a randomized, double-blind, placebo-controlled study ($n = 34$) (Amsterdam et al. 2004). Significant reductions in mean HAM-D and MADRS ratings, without an increase in YMRS scores, were reported in the active and placebo groups. These investigators also reported that in patients with bipolar depression who received eight weeks of open-label fluoxetine treatment, 48 % showed a HAM-D reduction greater than 50 %, while 7.3 % developed TEAS (YMRS ≥ 8) (Amsterdam et al. 1998). Other small double-blind studies of standard antidepressant use in bipolar depression include those of add-on tranylcypromine (Himmelhoch et al. 1991; Nolen et al. 2007) and desipramine versus bupropion (Sachs et al. 1994).

9.6.3   Studies Without Randomization

Altshuler and colleagues reported quasi-experimental results from the SFBN (Altshuler et al. 2003). Among the 1078 patients with BD, about 50 % became depressed and had a standard antidepressant added to their treatment regimen. Fifteen percent of these patients, for whom there was a clinical intent-to-treat with a standard antidepressant, achieved remission. A comparison of remitted patients (depending on whether antidepressants were continued for more than six months or discontinued before six months) revealed that 20–25 % experienced a relapse into depression over the first four months regardless of whether or not antidepressants were continued. Significantly, lower rates of relapse into depression over one year were observed for those who remained on antidepressants (36 %) compared with those who discontinued (70 %). However, when interpreting this finding, one must consider that the reasons for antidepressant discontinuation are not apparent; for example, treatments may have been discontinued due to lack of efficacy. A similarly designed study (Joffe et al. 2004) reported the one-year outcome of antidepressant treatment in 59 patients with bipolar depression and found comparable results, which are also subject to the same limitations.

The STEP-BD study reported a quasi-experimental comparison of outcomes for 1000 patients with BD treated openly and observed prospectively over one year of follow-up (Sachs 2004a). In this sample, 18 % of patients experienced the onset of a new depressive episode, including 5 % who had multiple depressive episodes. Outcome analysis for the first depressive episode revealed no statistically significant advantage for adding standard antidepressant medications compared to subjects managed without a standard antidepressant. Rates of TEAS were 14.6 % regardless of the use of standard antidepressants.

9.6.4   Meta-Analyses and Systematic Reviews

Gijsman and colleagues conducted a meta-analysis of 12 randomized, controlled trials on the efficacy and safety of antidepressants for the short-term treatment of bipolar depression and found that antidepressants were significantly more effective than placebo (Gijsman et al. 2004). While there was no increased risk of TEAS associated with standard antidepressants in general, the authors found a significant association for TEAS in the subgroup treated with tricyclic antidepressants.

In contrast, another meta-analysis (Sidor and Macqueen 2011) concluded that antidepressants were not statistically superior to placebo or other current standard treatments for bipolar depression. The above reports also included meta-analyses examining the TEAS data from these studies and found no evidence of an association between standard antidepressants and TEAS overall, but the studies that employed more sensitive criteria did find elevated switch rates among subjects receiving standard antidepressants. Notably, Sidor and McQueen included six additional randomized controlled trials ($N = 1034$), which assessed antidepressant use in the acute treatment of bipolar depression and brings the combined total to 15 studies containing 2373 patients. It seems unlikely that a study with a sample of this size would miss clinically meaningful effects related to efficacy or TEAS.

Several caveats pertain to translating these meta-analysis findings for clinical use. First, the results do not support the use of any one antidepressant agent. The "response rates" used in meta-analyses tend to overestimate drug effectiveness because the original studies defined response solely on depression outcomes; therefore, cases in which switch to mania occurred are included in the count of successful responses. For instance, about 25 % of the responders in tranylcypromine and imipramine studies are subjects who became manic (Himmelhoch et al. 1991). The TEAS findings are also limited by the lack of any formal assessment for TEAS in most studies. Meta-analysis is most appropriate and helpful when employed as a means of pooling results from underpowered homogeneous studies and is less desirable when high-quality, fully powered studies are available or when important differences between studies are overlooked.

9.6.5   Lithium

The American Psychiatric Association (APA) Guidelines recommend lithium as an initial treatment for bipolar depression of mild to moderate severity. There is, however, little statistical evidence comparing the antidepressant efficacy of lithium to placebo. Prior to 2008, relevant studies were limited to small placebo-controlled studies, crossover studies, or nonrandomized samples. The recently completed AstraZeneca-sponsored comparison of lithium, quetiapine, and placebo found no benefit for lithium compared to placebo in subjects with BD-I or BD-II depression. While the assay sensitivity of this trial was ostensibly established based on its success in detecting a substantial benefit for the groups treated with quetiapine 300 mg and 600 mg, the trial design may have inadvertently disadvantaged lithium relative to quetiapine, since lithium is widely used by the eligible population and lithium-responsive subjects were unlikely to enroll. In addition, lithium and placebo would be disadvantaged to the extent that quetiapine's sedative qualities and other adverse effects may have unblinded raters.

Studies intended to test other adjunctive treatments in which subjects in all treatment groups receive lithium cannot prove lithium's efficacy, but may nonetheless be instructive for clinical practice.

The efficacy of lithium may be estimated to some extent based on the aforementioned double-blind trial by Nemeroff and colleagues in which no overall benefit was found for adding standard antidepressants to lithium compared with lithium monotherapy at lithium levels ≥ 0.8 mEq/L (Nemeroff et al. 2001). Where lithium was dosed at levels ≤ 0.8 mEq/L, add-on antidepressant treatment was more beneficial than lithium monotherapy (Nemeroff et al. 2001). This finding suggests that subtherapeutic lithium levels are less effective than therapeutic lithium levels, but does not allow any conclusions to be drawn about the relative efficacy of lithium compared with antidepressants or placebo.

9.6.6   Valproate

Evidence supporting the antidepressant properties of valproate is limited. Two small placebo-controlled trials suggest that valproate may have beneficial effects for bipolar depression (Davis et al. 2005; Sachs et al. 2002). Bond and colleagues conducted a meta-analysis of four double-blind, controlled trials comparing valproate to placebo (total $N = 142$ subjects) (Bond et al. 2010). The relative risks of response ($RR = 2.10$, $p = 0.02$) and remission ($RR = 1.61$, $p = 0.04$) were significantly greater for divalproex than placebo. In addition, average response rates were 39.3 % for divalproex and 17.5 % for placebo. Since the total sample size in the four trials was relatively small, the evidence that divalproex is efficacious in the treatment of BD depression should be considered preliminary.

As with lithium, enthusiasm for use of valproate in bipolar depression is often muted by historical perspectives that may not apply to contemporary clinical nomenclature. When analyzing treatment effect in a large, open case series, Lambert's 1966 report finding moderate improvement in only 22 % of 103 "manic-depressive" patients treated with valpromide likely discouraged use of valproate in a manner similar to Cade's reported impression that lithium was ineffective for depression (Lambert et al. 1966). Since Lambert's series also largely comprised subjects who would be classified as having MDD according to DSM-5 criteria, the question of antidepressant efficacy for bipolar depression remains open.

9.6.7   Carbamazepine

Controlled studies supporting the efficacy of carbamazepine in bipolar depression are scarce. Post and colleagues published a randomized, double-blind study of 35 patients with bipolar depression that found at least mild improvement in symptoms (CGI ratings) in 57 % of patients and a more substantial improvement in 34.3 % (Post et al. 1986). This group also conducted a meta-analysis of carbamazepine treatment in individuals with MDD and bipolar depression (including several small, open-label, and controlled studies); response rates to carbamazepine treatment were observed in 56 % of open-label trials and 44 % of controlled studies.

9.7 Novel Treatment Options Supported by Proof-of-Concept Trials

Proof-of-concept trials are often conducted to evaluate clinical efficacy in small samples. Although systematic clinical trials can help advance clinical science, caution is necessary in applying results from proof-of-concept studies to clinical practice. When the evidence supporting innovative interventions is limited to studies of small sample size, both positive and negative results should always be considered preliminary.

9.7.1 Ketamine

Ketamine is an N-methyl-D-aspartate (NMDA) antagonist commonly used in medicine as a preanesthetic and, illicitly, as a party drug. Rapid resolution of depression and suicidal ideation following single intravenous infusions of low doses of ketamine have been reported for patients with BD as well as MDD (Diazgranados et al. 2010; Ionescu et al. 2015; Lee et al. 2015; Zarate et al. 2012).

Two small proof-of-concept studies provide further encouragement for this innovative approach to treatment-resistant bipolar depression. Both were randomized, placebo-controlled, double-blind, crossover, add-on studies in which subjects maintained at therapeutic levels of lithium or valproate received an intravenous infusion of either ketamine hydrochloride (0.5 mg/kg) or placebo on two treatment days two weeks apart. The onset of antidepressant effects was observed within 40 minutes and was sustained for several days.

Ionescu and colleagues found equally robust outcomes in response to open ketamine treatment for depressed BD patients with high levels of anxiety compared to those without high anxiety. Of course, this finding needs replication in a controlled trial, but is noteworthy since high levels of anxiety predict poor response to most treatments for bipolar depression (Coryell et al. 2009).

The results for ketamine are very encouraging, but there remains substantial potential that some or all of the putative antidepressant effect could be due to the small sample sizes and crossover design and that adverse effects such as dissociative symptoms often associated with ketamine use influence outcomes via functional unblinding of subjects and/or raters.

9.7.2 N-acetyl cysteine (NAC)

NAC may exert psychotropic effects through multiple mechanisms. These include being a precursor to the antioxidant glutathione, modulating glutamatergic transmission, and interacting with anti-inflammatory and neurotropic pathways. An intriguing report by Berk and colleagues sparked interest in NAC for bipolar depression (Berk et al. 2008). These investigators randomly assigned 75 BD patients to placebo or NAC (1 g twice daily) as an add-on to their ongoing maintenance treatment regimen. Although only a third of the subjects met criteria for a major depressive episode at baseline, the primary outcome—improvement from baseline MADRS score—showed a robust advantage for NAC over placebo. At end point, the study produced impressive effect sizes (medium to high) for improvements in MADRS and nine of the 12 secondary measures. Furthermore, loss of benefit was observed when NAC was replaced with placebo after the double-blind treatment phase. This finding needs replication in a trial specifically designed to treat acute depressive episodes.

9.7.3   Inositol

The controlled evidence supporting the use of inositol for bipolar depression consists of three small, double-blind placebo trials that found advantages over placebo of at least marginal statistical significance (Levine et al. 1995; Chengappa et al. 2000; Eden Evins et al. 2006). These studies in total, however, report results for only 34 inositol-treated subjects and 33 placebo-treated subjects. As noted above, in an open, randomized effectiveness study using durable recovery (a much more stringent outcome criterion defined as eight consecutive weeks euthymic), Nierenberg and colleagues found that the response of treatment-refractory bipolar depressed patients to adjunctive inositol (17.4 %) did not significantly differ from response to risperidone (4.6 %) or lamotrigine (23.8 %) (Nierenberg et al. 2006). In these studies, inositol was administered at doses ranging from 10 to 25 g per day.

9.7.4   Pramipexole

Two small, randomized, placebo-controlled trials have examined the effectiveness of pramipexole for treatment of bipolar depression. Goldberg and colleagues randomized 22 depressed BD patients to pramipexole (1.0–2.5 mg/day) plus a mood stabilizer or placebo plus a mood stabilizer (Goldberg et al. 2004). Pramipexole was well tolerated and more patients in the pramipexole group achieved a $\geq 50\%$ reduction from their baseline HAM-D scores and a greater mean change in HAM-D scores over the six-week study duration. In addition, the switch rates into mania and hypomania were not higher than in the placebo group and the side effects were mild.

Zarate and colleagues randomized 21 BD-II patients to either pramipexole (1.0 to 3.0 mg/day) plus a mood stabilizer or placebo plus a mood stabilizer (Zarate et al. 2004). They found that 60 % of the participants in the pramipexole group had a therapeutic response (as defined by a 50 % reduction in MADRS scores) compared to 9 % in the placebo group. Additionally, pramipexole appeared to be well tolerated as every participant completed the six-week study, except one from each group. These encouraging results need confirmation in a Category A study.

9.7.5 Riluzole

An open study of riluzole, a glutamate-modulating agent, in combination with lithium has been reported by Zarate and colleagues (2005). Of the 19 BD-I depressed patients who received riluzole (100–200 mg/day for six weeks), 43 % dropped out due to adverse events or lack of improvement. Among those remaining in the study, improvement in MADRS scores was noted in weeks three through six for all subjects. There were no switches to hypomania or mania. Future research is needed to determine whether riluzole has true antidepressant effects.

9.7.6 Other Agents

Beneficial effects have also been reported from small and often less rigorous studies involving a variety of newer treatment interventions including the use of stimulants, omega-3 fatty acids, dextromethorphan in combination with quinidine (DMQ), phototherapy, transcranial magnetic stimulation (TMS), and sleep deprivation.

Stimulants such as methylphenidate appear to have short-lived beneficial effects in bipolar depression, but are associated with mania, irritability, and psychosis (El-Mallakh 2000; Zarate et al. 2004). SFBN studies found no benefit for omega-3 fatty acids over placebo (Keck et al. 2006). In contrast, one open trial and a small, double-blind trial in children (Osher et al. 2005; Nemets et al. 2006) reported a beneficial effect from the use of omega-3 fatty acids in depression. However, these results have not been replicated and there remains no positive result for any form of omega-3 in a Category A study.

Positive findings have been reported for repetitive transcranial magnetic stimulation (Speer et al. 2000). Sleep deprivation has shown a transient mood-elevating effect in two studies (Riemann et al. 2002; Benedetti et al. 2001), but also carries a risk of precipitating mania.

Dextromethorphan is an NMDA antagonist that, unlike ketamine, can be given orally. When combined with a low dose of quinidine (DMQ), the problem of dextromethorphan's rapid metabolism is overcome. A retrospective chart review ($N = 77$) of chronically depressed BD-II and BD NOS patients found rapid improvement in response to DMQ (Kelly and Lieberman 2014). Although the combination of dextromethorphan/quinidine sulfate has FDA approval for treatment of pseudobulbar affect, there are no controlled data supporting its use for the treatment of bipolar depression.

9.8 Conclusion

The care model described here comprises a set of principles that clinicians can apply to current and emerging data. The model encourages clinicians to formulate a menu of reasonable choices personalized to each patient by drawing on two knowledge bases: data from randomized controlled trials and individual factor data acquired through clinical empiricism. This contrasts with usual published guidelines and algorithms, since knowledge bases are dynamic and allow the clinician to apply their judgment to all available information.

The right of patients to be made aware of evidence-based treatment options requires that practicing clinicians maintain awareness of the treatment options supported by Category A evidence. Currently, five medications are supported by Category A criteria. In order of the strength of support evidence, these are quetiapine, lurasidone, olanzapine, OFC, and lamotrigine. Regardless of whether patients accept these treatments or begin with alternatives, a measurement-based approach to treatment provides a systematic means of working toward an optimized individual treatment plan.

References

Altshuler L, Suppes T, Black D, Nolen WA, Keck PE Jr, Frye MA, McElroy S, Kupka R, Grunze H, Walden J, Leverich G, Denicoff K, Luckenbaugh D, Post R (2003) Impact of antidepressant discontinuation after acute bipolar depression remission on rates of depressive relapse at 1-year follow-up. Am J Psychiatry 160(7):1252–1262

Altshuler LL, Post RM, Hellemann G, Leverich GS, Nolen WA, Frye MA, Keck PE Jr, Kupka RW, Grunze H, McElroy SL, Sugar CA, Suppes T (2009) Impact of antidepressant continuation after acute positive or partial treatment response for bipolar depression: a blinded, randomized study. J Clin Psychiatry 70(4):450–457

Amsterdam JD, Shults J (2010) Efficacy and Safety of Long-Term Fluoxetine Versus Lithium Monotherapy of Bipolar II Disorder: A Randomized, Double-Blind, Placebo-Substitution Study. Am J Psychiatry 167:792–800

Amsterdam JD, Garcia-Espana F, Fawcett J, Quitkin FM, Reimherr FW, Rosenbaum JF, Schweizer E, Beasley C (1998) Efficacy and safety of fluoxetine in treating bipolar II major depressive episode. J Clin Psychopharmacol 18(6):435–440

Amsterdam JD, Shults J, Brunswick DJ, Hundert M (2004) Short-term fluoxetine monotherapy for bipolar type II or bipolar NOS major depression – low manic switch rate. Bipolar Disord 6 (1):75–81

Benedetti F, Barbini B, Campori E, Fulgosi MC, Pontiggia A, Colombo C (2001) Sleep phase advance and lithium to sustain the antidepressant effect of total sleep deprivation in bipolar depression: new findings supporting the internal coincidence model? J Psychiatr Res 35 (6):323–329

Berk M, Copolov DL, Dean O, Lu K, Jeavons S, Schapkaitz I, Anderson-Hunt M, Bush AI (2008) N-acetyl cysteine for depressive symptoms in bipolar disorder – a double-blind randomized placebo-controlled trial. Biol Psychiatry 64(6):468–475

Bond DJ, Lam RW, Yatham LN (2010) Divalproex sodium versus placebo in the treatment of acute bipolar depression: A systematic review and meta-analysis. J Affect Disord 124 (3):228–34

Bowden CL (2001) Strategies to reduce misdiagnosis of bipolar depression. Psychiatr Serv 52 (1):51–55

Brown EB, McElroy SL, Keck PE Jr, Deldar A, Adams DH, Tohen M, Williamson DJ (2006) A 7-week, randomized, double-blind trial of olanzapine/fluoxetine combination versus lamotrigine in the treatment of bipolar I depression. J Clin Psychiatry 67(7):1025–1033

Calabrese JR, Bowden CL, Sachs GS, Ascher JA, Monaghan E, Rudd GD (1999) A double-blind placebo-controlled study of lamotrigine monotherapy in outpatients with bipolar I depression. Lamictal 602 Study Group. J Clin Psychiatry 60(2):79–88

Calabrese JR, Keck PE Jr, Macfadden W, Minkwitz M, Ketter TA, Weisler RH, Cutler AJ, McCoy R, Wilson E, Mullen J (2005) A randomized, double-blind, placebo-controlled trial of quetiapine in the treatment of bipolar I or II depression. Am J Psychiatry 162(7):1351–1360

Calabrese JR, Kemp DE, Ganocy SJ, Brecher M, Carlson BX, Edwards S, Eudicone J, Evoniuk G, Jansen WT, McQuade R, Millen B, Minkwitz M, Owen R, Pikalov A, Szegedi A, Tohen M, Vester-Blokland E, Willigenburg AP (2010a) Knowing when a drug is not going to work in bipolar depression: absence of early improvement as a predictor of later non-response in 3,369 patients from 10 placebo-controlled acute trials. Paper presented at the International Society for Clinical Trial Methodology, Washington, DC

Calabrese JR, Ketter TA, Youakim JM, Tiller JM, Yang R, Frye MA (2010b) Adjunctive armodafinil for major depressive episodes associated with bipolar I disorder: a randomized, multicenter, double-blind, placebo-controlled, proof-of-concept study. J Clin Psychiatry 71:1363–1370

Calabrese JR, Frye MA, Yang R, Ketter TA, Armodafinil Treatment Trial Study N (2014) Efficacy and safety of adjunctive armodafinil in adults with major depressive episodes associated with bipolar I disorder: a randomized, double-blind, placebo-controlled, multicenter trial. J Clin Psychiatry 75(10):1054–1061

Chengappa KN, Levine J, Gershon S, Mallinger AG, Hardan A, Vagnucci A, Pollock B, Luther J, Buttenfield J, Verfaille S, Kupfer DJ (2000) Inositol as an add-on treatment for bipolar depression. Bipolar Disord 2(1):47–55

Coryell W, Solomon DA, Fiedorowicz JG, Endicott J, Schettler PJ, Judd LL (2009) Anxiety and outcome in bipolar disorder. Am J Psychiatry 166(11):1238–1243

Davis LL, Bartolucci A, Petty F (2005) Divalproex in the treatment of bipolar depression: a placebo-controlled study. J Affect Disord 85(3):259–266

Diazgranados N, Ibrahim L, Brutsche NE, Newberg A, Kronstein P, Khalife S, Kammerer WA, Quezado Z, Luckenbaugh DA, Salvadore G, Machado-Vieira R, Manji HK, Zarate CA Jr (2010) A randomized add-on trial of an N-methyl-D-aspartate antagonist in treatment-resistant bipolar depression. Arch Gen Psychiatry 67(8):793–802

Eden Evins A, Demopulos C, Yovel I, Culhane M, Ogutha J, Grandin LD, Nierenberg AA, Sachs GS (2006) Inositol augmentation of lithium or valproate for bipolar depression. Bipolar Disord 8(2):168–174

El-Mallakh RS (2000) An open study of methylphenidate in bipolar depression. Bipolar Disord 2 (1):56–59

Frye MA, Grunze H, Suppes T, McElroy SL, Keck PE Jr, Walden J, Leverich GS, Altshuler LL, Nakelsky S, Hwang S, Mintz J, Post RM (2007) A placebo-controlled evaluation of adjunctive modafinil in the treatment of bipolar depression. Am J Psychiatry 164(8):1242–1249

Gao K, Wu R, Kemp DE, Chen J, Karberg E, Conroy C, Chan P, Ren M, Serrano MB, Ganocy SJ, Calabrese JR (2014) Efficacy and safety of quetiapine-XR as monotherapy or adjunctive therapy to a mood stabilizer in acute bipolar depression with generalized anxiety disorder and other comorbidities: a randomized, placebo-controlled trial. J Clin Psychiatry 75 (10):1062–1068

Geddes JR, Calabrese JR, Goodwin GM (2009) Lamotrigine for treatment of bipolar depression: independent meta-analysis and meta-regression of individual patient data from five randomised trials. Br J Psychiatry 194(1):4–9

Gijsman HJ, Geddes JR, Rendell JM, Nolen WA, Goodwin GM (2004) Antidepressants for bipolar depression: a systematic review of randomized, controlled trials. Am J Psychiatry 161 (9):1537–1547

Goldberg JF, Burdick KE, Endick CJ (2004) Preliminary randomized, double-blind, placebo-controlled trial of pramipexole added to mood stabilizers for treatment-resistant bipolar depression. Am J Psychiatry 161(3):564–566

Himmelhoch JM, Thase ME, Mallinger AG, Houck P (1991) Tranylcypromine versus imipramine in anergic bipolar depression. Am J Psychiatry 148(7):910–916

Ionescu DF, Luckenbaugh DA, Niciu MJ, Richards EM, Zarate CA Jr (2015) A single infusion of ketamine improves depression scores in patients with anxious bipolar depression. Bipolar Disord 17(4):438–443

Joffe RT, MacQueen GM, Marriott M, Trevor Young L (2004) A prospective, longitudinal study of percentage of time spent ill in patients with bipolar I or bipolar II disorders. Bipolar Disord 6 (1):62–66

Johnson SL, Brickman AL (2006) Diagnostic inconsistency: a marker of service utilization in bipolar disorder. Manag Care Interface 19(4):41–45

Judd LL, Schettler PJ, Akiskal HS, Maser J, Coryell W, Solomon D, Endicott J, Keller M (2003) Long-term symptomatic status of bipolar I vs. bipolar II disorders. Int J Neuropsychopharmacol 6(2):127–137

Keck PE Jr, Mintz J, McElroy SL, Freeman MP, Suppes T, Frye MA, Altshuler LL, Kupka R, Nolen WA, Leverich GS, Denicoff KD, Grunze H, Duan N, Post RM (2006) Double-blind, randomized, placebo-controlled trials of ethyl-eicosapentanoate in the treatment of bipolar depression and rapid cycling bipolar disorder. Biol Psychiatry 60(9):1020–1022

Keller MB, Hanks DL, Klein DN (1996) Summary of the DSM-IV mood disorders field trial and issue overview. Psychiatr Clin North Am 19(1):1–28

Kelly TF, Lieberman DZ (2014) The utility of the combination of dextromethorphan and quinidine in the treatment of bipolar II and bipolar NOS. J Affect Disord 167:333–335

Kemp DE, Ganocy SJ, Brecher M, Carlson BX, Edwards S, Eudicone JM, Evoniuk G, Jansen W, Leon AC, Minkwitz M, Pikalov A, Stassen HH, Szegedi A, Tohen M, Van Willigenburg AP, Calabrese JR (2011a) Clinical value of early partial symptomatic improvement in the prediction of response and remission during short-term treatment trials in 3369 subjects with bipolar I or II depression. J Affect Disord 130(1–2):171–179

Kemp DE, Johnson E, Wang WV, Tohen M, Calabrese JR (2011b) Clinical utility of early improvement to predict response or remission in acute mania: focus on olanzapine and risperidone. J Clin Psychiatry 72:1236–1241

Klerman GL (1990) The psychiatric patient's right to effective treatment: implications of Osheroff v. Chestnut Lodge. Am J Psychiatry 147(4):409–418

Lambert PA, Carraz G, Borselli S (1966) Action neuropsychotrope d'un novel anti-epileptique: le Depamide [French]. Ann Med Psychol 1:707–710

Lee EE, Della Selva MP, Liu A, Himelhoch S (2015) Ketamine as a novel treatment for major depressive disorder and bipolar depression: a systematic review and quantitative meta-analysis. Gen Hosp Psychiatry 37(2):178–184

Leverich GS, Altshuler LL, Frye MA, Suppes T, McElroy SL, Keck PE Jr, Kupka RW, Denicoff KD, Nolen WA, Grunze H, Martinez MI, Post RM (2006) Risk of switch in mood polarity to hypomania or mania in patients with bipolar depression during acute and continuation trials of venlafaxine, sertraline, and bupropion as adjuncts to mood stabilizers. Am J Psychiatry 163 (2):232–239

Levine J, Barak Y, Gonzalves M, Szor H, Elizur A, Kofman O, Belmaker RH (1995) Double-blind, controlled trial of inositol treatment of depression. Am J Psychiatry 152(5):792–794

Loebel A, Cucchiaro J, Silva R, Kroger H, Hsu J, Sarma K, Sachs G (2014a) Lurasidone monotherapy in the treatment of bipolar I depression: a randomized, double-blind, placebo-controlled study. Am J Psychiatry 171(2):160–168

Loebel A, Cucchiaro J, Silva R, Kroger H, Sarma K, Xu J, Calabrese JR (2014b) Lurasidone as adjunctive therapy with lithium or valproate for the treatment of bipolar I depression: a randomized, double-blind, placebo-controlled study. Am J Psychiatry 171(2):169–177

Loebel A, Cucchiaro J, Silva R, Hsu J, Sarma K (2015) Double blind placebo controlled trial of Lurasidone vs placebo for treatment of unipolar major depressive episode with mixed features. In: Annual meeting American Psychaitric Association, Toronto

McElroy SL, Weisler RH, Chang W, Olausson B, Paulsson B, Brecher M, Agambaram V, Merideth C, Nordenhem A, Young AH (2010) A double-blind, placebo-controlled study of quetiapine and paroxetine as monotherapy in adults with bipolar depression (EMBOLDEN II). J Clin Psychiatry 71(2):163–174

Nemeroff CB, Evans DL, Gyulai L, Sachs GS, Bowden CL, Gergel IP, Oakes R, Pitts CD (2001) Double-blind, placebo-controlled comparison of imipramine and paroxetine in the treatment of bipolar depression. Am J Psychiatry 158(6):906–912

Nemets H, Nemets B, Apter A, Bracha Z, Belmaker RH (2006) Omega-3 treatment of childhood depression: a controlled, double-blind pilot study. Am J Psychiatry 163(6):1098–1100

Nierenberg AA, Farabaugh AH, Alpert JE, Gordon J, Worthington JJ, Rosenbaum JF, Fava M (2000) Timing of onset of antidepressant response with fluoxetine treatment. Am J Psychiatry 157(9):1423–1428

Nierenberg AA, Ostacher MJ, Calabrese JR, Ketter TA, Marangell LB, Miklowitz DJ, Miyahara S, Bauer MS, Thase ME, Wisniewski SR, Sachs GS (2006) Treatment-resistant bipolar depression: a STEP-BD equipoise randomized effectiveness trial of antidepressant augmentation with lamotrigine, inositol, or risperidone. Am J Psychiatry 163(2):210–216

Nolen WA, Kupka RW, Hellemann G, Frye MA, Altshuler LL, Leverich GS, Suppes T, Keck PE Jr, McElroy S, Grunze H, Mintz J, Post RM (2007) Tranylcypromine vs. lamotrigine in the treatment of refractory bipolar depression: a failed but clinically useful study. Acta Psychiatr Scand 115(5):360–365

Osher Y, Bersudsky Y, Belmaker RH (2005) Omega-3 eicosapentaenoic acid in bipolar depression: report of a small open-label study. J Clin Psychiatry 66(6):726–729

Ostacher MJ (2014) When positive isn't positive: the hopes and disappointments of clinical trials. J Clin Psychiatry 75(10):e1186–e1187

Pacchiarotti I, Bond DJ, Baldessarini RJ, Nolen WA, Grunze H, Licht RW, Post RM, Berk M, Goodwin GM, Sachs GS, Tondo L, Findling RL, Youngstrom EA, Tohen M, Undurraga J, Gonzalez-Pinto A, Goldberg JF, Yildiz A, Altshuler LL, Calabrese JR, Mitchell PB, Thase ME, Koukopoulos A, Colom F, Frye MA, Malhi GS, Fountoulakis KN, Vazquez G, Perlis RH, Ketter TA, Cassidy F, Akiskal H, Azorin JM, Valenti M, Mazzei DH, Lafer B, Kato T, Mazzarini L, Martinez-Aran A, Parker G, Souery D, Ozerdem A, McElroy SL, Girardi P, Bauer M, Yatham LN, Zarate CA, Nierenberg AA, Birmaher B, Kanba S, El-Mallakh RS, Serretti A, Rihmer Z, Young AH, Kotzalidis GD, MacQueen GM, Bowden CL, Ghaemi SN, Lopez-Jaramillo C, Rybakowski J, Ha K, Perugi G, Kasper S, Amsterdam JD, Hirschfeld RM, Kapczinski F, Vieta E (2013) The International Society for Bipolar Disorders (ISBD) task force report on antidepressant use in bipolar disorders. Am J Psychiatry 170(11):1249–1262

Post RM, Uhde TW, Roy-Byrne PP, Joffe RT (1986) Antidepressant effects of carbamazepine. Am J Psychiatry 143(1):29–34

Post RM, Altshuler LL, Leverich GS, Frye MA, Nolen WA, Kupka RW, Suppes T, McElroy S, Keck PE, Denicoff KD, Grunze H, Walden J, Kitchen CM, Mintz J (2006) Mood switch in bipolar depression: comparison of adjunctive venlafaxine, bupropion and sertraline. Br J Psychiatry 189:124–131

Revicki DA, Hanlon J, Martin S, Gyulai L, Nassir Ghaemi S, Lynch F, Mannix S, Kleinman L (2005) Patient-based utilities for bipolar disorder-related health states. J Affect Disord 87 (2–3):203–210

Riemann D, Voderholzer U, Berger M (2002) Sleep and sleep-wake manipulations in bipolar depression. Neuropsychobiology 45(Suppl 1):7–12

Rush AJ, Trivedi MH, Ibrahim HM, Carmody TJ, Arnow B, Klein DN, Markowitz JC, Ninan PT, Kornstein S, Manber R, Thase ME, Kocsis JH, Keller MB (2003) The 16-Item Quick Inventory of Depressive Symptomatology (QIDS), clinician rating (QIDS-C), and self-report (QIDS-SR): a psychometric evaluation in patients with chronic major depression. Biol Psychiatry 54 (5):573–583

Rush AJ, Kraemer HC, Sackeim HA, Fava M, Trivedi MH, Frank E, Ninan PT, Thase ME, Gelenberg AJ, Kupfer DJ, Regier DA, Rosenbaum JF, Ray O, Schatzberg AF (2006) Report by the ACNP Task Force on response and remission in major depressive disorder. Neuropsychopharmacology 31(9):1841–1853

Sachs GS (2004a) Effectiveness study design and preliminary finding from STEP-BD. In: 59th Annual Convention Society for Biological Psychiatry, New York, NY

Sachs GS (2004b) Strategies for improving treatment of bipolar disorder: integration of measurement and management. Acta Psychiatr Scand Suppl (422):7–17

Sachs GS (2007) Bipolar disorder clinical synthesis: Where does the evidence lead? Focus V (1):1–11

Sachs GS, Lafer B, Stoll AL, Banov M, Thibault AB, Tohen M, Rosenbaum JF (1994) A double-blind trial of bupropion versus desipramine for bipolar depression. J Clin Psychiatry 55 (9):391–393

Sachs GA, Ketter L, Suppes T, Rasgon N, Frey M, Collins M (2002) Divalproex versus placebo for treatment of Bipolar depression. In: 155th Annual Meeting of the American Psychiatric Association, Philadelphia, PA

Sidor MM, Macqueen GM (2011) Antidepressants for the acute treatment of bipolar depression: a systematic review and meta-analysis. J Clin Psychiatry 72(2):156–167

Silverstone T (2001) Moclobemide vs. imipramine in bipolar depression: a multicentre double-blind clinical trial. Acta Psychiatr Scand 104(2):104–109

Speer AM, Kimbrell TA, Wassermann EM, D Repella J, Willis MW, Herscovitch P, Post RM (2000) Opposite effects of high and low frequency rTMS on regional brain activity in depressed patients. Biol Psychiatry 48(12):1133–1141

Suppes T, Datto C, Minkwitz M, Nordenhem A, Walker C, Darko D (2010) Effectiveness of the extended release formulation of quetiapine as monotherapy for the treatment of acute bipolar depression. J Affect Disord 121(1–2):106–115

Szegedi A, Jansen WT, van Willigenburg AP, van der Meulen E, Stassen HH, Thase ME (2009) Early improvement in the first 2 weeks as a predictor of treatment outcome in patients with major depressive disorder: a meta-analysis including 6562 patients. J Clin Psychiatry 70 (3):344–353

Takayanagi Y, Spira AP, Roth KB, Gallo JJ, Eaton WW, Mojtabai R (2014) Accuracy of reports of lifetime mental and physical disorders: results from the Baltimore Epidemiological Catchment Area study. JAMA Psychiatry 71(3):273–280

Thase ME, Macfadden W, Weisler RH, Chang W, Paulsson B, Khan A, Calabrese JR, Group BIS (2006) Efficacy of quetiapine monotherapy in bipolar I and II depression: a double-blind, placebo-controlled study (the BOLDER II study). J Clin Psychopharmacol 26(6):600–609

Tohen M, Vieta E, Calabrese J, Ketter TA, Sachs G, Bowden C, Mitchell PB, Centorrino F, Risser R, Baker RW, Evans AR, Beymer K, Dube S, Tollefson GD, Breier A (2003) Efficacy of olanzapine and olanzapine-fluoxetine combination in the treatment of bipolar I depression. Arch Gen Psychiatry 60(11):1079–1088

Tohen M, Calabrese JR, Sachs GS, Banov MD, Detke HC, Risser R, Baker RW, Chou JC, Bowden CL (2006) Randomized, placebo-controlled trial of olanzapine as maintenance therapy in patients with bipolar I disorder responding to acute treatment with olanzapine. Am J Psychiatry 163(2):247–256

Tohen M, McDonnell DP, Case M, Kanba S, Ha K, Fang YR, Katagiri H, Gomez JC (2012) Randomised, double-blind, placebo-controlled study of olanzapine in patients with bipolar I depression. Br J Psychiatry 201(5):376–382 van der Loos ML, Mulder PG, Hartong EG, Blom MB, Vergouwen AC, de Keyzer HJ, Notten PJ, Luteijn ML, Timmermans MA, Vieta E, Nolen WA (2009) Efficacy and safety of lamotrigine as add-on treatment to lithium in bipolar depression: a multicenter, double-blind, placebo-controlled trial. J Clin Psychiatry 70(2):223–231

Vieta E, Martinez-Aran A, Goikolea JM, Torrent C, Colom F, Benabarre A, Reinares M (2002) A randomized trial comparing paroxetine and venlafaxine in the treatment of bipolar depressed patients taking mood stabilizers. J Clin Psychiatry 63(6):508–512

Young AH, McElroy SL, Bauer M, Philips N, Chang W, Olausson B, Paulsson B, Brecher M (2010) A double-blind, placebo-controlled study of quetiapine and lithium monotherapy in adults in the acute phase of bipolar depression (EMBOLDEN I). J Clin Psychiatry 71 (2):150–162

Young AH, Calabrese JR, Gustafsson U, Berk M, McElroy SL, Thase ME, Suppes T, Earley W (2013) Quetiapine monotherapy in bipolar II depression: combined data from four large, randomized studies. Int J Bipolar Disord 1:10

Zarate CA Jr, Payne JL, Singh J, Quiroz JA, Luckenbaugh DA, Denicoff KD, Charney DS, Manji HK (2004) Pramipexole for bipolar II depression: a placebo-controlled proof of concept study. Biol Psychiatry 56(1):54–60

Zarate CA Jr, Quiroz JA, Singh JB, Denicoff KD, De Jesus G, Luckenbaugh DA, Charney DS, Manji HK (2005) An open-label trial of the glutamate-modulating agent riluzole in combination with lithium for the treatment of bipolar depression. Biol Psychiatry 57(4):430–432

Zarate CA Jr, Brutsche NE, Ibrahim L, Franco-Chaves J, Diazgranados N, Cravchik A, Selter J, Marquardt CA, Liberty V, Luckenbaugh DA (2012) Replication of ketamine's antidepressant efficacy in bipolar depression: a randomized controlled add-on trial. Biol Psychiatry 71 (11):939–946

Chapter 10
Pharmacological Treatment of the Maintenance Phase of Bipolar Depression: Efficacy and Side Effects

Keming Gao, Renrong Wu, and Joseph R. Calabrese

Abstract To prevent relapse/recurrence of mood episodes in bipolar disorder (BD), medications must demonstrate efficacy in preventing both manic/mixed/hypomanic and depressive relapses/recurrences. Most maintenance treatment studies of BD have continued using relapse-prevention designs to preserve assay sensitivity despite their limited generalizability. This chapter focuses on all traditional efficacy measures including time to intervention for relapse/recurrence of any mood episodes, manic/mixed episodes, and depressed episodes. For safety, we have focused on short- and long-term side effects, especially $\geq 7\%$ weight gain. To date, lithium, lamotrigine, and quetiapine have been investigated in both manic and depressive index episodes. Divalproex, olanzapine, aripiprazole, paliperidone, and risperidone long-acting injections have been evaluated in manic/mixed index episodes. Both time-to-event results and cumulative risk for relapse/recurrence showed that only quetiapine demonstrated bimodal efficacy in preventing both depressive and manic/mixed/hypomanic relapses. The other medications were only efficacious for preventing manic/mixed relapses or more effective in preventing manic/mixed relapses than depressive relapses. Short-term side effects varied widely among the aforementioned medications, but newly emergent side effects during long-term treatment were relatively rare. With the exception of ziprasidone, significantly higher rates of $\geq 7\%$ weight gain were observed with all other antipsychotics. The short- and long-term side effects suggest that relapse-prevention studies also inflate the safety and tolerability of active treatments.

K. Gao, MD, PhD (✉) • J.R. Calabrese, MD
Department of Psychiatry, Mood and Anxiety Clinic in Mood Disorders Program, University Hospitals Case Medical Center, Case Western Reserve University School of Medicine, 10524 Euclid Avenue, 12th Floor, Cleveland, OH 44106, USA
e-mail: keming.gao@uhhospitals.org

R. Wu, MD, PhD
Institute of Mental Health of Second Xiangya Hospitals of Central South University, Hunan, P.R. China

© Springer International Publishing Switzerland 2016
C.A. Zarate Jr., H.K. Manji (eds.), *Bipolar Depression: Molecular Neurobiology, Clinical Diagnosis, and Pharmacotherapy*, Milestones in Drug Therapy,
DOI 10.1007/978-3-319-31689-5_10

Keywords Bipolar depression • Efficacy • Relapse-prevention design • Time-to-event analysis • Maintenance • Common side effects

10.1 Introduction

Symptomatic patients with bipolar I disorder (BD-I) experience depressive symptoms three to four times as often as manic symptoms (Judd et al. 2002), while symptomatic patients with bipolar II disorder (BD-II) experience depressive symptoms approximately 39 times more commonly than hypomanic symptoms (Judd et al. 2003). Moderate to extreme impairment in work, social, and family life is known to occur significantly more often with depressive than manic symptoms in patients with BD (Calabrese et al. 2004a). In addition, significant psychosocial impairment during illness-free periods is more strongly predicted by the number of past depressive episodes than past manias (MacQueen et al. 2000). Moreover, there is a higher risk for suicide in BD than in other major psychiatric disorders, particularly during depressive episodes (Dilsaver et al. 1997; Goodwin and Jamison 2007; Isometsa et al. 1994).

These findings suggest that the prevention of acute major depressive episodes is critical to reducing the morbidity and mortality associated with BD. However, preventing bipolar depression during maintenance treatment is less well studied compared to preventing mania, although some progress has been made since the publication of the first edition of this chapter (Gao et al. 2009). Most maintenance studies have continued using a relapse-prevention design instead of a prophylaxis design (see the first edition of this chapter (Gao et al. 2009) for details).

Although results from relapse-prevention design studies are not generalizable, such studies will likely play a more important role in identifying subgroups of patients with certain biomarkers for efficacy and/or safety, given the more recent emphasis on "personalized" medicine. One example is a multicenter study of the pharmacogenomics of mood stabilizer response in bipolar disorder (PGBD) which was funded by the National Institute of Mental Health (NIMH, NCT01272531). The indispensable role of the relapse-prevention design in bipolar maintenance studies highlights the importance of understanding the results of such studies. Here, we focus on the efficacy and safety of all drugs studied with modern methodology for maintenance research in BD based on DSM-III or IV diagnostic criteria.

10.2 Evolution of Methodologies in Bipolar Maintenance Research

Since the publication of the first edition of this book, little consensus on a "standard" methodology for bipolar maintenance studies has been achieved. Over the years, the methods that have evolved the most include study enrollment, study design, outcome measures, index episode, study duration, and statistical analysis

(Calabrese et al. 2001; Goodwin and Jamison 2007; Gao et al. 2009). All these components can directly affect the interpretation of study results.

10.2.1 Study Enrollment

Due to different inclusion and exclusion criteria, selection biases stemming from study enrollment are unavoidable. Although modern maintenance studies have used standard diagnostic instruments and symptom severity measures, the eligibility requirements for studies vary widely. Therefore, understanding each study's enrollment requirements will help determine the relevance of study findings to routine clinical practice.

10.2.2 Study Design (Crossover Versus Parallel)

Crossover designs were used in some early lithium maintenance studies (Calabrese et al. 2001). However, most modern maintenance studies have used an open-label stabilization phase in which all patients receive active drug(s) followed by randomization of those who meet certain criteria to different treatment arms in a parallel fashion. The "enriched" nature of the studied sample tends to increase homogeneity, similar to the crossover design, and to inflate therapeutic effect size. Meanwhile, the parallel discontinuation from a studied drug might also inflate response to the studied drug because patients who were assigned to placebo or a new active treatment might have early relapse due to withdrawal from the open-label active treatment(s).

10.2.3 Outcome Measures

Most early lithium studies did not measure mood severity with standardized rating scales, such as the Hamilton Rating Scale for Depression or the Young Mania Rating Scale. More recent maintenance studies have used standardized rating scales to quantify the minimum severity for an index episode or DSM-IV criteria for a mood episode. However, rating scales are limited by their cross-sectional assessment of a period of seven days prior to completing the rating scale. Time to intervention for additional episodes has been used as a primary outcome measure, but each individual study has its own criteria for intervention (Calabrese et al. 2001; Gao et al. 2009).

10.2.4 Study Index Episode

Evidence suggests that index mood episode often not only predicts the polarity of future relapse at different times after remission (Calabrese et al. 2004b) but may also predict treatment response (Shapiro et al. 1989). It is likely that results from patients presenting with mania may not apply to those presenting with depression or vice versa, suggesting: (1) that a drug must be studied with different index mood episodes, and (2) that the study must last long enough to show a spectrum of efficacy in the prevention of early and late mood relapses.

10.2.5 Study Duration

Since early relapse and late relapse may represent different pathological processes (Calabrese et al. 2004b), the length of study duration is also important for a maintenance study. However, no consensus has been achieved for the ideal duration of a bipolar maintenance study. Investigators have purported that "pure" maintenance efficacy cannot be established if the study lasts less than six months (Goodwin and Jamison 2007; Ghaemi et al. 2004).

10.2.6 Statistical Analyses

The early maintenance studies used responder analyses with little or no distinction between primary and secondary outcome measures (Gao et al. 2009). The use of survival analysis has become the standard method of examining data from BD maintenance studies. Relapse-prevention trials typically assess length of time to the first event (starting at the point of randomization and ending when relapse or recurrence has occurred), which not only records how many events occur during the study period but also records when the events occur. For this reason, time to intervention/relapse has commonly been used as a primary outcome measure in modern BD maintenance studies, and hazard ratio (HR) is commonly used to measure the magnitude of difference in time-event data between groups. Survival data are commonly depicted with a Kaplan–Meier curve, which can also be used to demonstrate median survival (the time at which 50 % of patients have a relapse or recurrence). The primary problems with survival analysis techniques derive from sample size and dropouts. As the sample size decreases over time due to dropouts, a survival analysis is most valid for the earlier portion of the curves, where there are a larger number of patients in the study (Goodwin and Jamison 2007).

10.3 Efficacy Based on Time to Relapse

HR is the ratio of the probability of an event in an experimental arm to the probability in a comparator arm and can be larger or smaller than one. The HR takes both events and time into account, which makes it more accurate than relative risk (which simply counts events cumulatively over time for time-to-event data). However, the HR is meaningful only when it is consistent over time. Looking at the survival curves of a study will help determine if an HR is consistent or not.

10.3.1 Time to Any Mood Relapse

All US FDA-approved agents for BD maintenance treatment, including lithium, lamotrigine, aripiprazole, olanzapine, quetiapine, ziprasidone, and risperidone long-acting injection had significantly longer time to any mood relapse compared to placebo (Table 10.1). The fact that the approval of all these drugs was based on placebo-controlled relapse-prevention studies supports previous observations that the relapse-prevention design had the best assay sensitivity to detect differences between active treatment(s) and placebo (Calabrese et al. 2001). Although lithium's approval as a maintenance treatment for BD was based on early maintenance studies, its efficacy in preventing mood relapses was replicated by most studies using modern methodology (Bowden et al. 2000, 2003, Calabrese et al. 2003; Weisler et al. 2011). However, in the first maintenance study using modern research methods, neither lithium nor divalproex was superior to placebo in delaying any mood relapse, and there was no significant difference between the two agents in delaying mood relapses (Bowden et al. 2000).

In the BALANCE study (2010), lithium was superior to valproate in delaying any mood relapse with an HR of 0.71 (95 % 0.51, 1.00), corresponding to 29 % risk reduction. The contradictory findings of lithium versus divalproex in delaying any mood relapse from the BALANCE study and that of Bowden and colleagues (Bowden et al. 2000) might be due to different study designs. In the BALANCE study, no specific medication was required for stabilization and all patients were euthymic. In the study by Bowden and colleagues, patients could be treated with valproate, lithium, valproate plus lithium, or with neither drug and were in euthymia or had differing severity of manic symptoms. Moreover, in the divalproex study, confounding factors included a higher dropout rate and fewer randomized patients in the lithium group, as well as milder symptom severity than in other studies.

The combination of lithium and valproate was superior to valproate alone in delaying mood relapse with an HR of 0.57 (95 % CI 0.40, 0.80), corresponding to a 43 % risk reduction (Table 10.1). This suggests that combination therapy should be considered when valproate alone is not effective. The run-in treatment with lithium and valproate to enrich a sample of patients who tolerated these two medications made this study similar to a study with relapse-prevention design. However, randomization in this study was not blinded, which limits its generalizability.

Quetiapine monotherapy (HR = 0.29) (Weisler et al. 2011) and quetiapine combination therapy relative to placebo (HR = 0.28 − 0.32) have similar HRs, suggesting that quetiapine is effective in preventing both manic/mixed and depressive relapses (Table 10.1), and that it may not be necessary to start quetiapine with a

Table 10.1 Summary of randomized controlled trials in bipolar maintenance and time to mood episode relapse of active treatment versus placebo or an active comparator

Study and sample size	Index episodes and open-label treatment	Duration of randomized phase	Treatments during randomization and number of patients	Time to any mood relapse or HR mean (95 % CI)	Time to mania relapse or HR mean (95 % CI)	Time to depression relapse or HR mean (95 % CI)
Placebo-controlled monotherapy maintenance treatments						
Bowden et al. (2000)[a] N = 571	Manic, partially recovered from mania, or euthymia after a manic episode; treated with VPA, Li, VPA + Li, or neither for 12 weeks	52 weeks	VPA 71–125 μg/ml, n = 187; Li 0.8–1.2 mmol/l, n = 90; PBO, n = 92	275 days, NS; 189 days, NS; 173 days	>365 days, NS; 293 days, NS; 189 days	126 days, NS; 81 days, NS; 101 days
Bowden et al. (2003) N = 349	Current manic/hypomanic or manic/hypomanic within 60 days; treated with LTG 100–200 mg/day monotherapy or adjunctive therapy for 8–16 weeks	18 months	LTG 100–400 mg/day, n = 58; Li 0.8–1.1 mEq/l, n = 44; PBO, n = 69	141 days, p = 0.02; 292 days, p = 0.003; 85 days	NC, NS; NC, p = 0.006; 203 days	NC, p = 0.02; NC, NS; 269 days
Calabrese et al. (2003) N = 966	BD-I depression or major depressive episode within six months; treated with LTG 100–200 mg/day monotherapy or adjunctive therapy for eight to 16 weeks	18 months	LTG 200/400 mg/day, n = 221; Li 0.8–1.1 mEq/l, n = 121; PBO, n = 121	200 days, p = 0.029; 170 days, p = 0.029; 93 days	NC, NS; NC, p = 0.026; NC	NC, p = 0.047; NC, NS; NC
Goodwin et al. (2004) N = 1315	Manic, mixed, or depressed; treated with LTG 100–200 mg as adjunctive therapy or monotherapy for eight to 16 weeks	18 months	LTG 100–400 mg/days, n = 223; Li 0.8–1.1 mEq/l, n = 188	197 days, p < 0.001; 184 days, p < 0.001; 86 days	HR 0.64 (0.43, 0.97); HR 0.35 (0.21, 0.61)	HR 0.64 (0.45, 0.90); HR 0.76 (0.53, 1.08)
Tohen et al. (2006)[b] N = 731	Manic or mixed; treated with OLZ 5–20 mg/day for six to 12 weeks with two consecutive weeks of mood stability	48 weeks	OLZ 5–20 mg/days, n = 225; PBO, n = 136	174 days, p < 0.001; 22 days	NC, p < 0.001	49 days, p < 0.001; 18 days
Keck et al. (2006, 2007) N = 567	Manic or mixed; treated with APZ 15 or 30 mg/day for six to 18 weeks	26 weeks; 100 weeks	APZ 15 or 30 mg/days, n = 78; PBO, n = 83; APZ 15 or 30 mg/days, n = 77; PBO, n = 83	HR 0.52 (0.31, 0.91); HR 0.53 (0.32, 0.87)	HR 0.31 (0.12, 0.77); HR 0.35 (0.16, 0.75)	HR 0.83 (0.35, 2.01); HR 0.81 (0.36, 1.81)

Study	Description	Duration	Treatment			
Quiroz et al. (2010) N = 559	Recent/current manic or mixed; treated with oral risperidone 1–6 mg/day for three weeks +RLAI 12.5–50 mg/two weeks for 26 weeks	24 months	RLAI 12.5–50 mg/2 weeks, n = 154 PBO, n = 149	HR 0.40 (0.27, 0.59)	HR 0.25 (0.15, 0.41)	HR 1.09 (0.55, 2.19)
Weisler et al. (2011) N = 2438	Manic, mixed, or depressed; treated with QTP 300–800 mg/day for four to 24 weeks with four weeks of clinical stability	104 weeks	QTP 300–800 mg/day, n = 403 Li 600–1800 mg/day, n = 364 PBO, n = 404	HR 0.29 (0.23, 0.38) HR 0.46 (0.36, 0.59)	HR 0.29 (0.21, 0.40) HR 0.37 (0.27, 0.53)	HR 0.30 (0.20, 0.40) HR 0.59 (0.42, 0.84)
Young et al. (2014) N = 584	Depressed; treated with QTP 300 or 600 mg/day double-blind acute treatment for eight weeks	52 weeks	QTP 300–600 mg/day, n = 290 PBO, n = 294	HR 0.51 (0.38, 0.69)	HR 0.75 (0.45, 1.24)	HR 0.43 (0.30, 0.62)
Berwaerts et al. (2012) N = 766	Manic or mixed; treated with PLAI-ER 3–12 mg/day or OLZ 5–20 mg/day for three weeks +12 week continuation for responders	>24 months	PLAI 3–12 mg/days, n = 152 OLZ, 5–20 mg/days, n = 152 PBO, n = 148	558 days, p = 0.0017; NC, p ≤ 0.0001; 283 days	HR 2.06 (1.32, 3.22) SL, ≤0.001	HR 0.88 (0.53, 1.46) N/A
Vieta et al. (2012)[c] N = 560	Acute or non-acute manic or mixed; treated with RLAI 25–50 mg/two weeks for 12 weeks	18 months	RLAI 25–50 mg/2 weeks + PBO, n = 132; OLZ 10 mg/day + PBO injection, n = 131; PBO + PBO injection, n = 135	NC, p = 0.057 NC, p < 0.0001 198 days	323 days, p = 0.005 NC, p < 0.0001 323 days	NC, p = 0.655 NC, p = 0.011 NC
Placebo-controlled combination treatments of antipsychotics and mood stabilizers						
Tohen et al. (2004) N = 160	Manic or mixed; treated with Li 0.6–1.2 mEq/l or VPA 50–125 μg/ml for ≥ two weeks + OLZ 5–20 mg/day for six weeks	18 months	OLZ 5–20 mg/day + Li or VPA, n = 51 PBO + Li or VPA, n = 48	163 days, p = 0.023 42 days	171.5 days, NS 59 days	163 days, NS 55 days
Vieta et al. (2008b) N = 1461	Manic, mixed, or depressed; treated with QTP 400–800 mg/day + Li 0.5–1.2 mEq/l or VPA 50–125 μg/ml for up to 36 weeks with at least 12 weeks of clinical stability	104 weeks	QTP 400–800 mg/day + Li or VPA, n = 336 PBO + Li or VPA, n = 367	HR 0.28 (0.21, 0.37)	HR 0.30 (0.20, 0.44)	HR 0.26 (0.17, 0.41)

(continued)

Table 10.1 (continued)

Study and sample size	Index episodes and open-label treatment	Duration of randomized phase	Treatments during randomization and number of patients	Time to any mood relapse or HR mean (95 % CI)	Time to mania relapse or HR mean (95 % CI)	Time to depression relapse or HR mean (95 % CI)
Vieta et al., (2008a)[d] N = 58	Euthymia or mild symptoms, treated with Li ≥ 0.6 mEq/l during the last year with one episode in six months	52 weeks	OCBZ 1200 mg/d + Li; n = 26 PBO + Li, n = 29	19.2 weeks, NS 18.6 weeks	N/A	N/A
Suppes et al. (2009) N = 1953	Manic, mixed, or depressed; treated with QTP 400–800 mg/day + Li 0.5–1.2 mEq/l or VPA 50–125 µg/ml for up to 36 weeks with at least 12 weeks of clinical stability	104 weeks	QTP 400–800 mg/d + Li or VPA, n = 310 PBO + Li or VPA, n = 313	HR 0.32 (0.24, 0.42)	HR 0.3 (0.18, 0.49)	HR 0.33 (0.23, 0.48)
Macfadden et al. (2009) N = 240	Manic, hypomanic, depressed, mixed, or euthymic; treated with RLAI 25–50 mg/two weeks + TAU for 16 weeks	52 weeks	RLAI 25–50 mg/2 weeks + TAU, n = 65 PBO + TAU, n = 59	SL, p = 0.010	N/A	N/A
Bowden et al. (2010) N = 584	Recent/current manic/mixed; treated with Li 0.6–1.2 mEq/l or VPA 50–125 µg/ml for ≥ 2 weeks + ZPA 80–160 mg/day for up to 16 weeks	6 months	ZPA 80–160 mg/day + Li or VPA, n = 127 PBO + Li or VPA, n = 127	43 days, p = 0.0104 26.5 days	SL, p = 0.0035	NS
Marcus et al. (2011) N = 686	Current manic/mixed, inadequate response to Li or VPA; treated with Li 0.6–1.0 mEq/l or VPA 50–125 µg/ml for ≥ two weeks + APZ 10–30 mg/day for 12 weeks	52 weeks	Li or VPA + APZ 10–30 mg/day, n = 168 Li or VPA + PBO, n = 169	HR 0.54 (0.33, 0.89)	HR 0.35 (0.15, 0.83)	HR 0.73 (0.36, 1.48)
Woo et al. (2011) N = 175	Manic or mixed: treated with APZ 10–30 mg/day + VPA 50–120 µg/ml for six weeks	24 weeks	APZ 10–30 mg/day + VPA, n = 40 PBO + VPA, n = 43	Longer, NS	NS	Longer, NS
Carlson et al. (2012) N = 787	Manic or mixed: treated with single-blind APZ 10–30 mg/day + open-label LTG 100 or 200 mg/day for nine to 24 weeks	52 weeks	APZ 10–30 mg/day + LTG, n = 178 PBO + LTG, n = 173	HR 0.7 (0.5, 1.0) + 98	HR 0.6 (0.3, 1.0)	HR 0.8 (0.5, 1.4)

Non-placebo-controlled head-to-head comparison treatments

Tohen et al. (2005) N = 543	Manic or mixed; treated with OLZ 15 mg/day + Li 600 mg/day for six to 12 weeks	48 weeks	OLZ 5–20 mg/d, $n = 217$ Li 0.6–1.2 mEq/l, $n = 214$	SR, longer, $p = 0.07$ SYR, SL, $p = 0.04$	N/A	N/A
Calabrese et al. (2005a) N = 254	RCBD; treated with Li $\geq$ 0.8 mEq/l + VPA $\geq$ 50 µg/ml for 12–24 weeks for at least four consecutive weeks of mood stability	20 months	Li $\geq$ 0.8 mEq/l, $n = 32$ VPA $\geq$ 50 µg/ml, $n = 28$	18 weeks, NS 45 weeks	NC, NS NC	36 weeks, NS NC
BALANCE (2010) N = 459	Bipolar I euthymia; run-in period with both Li 0.4–1.0 mEq/l and VPA 750–1250 mg/day for four to eight weeks	24 months	Li 0.4–1.0 mEq/l, $n = 110$ VPA $\geq$ 50 µg/ml, $n = 110$ Li + VPA, $n = 110$	HR 0.8 (0.57, 1.15) HR 0.57 (0.40, 0.80)	HR 0.66 (0.41,1.07) HR 0.50 (0.31,0.79)	HR 1.06 (0.67, 1.67) HR 0.71 (0.46, 1.08)
Licht et al. (2010)	Manic/mixed or depressed; treated with various psychotropics for $\leq$ 12 months without operational criteria, but through clinical judgment	5.8 years	Li 0.5–1.0 mEq/l, $n = 78$ LTG up to 400 mg/days, $n = 77$	232 days, NS 202 days	HR 1.91 (0.73, 5.04)	HR 0.69 (0.41, 1.22)

Abbreviations: APZ, aripiprazole; CI, confidence interval; HR, hazard ratio; LTG, lamotrigine; Li, lithium; N/A, not available; NC, not calculable; NS, not significantly different from placebo or comparator; OCBZ: oxcarbamazepine; OLZ, olanzapine; PBO, placebo; PLAI, paliperidone long-acting injection; QTP, quetiapine; RCBD, rapid cycling bipolar disorder; RLAI, risperidone long-acting injection; SL, significantly longer; SR, symptomatic relapse; SYR, syndromic relapse; TAU, treatment as usual; VPA, valproate; wks, weeks; ZPA, ziprasidone

[a] The time to relapse was the median time to 50 % relapse for any mood and 25 % relapse for manic/mixed or depressive episodes

[b] The time to relapse was the median time to 50 % relapse for any symptomatic mood episode and 25 % relapse for manic/mixed or depressive symptomatic episodes

[c] The time to relapse was the median time to 25 % relapse for any mood, manic/mixed, or depressive episodes

[d] Average time to relapse

mood stabilizer for maintenance treatment of BD-I (Suppes et al. 2009; Vieta et al. 2008b). The rates from open-label treatment to randomization of quetiapine monotherapy and combination therapy were 48.1 % (Weisler et al. 2011) and 38.8 % (Suppes et al. 2009; Vieta et al. 2008b), respectively. Discontinuation due to adverse events during the open-label phase was 14.2 % for monotherapy and 18.2 % for combination therapy, suggesting that combination therapy increased risk of non-randomization due to adverse events.

Another study found a trend for olanzapine to delay symptomatic relapse compared to lithium (Tohen et al. 2005). The same study found that olanzapine delayed syndromic relapse for significantly longer than lithium, underscoring the notion that the definition of relapse can affect study results. Quetiapine was superior to lithium with an HR of 0.66 (95% CI 0.49, 0.88) for any mood relapse (Weisler et al. 2011), and olanzapine was superior to paliperidone in delaying any mood relapse (Berwaerts et al. 2012), indicating that some medications might be superior to others in delaying any mood relapse and/or depressive relapse.

10.3.2   Time to Manic/Mixed Relapse

With the exception of quetiapine in patients with a depressive index episode (Young et al. 2014) and olanzapine therapy adjunctive to a mood stabilizer (Tohen et al. 2004), active treatments in all other studies were superior to placebo in delaying mood relapse as well as manic/mixed relapses (Table 10.1). The non-superiority of quetiapine in this study might be due to the study index episode (Young et al. 2014). The rate of manic/mixed relapses in the placebo arm of this study was 11 % (Young et al. 2014), and in the study with a manic, mixed, or depressive index episode it was 30 % (Weisler et al. 2011). This supports the possibility that a study that includes only patients with a depressive index episode might not have enough manic/mixed episodes to compare between treatment arms.

The importance of sample size in detecting a difference between active treatments and placebo was supported by lamotrigine studies and the olanzapine combination therapy study (Table 10.1). In original separate analyses, lamotrigine was not superior to placebo in delaying manic/mixed relapses in patients with a manic/mixed index episode (Bowden et al. 2003) or with a depressive index episode (Calabrese et al. 2003). In a pooled analysis of these two studies, both lamotrigine and lithium were significantly superior to placebo in delaying manic/mixed relapses (Goodwin et al. 2004) although the effect of lamotrigine was small (HR for manic/mixed relapses was 0.64, corresponding to a 36 % risk reduction). Both separate and pooled analyses showed that lithium was more effective than lamotrigine in delaying manic/mixed relapses with an HR of 0.55 (95% CI 0.319, 0.949) (Goodwin et al. 2004).

Among active treatment head-to-head comparisons, only olanzapine was superior to paliperidone in delaying manic/mixed relapses. There were no significant differences between lithium and valproate or between quetiapine and lithium in delaying manic/mixed relapses (Table 10.1). Lithium was superior to lamotrigine in delaying manic/mixed relapse in "enriched" relapse-prevention studies (Goodwin et al. 2004), but not in an open-label effectiveness trial mimicking clinical practice (Licht et al. 2010). These findings highlight the impact of enrollment and study design on results from different studies.

10.3.3 Time to Depressive Relapse

In contrast to delaying the time to relapse of manic/mixed episodes (Table 10.1), only quetiapine, lamotrigine, olanzapine, and lithium were significantly superior to placebo in delaying time to depressive relapse. For lithium, this was only observed in one of three studies (Bowden et al. 2003; Calabrese et al. 2003; Weisler et al. 2011). The risk reduction for depressive relapses with lithium was smaller than that for manic/mixed relapses, with an HR of 0.59, corresponding to a 41 % risk reduction for depressive relapses versus an HR of 0.37, or a 63 % risk reduction for manic/mixed relapses (Weisler et al. 2011).

Olanzapine also appeared to more effectively prevent manic/mixed relapses than depressive relapses (Tohen et al. 2006) although olanzapine was significantly superior to placebo in delaying depressive relapses in two studies (Tohen et al. 2006; Vieta et al. 2012). However, only quetiapine demonstrated robust results of equal or better efficacy for preventing depressive relapses than for manic/mixed relapses (Table 10.1). Among the combination therapies, only quetiapine with a mood stabilizer was superior to placebo in delaying depressive relapses. Quetiapine was superior to lithium in delaying depressive relapses with an HR of 0.54 (95 % CI 0.35, 0.84), corresponding to a 46 % risk reduction (Weisler et al. 2011). In terms of lithium versus divalproex in preventing depressive relapses, the results were inconsistent (Bowden et al. 2000, BALANCE 2010).

10.3.4 Effect of Early Relapse

Early relapse might occur as a result of acute withdrawal of an active open-label treatment in patients who were randomized to placebo and/or a different active treatment. Such an effect might distort the results of a relapse-prevention study, i.e., inflate the effect of a blind treatment with the same treatment as that in the open-label phase. However, post hoc analyses of lithium and lamotrigine maintenance studies—after excluding all events in the first 28 days—found that lamotrigine and lithium were still significantly superior to placebo in delaying intervention for any mood relapse (Goodwin et al. 2004).

Similarly, after excluding events in the first four weeks post-randomization, quetiapine and lithium were still significantly superior to placebo in delaying any mood recurrences, with HRs of 0.27 and 0.41, respectively. Quetiapine was also superior to lithium for delaying any mood relapses with an HR of 0.70 (Weisler et al. 2011). All these differences in HRs were similar to those that included all events in the primary analyses (Table 10.1). Similar findings were also observed in quetiapine therapy adjunctive to a mood stabilizer after excluding events occurring in the first 28 days (Vieta et al. 2008b; Suppes et al. 2009). These data suggest that early relapses had little impact on the overall efficacy of a medication for long-term prevention of any mood recurrence.

10.3.5 Relapse Versus Recurrence

Although relapse and recurrence are interchangeably used in most maintenance studies, some researchers believe that there is a difference between the terms. Mood episodes that have the same polarity as the index episode and that occur within the first few months are considered to be relapses instead of recurrences (Ghaemi et al. 2004). Prevention of recurrences should be a main outcome measure of true maintenance efficacy (Calabrese et al. 2006). However, most maintenance studies have not separated relapses from recurrences. A post hoc analysis of lithium and lamotrigine maintenance trials found that after excluding patients who relapsed to the same polarity as the index episode within 90 days, both agents were still significantly superior to placebo in delaying any mood relapse (Calabrese et al. 2006). Similar results were observed after excluding patients who relapsed into the same polarity as the index episode within 180 days.

These time-to-event data suggest that among all studied drugs in bipolar maintenance, only quetiapine monotherapy or adjunctive therapy to lithium or valproate were equally effective in preventing manic and depressive relapses. The rest of the studied drugs were only effective in preventing manic/mixed relapses or more effective in preventing manic/mixed relapses than preventing depressive relapses. Limited data suggest that early relapse had little effect on the overall maintenance efficacy of the studied drugs.

However, it should be kept in mind that different studies used different study index episodes; most studies used a manic/mixed episode as the index episode (Table 10.1). Only one study included patients with BD-II (Young et al. 2014). Among the larger sample studies, only the BALANCE study used a design close to a prophylactic study. Because of these differences, it is inappropriate to compare the results from different studies. Even in the same study, true differences between a drug, placebo, or comparator cannot be defined because remitters were "enriched" in the open-label phase with medications that the study sponsor selected. The only "fair" comparison from these relapse-prevention studies probably occurred in the olanzapine versus lithium study (Tohen et al. 2005), where the patients who entered the randomization phase were those who responded and tolerated both lithium and olanzapine.

10.4   Efficacy Based on Relapse Rate

In contrast to HR, the relapse risk can also be estimated without considering the length of time when a recurrence occurs. Most recent studies used relapse/recurrence rates of any mood, manic/mixed, or depressive episode as secondary outcome measures. Comparing recurrence rates between active treatment(s) and placebo or a comparator can also provide clinically relevant information, especially when using evidence-based measures such as the number needed to treat (NNT). Our previous analyses showed that NNTs for preventing any mood relapse ranged widely (Gao et al. 2015). Below, we use NNT to estimate the magnitude of efficacy differences between active treatments and their comparators.

10.4.1   Any Mood Relapse

With the exception of a few studies with smaller sample sizes, all studies with larger sample sizes showed that active treatments were superior to placebo in preventing any mood relapse, with NNTs from three to 13 (Table 10.2). The results of olanzapine in three different studies suggest that the effect of open-label treatment on post-randomization might depend on the spectrum efficacy of open-label medications and post-randomization active treatments (Berwaerts et al. 2012; Tohen et al. 2006; Vieta et al. 2012). All three studies enrolled patients in manic/mixed states. Patients were treated with olanzapine during the open-label phase in two studies (Berwaerts et al. 2012; Tohen et al. 2006), but in the third study (Vieta et al. 2012) patients were treated with risperidone long-acting injection during the open-label phase. However, the NNTs for preventing any mood relapse with olanzapine relative to placebo in these three studies were the same, with a mean NNT of three (Table 10.2). It is possible that risperidone long-acting injection had no "withdrawal" effect on mood relapse in patients who switched to olanzapine. Meanwhile, both risperidone and olanzapine were more effective for preventing manic than depressive relapses (Table 10.1), which might be the reason that olanzapine had the same efficacy in three studies regardless of the difference in treatments during the open-label phase.

The index episode appeared to play a more important role than open-label and post-randomization treatments in study outcomes (Weisler et al. 2011; Young et al. 2014). In both quetiapine monotherapy studies, patients were treated with quetiapine during the open-label phase for stabilization, but one study included patients with a manic/mixed or depressive index episode (Weisler et al. 2011) and another included only patients with a depressive index episode (Young et al. 2014). The NNTs for preventing any mood relapse with quetiapine relative to placebo were three (95 % CI 3, 4) in the study with the manic/mixed or depressive index episode and six (95 % CI 4, 12) in the study with the depressive episode only (Table 10.2). These support an early observation that an index episode could also affect the efficacy of a drug in maintenance treatment of BD (Shapiro et al. 1989).

Table 10.2 Summary of randomized, controlled trials in bipolar maintenance and mood relapse rates of active treatment versus placebo or an active comparator

Study and sample size	Index episodes and open-label treatment	Duration of randomized phase	Treatments during randomization and number of patients	Any mood relapse NNT Mean (95 % CI)	Mania/Mixed relapse NNT Mean (95 % CI)	Depressive relapse NNT Mean (95 % CI)
Placebo-controlled monotherapy maintenance treatments						
Bowden et al. (2000) N = 571	Manic, partially recovered from mania, or euthymia after a manic episode; treated with VPA, Li, VPA + Li, or neither for 12 weeks	52 weeks	VPA 71–125 µg/ml, n = 187; Li 0.8–1.2 mmol/l, n = 90; PBO, n = 92	7 (4, 34)[a] 13 (5, ∞, −16)	21 (92, ∞, −21)[a] 68 (8, ∞, −10)	10 (5, 49)[a] 16 (6, ∞, −26)
Bowden et al. (2003) N = 349	Current manic/hypomanic or manic/hypomanic within 60 days; treated with LTG 100–200 mg/day monotherapy or adjunctive therapy for eight to 16 weeks	18 months	LTG 100–400 mg/days, n = 58; Li 0.8–1.1 mEq/l, n = 44; PBO, n = 69	4 (3, 8) 3 (2, 9)	16 (5, ∞, −9) 4 (3, 20)	6 (3, 51) 13 (4, ∞, −11)
Calabrese et al. (2003)[b] N = 966	BD-I depression or major depressive episode within six months; treated with LTG 100–200 mg/day monotherapy or adjunctive therapy for eight to 16 weeks	18 months	LTG 200/400 mg/days, n = 221; Li 0.8–1.1 mEq/l, n = 121; PBO, n = 121	19 (6, ∞, −15) 11 (5, ∞, −26)	479 (11, ∞, −12) 13 (6, ∞, −133)	20 (6, ∞, −16) 86 (8, ∞, −9)
Goodwin et al. (2004) N = 1315	Manic, mixed, or depressed; treated with LTG 100–200 mg as adjunctive therapy or monotherapy for eight to 16 weeks	18 months	LTG 100–400 mg/days, n = 223; Li 0.8–1.1 mEq/l, n = 164; PBO, n = 188	10 (5, 122)[c] 6 (4, 18)	33 (9, ∞, −19)[c] 8 (5, 22)	14 (6, ∞, −55)[c] 53 (9, ∞, −13)
Tohen et al. (2006) N = 731	Manic or mixed; treated with OLZ 5–20 mg/day for six to 12 weeks with two consecutive weeks of mood stability	48 weeks	OLZ 5–20 mg/days, n = 225; PBO, n = 136	3 (2, 4)	4 (3, 7)	11 (5, ∞, −79)
Keck et al. (2006) N = 567	Manic or mixed; treated with APZ 15 or 30 mg/day for six to 18 weeks	26 weeks	APZ 15 or 30 mg/days, n = 78; PBO, n = 83	5 (3, 25) 5 (3, 25)[d]	6 (4, 31) 6 (3, 30)[d]	64 (8, ∞, −11) 73 (8, ∞, −10)[d]

Study	Description	Duration	Treatments (n)			
Quiroz et al. (2010) N = 559	Recent/current manic or mixed; treated with oral risperidone 1–6 mg/day for three weeks + RLAI 12.5–50 mg/two weeks for 26 weeks	24 months	RLAI 12.5–50 mg/2 weeks, n = 154 PBO, n = 149	4 (3, 6)	3 (2, 5)	−26 (25, ∞, −8)
Weisler et al. (2011) N = 2438	Manic, mixed, or depressed; treated with QTP 300–800 mg/day for four to 24 weeks with four weeks of clinical stability	104 weeks	QTP 300–800 mg/day, n = 403 Li 600–1800 mg/day, n = 364 PBO, n = 404	3 (3, 4) 4 (3, 5)	6 (4, 9) 6 (4, 8)	8 (6, 14) 13 (8, 45)
Young et al. (2014) N = 584	Depressed; treated with QTP 300 or 600 mg/day double-blind acute treatment for eight weeks	52 weeks	QTP 300–600 mg/day, n = 290 PBO, n = 294	6 (4, 12)	81 (16, ∞, −27)	7 (5, 12)
Berwaerts et al. (2012) N = 766	Manic or mixed; treated with PLAI-ER 3–12 mg/day or OLZ 5–20 mg/day for three weeks + 12 week continuation for responders	>24 months	PLAI 3–12 mg/days, n = 152 OLZ, 5–20 mg/days, n = 152 PBO, n = 148	8 (4, 63)[e] 3 (2, 4)[e]	9 (5, 51)[f] 5 (3, 8)[f]	−55 (14, ∞, −9)[f] 14 (7, ∞, −34)[f]
Vieta et al. (2012) N = 560	Acute or non-acute manic or mixed; treated with RLAI 25–50 mg/two weeks for 12 weeks.	18 months	RLAI 25–50 mg/2 weeks + PBO, n = 132; OLZ 10 mg/day + PBO injection, n = 131; PBO + PBO injection, n = 135	6 (3, 8) 3 (2, 5)	6 (4, 17) 4 (3, 7)	−56 (13, ∞, −9) 12 (6, ∞, −446)
Placebo-controlled combination treatments of antipsychotics and mood stabilizers						
Tohen et al. (2004) N = 160	Manic or mixed; treated with Li 0.6–1.2 mEq/l or VPA 50–125 μg/ml for ≥ two weeks + OLZ 5–20 mg/day for six weeks	18 months	OLZ 5–20 mg/day + Li or VPA, n = 51 PBO + Li or VPA, n = 48	5 (3, ∞, −20)[g]	11 (4, ∞, −8)[g]	6 (3, ∞, −16)[g]
Vieta et al. (2008b) N = 1461	Manic, mixed, or depressed; treated with QTP 400–800 mg/day + Li 0.5–1.2 mEq/l or VPA 50–125 μg/ml for up to 36 weeks with at least 12 weeks of clinical stability	104 weeks	QTP 400–800 mg/day + Li or VPA, n = 336 PBO + Li or VPA, n = 367	3 (3, 4)	6 (5, 10)	7 (5, 10)

(continued)

Table 10.2 (continued)

Study and sample size	Index episodes and open-label treatment	Duration of randomized phase	Treatments during randomization and number of patients	Any mood relapse NNT Mean (95 % CI)	Mania/Mixed relapse NNT Mean (95 % CI)	Depressive relapse NNT Mean (95 % CI)
Vieta et al. (2008a) N = 58	Euthymia or mild symptoms, treated with Li ≥ 0.6 mEq/l during the last year with one episode in six months	52 weeks	OCBZ 1200 mg/d + Li; n = 26 PBO + Li, n = 29	5 (2, ∞, −17)	8 (3, ∞, −10)	5 (3, ∞, −37)
Suppes et al. (2009) N = 1953	Manic, mixed, or depressed; treated with QTP 400–800 mg/day + Li 0.5–1.2 mEq/l or VPA 50–125 µg/ml for up to 36 weeks with at least 12 weeks of clinical stability	104 weeks	QTP 400–800 mg/d + Li or VPA, n = 310 PBO + Li or VPA, n = 313	3 (3, 4)	8 (6, 14)	5 (4, 8)
Macfadden et al. (2009) N = 240	Manic, hypomanic, depressed, mixed, or euthymic; treated with RLAI 25–50 mg/two weeks + TAU for 16 weeks	52 weeks	RLAI 25–50 mg/2 weeks + TAU, n = 65 PBO + TAU, n = 59	4 (3, 17)	6 (3, 39)	16 (5, ∞, −15)
Bowden et al. (2010) N = 584	Recent/current manic/mixed; treated with Li 0.6–1.2 mEq/l or VPA 50–125 µg/ml for ≥ two weeks + ZPA 80–160 mg/day for up to 16 weeks	6 months	ZPA 80–160 mg/day + Li or VPA, n = 127 PBO + Li or VPA, n = 127	8 (4, 52)	N/A	N/A
Carlson et al. (2012) N = 787	Manic or mixed; treated with single-blind APZ 10–30 mg/day + open-label LTG 100 or 200 mg/day for nine to 24 weeks	52 weeks	APZ 10–30 mg/day + LTG, n = 178 PBO + LTG, n = 173	7 (4, 19)	8 (5, 25)	17 (7, ∞, −44)
Non-placebo-controlled head-to-head comparison treatments						
Tohen et al. (2005) N = 543	Manic or mixed; treated with OLZ 15 mg/day + Li 600 mg/day for six to 12 weeks	48 weeks	OLZ 5–20 mg/days, n = 217 Li 0.6–1.2 mEq/l, n = 214	11 (6, ∞, −755) 11 (5, 252)[h]	7 (5, 16) 7 (4, 14)[h]	−20 (65, ∞, −9) −18 (143, ∞, −8)[h]
Calabrese et al. (2005a) N = 254	RCBD; treated with Li ≥ 0.8 mEq/l + VPA ≥ 50 µg/ml for 12–24 weeks for at least four consecutive weeks of mood stability	20 months	Li ≥ 0.8 mEq/l, n = 32 VPA ≥ 50 µg/ml, n = 28	16 (3, ∞, −6)	224 (5, ∞, −5)	17 (4, ∞, −6)

Study	Inclusion criteria	Duration	Treatment			
BALANCE (2010) N = 459	Bipolar I euthymia; run-in period with both Li 0.4–1.0 mEq/l and VPA 750–1250 mg/day for four to eight weeks	24 months	Li + VPA, n = 110 Li 0.4–1.0 mEq/l, n = 110 VPA ≥ 50 µg/ml, n = 110	6 (4, 39) 10 (5, ∞, −38) 18 (5, ∞, −13)[i]	6 (3, 22) 12 (5, ∞, −21) 11 (5, ∞, −31)[i]	10 (4, ∞, −34) 7 (4, 129) 28 (6, ∞, −11)[i]
Licht et al. (2010)	Manic/mixed or depressed; treated with various psychotropics for ≤ 12 months without operational criteria, but through clinical judgment	5.8 years	Li 0.5–1.0 mEq/l, n = 78 LTG up to 400 mg/days, n = 77	−177 (7, ∞, −6)	4 (3, 12)	−4 (−11, −3)

Abbreviations: APZ, aripiprazole; BD, bipolar disorder; CI, confidence interval; LTG, lamotrigine; Li, lithium; N/A, not available; NNT, number needed to treat; benefit; OCBZ: oxcarbazepine; OLZ, olanzapine; PBO, placebo; PLAI, paliperidone long-acting injection; QTP, quetiapine; RLAI, risperidone long-acting injection; TAU, treatment as usual; VPA, valproate; ZPA, ziprasidone

[a]Calculated based on the reasons for premature termination

[b]In original analysis, lithium was superior to placebo in preventing manic relapses, and lamotrigine 200 mg/day was superior to placebo in preventing depressive relapses

[c]Only data for lamotrigine 200 mg/day and 400 mg/day were included. When including data for lamotrigine 50 mg/day, the difference between lamotrigine and placebo was not significant

[d]Based on data for 100 weeks

[e]The NNT was calculated based on the data at the end of the first year. At the end of the second year, the recurrence rates between paliperidone-ER and placebo were not significantly different

[f]The NNT was calculated based on the data at 400 days after randomization

[g]Based on 30 patients in combination therapy and 38 patients in monotherapy for symptomatic relapses

[h]Syndromic relapses of mood episodes based on DSM-IV criteria

[i]Comparison between lithium and the combination of lithium and valproate

∞ Indefinite

That the NNTs of quetiapine monotherapy and adjunctive therapy to lithium or valproate were the same or similar (Table 10.2) suggests that quetiapine adjunctive therapy might be as effective as monotherapy in preventing any mood relapse, which is consistent with the time-to-event analysis with HR (Table 10.1). The results of risperidone long-acting injection also support that risperidone long-acting injection plus treatment as usual was as effective as risperidone long-acting injection monotherapy in preventing any mood relapse (Table 10.2).

10.4.2 Manic/Mixed Relapse

As shown in Table 10.2, among monotherapy treatments, olanzapine had the smallest NNT of four (95 % CI 3, 7) for preventing manic/mixed relapses. The NNTs of the combination treatments ranged from six to 11. Results from head-to-head comparison studies suggest that some medications were superior to others in preventing manic/mixed relapse. These cumulative data suggest that some medications may be superior to others in preventing manic/mixed relapse as the time-to-event analyses reflected (Table 10.1).

10.4.3 Depressive Relapse

Among monotherapy treatments (Table 10.2), only quetiapine was repeatedly shown to be superior to placebo in preventing depressive relapse in two studies (Weisler et al. 2011; Young et al. 2014). Lamotrigine was superior to placebo in delaying depressive relapses only in studies with a manic index episode (Bowden et al. 2003). In contrast, only one of three studies showed that lithium was significantly superior to placebo in preventing depressive relapses with an NNT of 13 (95 % CI 8, 45). Valproate was also superior to placebo in preventing depressive relapse when using premature termination due to depression as an outcome measure, with an NNT of 10 (95 % CI 5, 49).

Among combination treatments, only quetiapine plus lithium or valproate were superior to placebo in preventing depressive relapse, with an NNT of five (95 % CI 4, 8) in one study and an NNT of seven (95 % CI 5, 10) in another (Table 10.2). The BALANCE study showed that lithium monotherapy was superior to valproate alone in preventing depressive relapses (BALANCE 2010). Lamotrigine was superior to lithium in preventing depressive relapse in a clinical setting (Table 10.2).

10.4.4 Effect of Early Relapse

In contrast to time to delay for any mood relapse, early relapse had some effect on the difference in cumulative risk for any mood relapse between active treatment and placebo based on a pooled analysis of maintenance studies of lamotrigine and lithium (Goodwin et al. 2004). In the first 28 days, the NNTs for preventing any mood relapse were 12 (95 % CI 6, 101) for lamotrigine and eight (95 % CI 5, 22) for lithium, respectively. During the rest of the study period, among those who did not relapse in the first 28 days, the NNT for preventing any mood relapse was six (95 % CI 4, 19) for lamotrigine and five (95 % CI 3, 14) for lithium. However, the overall NNTs for preventing any mood relapses were 10 for lamotrigine and six for lithium (Table 10.2), suggesting that inclusion of patients with early relapse might diminish the true magnitude of maintenance efficacy of an active treatment relative to placebo.

The larger NNTs in the first 28 days suggest that the active treatments were less effective in preventing early relapses regardless of what treatment patients received after randomization. High rates of relapse shortly after randomization could be due to the "fear" of receiving placebo, which made active treatments appear to be less effective. Increased rates of early relapse in all studied arms have been observed in all double-blind, placebo-controlled studies, as reflected by deeper slopes on survival curves of all treatments during the early study period. These across-the-board early relapses shortly after randomization made it difficult for active treatments to separate from placebo or comparators. This is commonly observed in survival curves for which there is no separation or a smaller separation between active treatment and placebo. The early relapses seemingly had no negative impact on the efficacy of late relapses although they may have diminished the magnitude of overall maintenance efficacy. In contrast, excluding patients with early relapses might increase the possibility of detecting a difference between an active treatment and placebo. These data also suggest that if an active treatment is superior to placebo in the early phase of a maintenance study, it will likely be superior to placebo in the late phase of the study.

10.4.5 Relapse Versus Recurrence

Early relapses seemed to have a negative impact on the overall maintenance efficacy of lamotrigine and lithium (see Sect. 10.4.4). Further analyses of lamotrigine and lithium in preventing recurrences of mood episodes by excluding patients who relapsed to the same polarity as the index episode within the first 90 or 180 days found that lithium and lamotrigine were superior to placebo in delaying any mood recurrence, with an NNT of five (95 % CI 3, 10) for lithium and eight (95 % CI 4, 91) for lamotrigine after excluding those who relapsed within the first 90 days. After excluding those who had relapsed within the first 180 days, the NNT was six (95 % CI 4, 20) for lithium and eight (95 % CI 4, 140) for lamotrigine (Calabrese et al. 2006), suggesting that lithium and lamotrigine had sustained maintenance efficacy. In contrast, the NNTs for preventing manic or depressive relapses were much larger, from 10 to 61. Only lamotrigine was still superior to placebo for preventing depressive recurrence with an NNT of 10 (95 % CI 5, 54) after excluding patients who relapsed within the first 180 days. Lithium was superior to placebo for manic recurrences, with an NNT of 10 (95 % CI 5, 154) after excluding those who relapsed within the first 90 days.

These data suggest that lamotrigine and lithium had "true" maintenance efficacy in preventing any mood recurrences although the prevention of manic and depressive recurrences was less robust. It remains unclear if other medications have similar sustained efficacy for preventing any mood recurrences.

10.5   Disagreement Between Time to Relapse and Rate of Relapse

In most cases, time-to-event data and rate data were consistent (Tables 10.1 and 10.2). However, in some instances, these two measures were not always parallel. For example, in the lamotrigine maintenance study with subjects who had an index depressive episode, both lamotrigine and lithium were superior to placebo in delaying any mood relapse (Table 10.1), but the rates of any mood relapse in the lamotrigine or lithium group were not significantly different from that in the placebo group (Table 10.2). The disagreement between these two measures could be due to the fact that there were so many relapses in all groups in the early phase of the study that the study could not detect differences in overall cumulative risk between treatments. The disagreement between the time-to-event results and rates of relapse also highlights the disadvantage of using relative risk instead of time-to-event analysis for relapse-prevention studies.

10.6   Side Effects in the Acute and Maintenance Phases

10.6.1   Self-Reported Side Effects in Acute Manic or Depressive Phases

As shown in Table 10.3, active drugs caused similar side effects in patients regardless of whether they were in a manic or depressive phase. However, some side effects might only occur during one phase. These differences could be due to true differences in the sensitivity of patients during different phases of illness (Gao et al. 2008a, b; Wang et al. 2011). Differences could also be due to study duration. Most mania studies only lasted three to four weeks, but most bipolar depression studies lasted six to eight weeks. In addition, antipsychotic adjunctive therapy appeared to cause more side effects than monotherapy with some antipsychotics, but not others. It remains unclear if the increased risk for side effects from adjunctive therapy was drug-specific.

10.6.2   Self-Reported Side Effects in the Maintenance Phase

Because only patients who tolerated open-labeled treatment(s) and met randomization criteria continued in a maintenance phase, the rates of side effects from an active treatment relative to placebo after randomization were much lower, with only a few new side effects emerging during the maintenance phase (Table 10.3). Clearly, a study with a relapse-prevention design also potentially inflates the safety and tolerability of a studied drug in the maintenance phase. Therefore, low rates of side effects in the maintenance phase should not be interpreted as the overall safety and tolerability of a studied drug.

10.6.3   ≥7 % Weight Gain During the Maintenance Phase

Weight gain was a common self-reported side effect during maintenance treatment studies of BD (Table 10.3). However, the magnitude of differences between active treatments and placebo differed. Olanzapine-induced $\geq 7\%$ weight gain was significantly higher relative to placebo in all studies. Other antipsychotic-induced $\geq 7\%$ weight gain was significantly higher compared to placebo in some studies, but not others (Table 10.4). One possibility for such inconsistency is study duration, which is supported by risperidone long-acting injection studies (Table 10.4). Aripiprazole did not cause significant weight gain in acute mania or bipolar depression (Gao et al. 2011), but caused significantly more $\geq 7\%$ weight gain than placebo in three of five long-term studies (Table 10.4). Significantly, more $\geq 7\%$ weight gain was observed with quetiapine 600 mg/day relative to placebo, but not with quetiapine 300 mg/day, suggesting that the magnitude of weight gain might be related to the dose of an active treatment (Table 10.4).

10.7   Clinical Implications

With regard to the efficacy of preventing different mood relapses, medications for bipolar maintenance can be divided into four categories: (1) equal/similar bimodal mood stabilization (quetiapine monotherapy and adjunctive therapy); (2) more effective in preventing depressive relapse (lamotrigine); (3) more effective in preventing manic/hypomanic or mixed relapse (lithium, olanzapine); and (4) only

Table 10.3 Self-reported side effects from medications studied in different phases of bipolar disorder

Medications	Depression ≥5 % and ≥ 1.5 times higher than placebo	Mania ≥5 % and ≥ 1.5 times higher than placebo in placebo-controlled studies or ≥ 5 % in open-label phase of maintenance studies	Maintenance >5 % and ≥ 1.5 times higher than placebo
Aripiprazole	Akathisia, insomnia, nausea, fatigue, restlessness, dry mouth, anxiety, increased appetite, sedation, attentional disturbances (Thase et al. 2008)	Akathisia, nausea, sedation, somnolence, dyspepsia, constipation, vomiting, tremor, anxiety, insomnia (Keck et al. 2003; Sachs et al. 2006; Kanba et al. 2014)	Nausea, tremor, akathisia (Keck et al. 2006)
Aripiprazole + a mood stabilizer	Not available	Akathisia, tremor, nausea, insomnia, headache, dizziness, somnolence, diarrhea, extrapyramidal side effects, vomiting, sedation, restlessness (Vieta et al. 2008c; Marcus et al. 2011)	Tremor (Marcus et al. 2011)
Olanzapine	Somnolence, weight gain, increased appetite, dry mouth, asthenia (Tohen et al. 2003)	Somnolence, dry mouth, dyspepsia, asthenia, dizziness (Tohen et al. 2000)	Weight increase, fatigue (Tohen et al. 2006)
Olanzapine + a mood stabilizer	Not available	Somnolence, dry mouth, weight gain, increased appetite, tremor, asthenia, dizziness, thirst, speech disorder (Tohen et al. 2002)	Somnolence, weight gain, tremor (Tohen et al. 2004)
Quetiapine	Dry mouth, sedation, somnolence, dizziness, fatigue, constipation, dyspepsia, increased appetite, tremor (Calabrese et al. 2005b)	Somnolence, dry mouth, weight gain, dizziness, asthenia, sedation, headache, constipation, weight increase (Vieta et al. 2005; Weisler et al. 2011)	Dry mouth (Young et al. 2014); Somnolence (Weisler et al. 2011)
Quetiapine + a mood stabilizer	Not available	Somnolence, dry mouth, asthenia, postural hypotension, weight gain, pharyngitis, dizziness, sedation, tremor, headache,	Sedation, somnolence, weight increase, hypothyroidism (Suppes et al. 2009; Vieta et al. 2008b)

(continued)

Table 10.3 (continued)

Medications	Depression ≥5 % and ≥ 1.5 times higher than placebo	Mania ≥5 % and ≥ 1.5 times higher than placebo in placebo-controlled studies or ≥ 5 % in open-label phase of maintenance studies	Maintenance >5 % and ≥ 1.5 times higher than placebo
		increased appetite, nausea, dizziness, fatigue, constipation, vomiting, nasopharyngitis (Sachs et al. 2004; Yatham et al. 2004; Suppes et al. 2009; Vieta et al. 2008b)	
Ziprasidone	Not available	Somnolence, extrapyramidal syndrome, dizziness, akathisia, tremor, nausea, asthenia (Potkin et al. 2005)	Not available
Ziprasidone + a mood stabilizer	Somnolence, sedation, dizziness, nausea, insomnia, anxiety, akathisia, restlessness (Sachs et al. 2011)	Sedation, dizziness, akathisia, somnolence, tremor, insomnia, fatigue, nausea, headache (Sachs et al. 2012; Bowden et al. 2010)	Tremor (Bowden et al. 2010)
Lamotrigine	None (Calabrese et al. 2008)	Headache, nausea, infection, rash, dizziness, somnolence, diarrhea (Goodwin et al. 2004)	None (Goodwin et al. 2004)
Lithium	Somnolence, dry mouth, nausea, diarrhea, insomnia, tremor (Young et al. 2010)	Tremor, headache, somnolence, weight gain, dry mouth, dizziness, asthenia, weight loss, anorexia, nausea, vomiting (Bowden et al. 2005)	Tremor[a], weight gain[a], nausea[b], somnolence[b], diarrhea[b], tremors[b], insomnia[c], worsening of mania[c] (Goodwin et al. 2004; Tohen et al. 2005; Weisler et al. 2011)
Paliperidone-extended release	Not available	Somnolence, akathisia, sedation, hypertonia, constipation, dyspepsia (Vieta et al. 2010)	None (Berwaerts et al. 2012)
Risperidone long-acting injection + treatment as usual	Not available	Not available	Tremor, muscle rigidity, weight gain, hypokinesia, sedation (Macfadden et al. 2009)

(continued)

Table 10.3 (continued)

Medications	Depression ≥5 % and ≥ 1.5 times higher than placebo	Mania ≥5 % and ≥ 1.5 times higher than placebo in placebo-controlled studies or ≥ 5 % in open-label phase of maintenance studies	Maintenance >5 % and ≥ 1.5 times higher than placebo
ivalproex-extended release	Nausea, increased appetite, diarrhea, dry mouth, and cramps (Muzina et al. 2011)	Nausea, increased appetite (Tohen et al 2008)	Not available

[a]Stabilized with quetiapine monotherapy
[b]Stabilized with lamotrigine monotherapy
[c]Stabilized with olanzapine and lithium

effective in preventing manic/mixed relapses (aripiprazole, risperidone long-acting injection, and ziprasidone adjunctive therapy). Clearly, choosing a medication for preventing manic relapses appears easy, but it is more challenging to choose a medication to prevent both manic and depressive relapses. Although it is a common practice, it remains unclear if the combination of two medications such as an antipsychotic with a mood stabilizer like lithium, lamotrigine, or valproate is more effective for preventing depressive relapses than one medication alone. Aripiprazole adjunctive to lamotrigine was not superior to aripiprazole alone in preventing depressive relapses, suggesting that combining two medications with different polarity efficacy may not be better than one medication alone.

Since patients with bipolar depression are more sensitive, but less tolerant, to psychotropic medications (Gao et al. 2008a, b, 2011; Wang et al. 2011), a slow titration schedule and a lower targeted dose might reduce the possibility of discontinuation due to adverse events in patients presenting with a depressive episode. The risk and benefit from a medication should be clearly communicated with patients before beginning a medication (Wu et al. 2011). It is important to keep in mind that acute side effects do not predict long-term side effects. Aripiprazole-induced weight gain is a good example.

Relapse-prevention designs will continue to be used to study the efficacy and safety of psychotropic drugs in the long-term treatment of BD because of regulatory requirements for approving new drugs or indications, commercial interests in BD research, and the feasibility of conducting a study. More importantly, the relapse-prevention design has the best assay sensitivity to detect significant differences between an active treatment and placebo. The generalizability of the results from a relapse-prevention study is limited not only because of differences inherent to each individual trial but also because of the universal exclusion of patients with comorbid conditions such as substance use disorders, anxiety disorders, severe medical problems, and those with severe suicidality.

Table 10.4   Summary of $\geq 7\%$ weight gain in bipolar maintenance studies

Agents and study	Study duration	Treatment arms	Total N	N of $\geq 7\%$ WG	$\%$ of $\geq 7\%$ WG	NNH¥ Mean (95 % CI)
Aripiprazole (APZ)						
APZ (Keck et al. 2006)	26 weeks	APZ 15–30 mg/day	56	7	12.5	8 (4, 26)
		Placebo	60	0	0.0	
APZ (Keck et al. 2007)	100 weeks	APZ 15–30 mg/day	60	12	20.0	7 (4, 31)
		Placebo	61	3	4.9	
APZ + MS (Marcus et al. 2011)	52 weeks	APZ 10–30 mg/day + MS	156	22	14.1	48 (10, ∞, −18)
		Placebo + MS	158	19	12.0	
APZ + LTG (Carlson et al. 2012)	52 weeks	APZ 10–30 mg/day + LTG 100 or 200 mg/day	108	13	12.0	12 (6, 81)
		Placebo + LTG 100 or 200 mg/day	111	4	3.6	
APZ + DIV (Woo et al. 2011)	26 weeks	APZ 10–30 mg/day + DIV	40	9	22.5	26 (5, ∞, −7)
		Placebo + DIV	43	8	18.6	
Olanzapine (OLZ)						
OLZ + MS (Tohen et al. 2004)	18 months	OLZ 5–20 mg/day + MS	51	14	27.5	5 (3, 16)
		Placebo + MS	48	3	6.3	
OLZ (Tohen et al. 2006)	48 weeks	OLZ 5–20 mg/day	224	36	16.1	7 (5, 13)
		Placebo	133	3	2.3	
OLZ (Tohen et al. 2005)	12 months	OLZ 5–20 mg/day	215	64	27.8	5 (4, 8)
		Lithium 0.6–1.2 mEq/l	214	21	9.8	
Paliperidone-ER						
Paliperidone-ER (Berwaerts et al. 2012)	24 months	Paliperidone-ER 3–12 mg/day	141	41	29.1	12 (6, ∞, −47) 5 (3, 14)
		OLZ 5–20 mg/day	78	32	41.0	
		Placebo	138	29	21.1	
Quetiapine (QTP)						
QTP (Weisler et al. 2011)	104 weeks	QTP 300–800 mg/day	404	43	10.6	13 (9, 22)
		Lithium 0.6–1.2 mEq/l	418	23	5.4	36 (8, 5336)
		Placebo	404	11	2.6	
QTP (Young et al. 2014)	52 weeks	QTP 300 mg/day	141	9	6.4	43 (13, ∞, −53)
		QTP 600 mg/day	150	14	9.3	19 (9, 183)
		Placebo	294	12	4.1	
QTP + MS (Suppes et al. 2009)	104 weeks	QTP 400–800 mg/day + MS	310	36	11.6	13 (8, 28)
		Placebo + MS	313	12	3.8	
QTP + MS (Vieta et al. 2008b)	104 weeks	QTP 400–800 mg/day + MS	336	25	7.4	18 (11, 40)
		Placebo + MS	367	7	1.9	

(continued)

Table 10.4 (continued)

Agents and study	Study duration	Treatment arms	≥7 % Weight gain			
			Total N	N of ≥ 7 % WG	% of ≥ 7 % WG	NNH¥ Mean (95 % CI)
Risperidone						
Risperidone-LAI (Vieta et al. 2012)	78 weeks	Risperidone LAI 25, 37.5, or 50	132	24	18.2	8 (5, 20)
		OLZ	135	37	27.4	5 (3, 7)
		Placebo	131	7	5.3	
Risperidone-LAI (Macfadden et al. 2009)	52 weeks	Risperidone LAI 25, 37.5, or 50 mg/ 2 weeks + TAU	65	18	27.7	−36 (8, ∞, −5)
		Placebo + TAU	59	18	30.5	
Risperidone-LAI (Quiroz et al. 2010)	24 months	Risperidone LAI 25, 37.5, or 50	135	16	11.9	11 (6, 39)
		Placebo	133	4	3.0	
Ziprasidone (ZPA)						
ZPA + MS (Bowden et al. 2010)	26 weeks	ZPA 80–160 mg/day + MS	127	7	5.5	647 (16, ∞, −16)
		Placebo + MS	112	6	5.4	

Abbreviations: APZ, aripiprazole; DIV, divalproex; LAI, long-acting injection; LTG, lamotrigine; MS, mood stabilizer; NNH, number needed to treat to harm; OLZ, olanzapine; QTP, quetiapine; TAU, treatment as usual; ZPA, ziprasidone

10.8   Conclusion

The generalizability of a relapse-prevention study can be increased by broadening the inclusion criteria of a study such as comorbid psychiatric and medical conditions and using less "enriched" treatments during the open-label phase. Finding drugs with bimodal stabilization is an urgent unmet need for BD maintenance treatment. Future research should focus on agents with both antimanic and antidepressant properties. Quetiapine's ability to equally prevent manic/mixed and depressive relapses suggests that acute antimanic and antidepressant efficacy may predict long-term treatment efficacy. Agents with both acute antimanic and antidepressant properties should be prioritized in future BD maintenance studies.

References

BALANCE Investigators and Collaborators, Geddes JR, Goodwin GM et al (2010) Lithium plus valproate combination therapy versus monotherapy for relapse prevention in bipolar I disorder (BALANCE): a randomized open-label trial. Lancet 375(9712):385–395

Berwaerts J, Melkote R, Nuamah I et al (2012) A randomized, placebo- and active controlled study of paliperidone extended-release as maintenance treatment in patients with bipolar I disorder after an acute manic or mixed episode. J Affect Disord 138(3):247–258

Bowden CL, Calabrese JR, McElroy SL et al (2000) A randomized, placebo-controlled 12-month trial of divalproex and lithium in treatment of outpatients with bipolar I disorder. Divalproex Maintenance Study Group. Arch Gen Psychiatry 57(5):481–489

Bowden CL, Calabrese JR, Sachs G et al (2003) A placebo-controlled 18-month trial of lamotrigine and lithium maintenance treatment in recently manic or hypomanic patients wth bipolar I disorder. Arch Gen Psychiatry 60(4):392–400

Bowden CL, Grunze H, Mullen J et al (2005) A randomized, double-blind, placebo controlled efficacy and safety study of quetiapine or lithium as monotherapy for mania in bipolar disorder. J Clin Psychiatry 66(1):111–121

Bowden CL, Vieta E, Ice KS et al (2010) Ziprasidone plus a mood stabilizer in subjects with bipolar I disorder: a 6-month, randomized, placebo-controlled, double-blind trial. J Clin Psychiatry 71(2):130–137

Calabrese JR, Rapport DJ, Shelton MD et al (2001) Evolving methodologies in bipolar disorder maintenance research. Br J Psychiatry Suppl 41:s157–s163

Calabrese JR, Bowden CL, Sachs G et al (2003) A placebo-controlled 18-month trial of lamotrigine and lithium maintenance treatment in recently depressed patients with bipolar I disorder. J Clin Psychiatry 64(9):1013–1024

Calabrese JR, Hirschfeld RM, Frye MA et al (2004a) Impact of depressive symptoms compared with manic symptoms in bipolar disorder: results of a U.S. community-based sample. J Clin Psychiatry 65(11):1499–1504

Calabrese JR, Vieta E, El-Mallakh R et al (2004b) Mood state at study entry as predictor of the polarity of relapse in bipolar disorder. Biol Psychiatry 56(12):957–963

Calabrese JR, Shelton MD, Rapport DJ et al (2005a) A 20-month, double-blind, maintenance trial of lithium versus divalproex in rapid-cycling bipolar disorder. Am J Psychiatry 162 (11):2152–2161

Calabrese JR, Keck PE Jr, Macfadden W et al (2005b) A randomized, double-blind, placebo-controlled trial of quetiapine in the treatment of bipolar I or II depression. Am J Psychiatry 162:1351–1360

Calabrese JR, Goldberg JF, Ketter TA et al (2006) Recurrence in bipolar I disorder: a post hoc analysis excluding relapses in two double-blind maintenance studies. Biol Psychiatry 59 (11):1061–1064

Calabrese JR, Huffman RF, White RL et al (2008) Lamotrigine in the acute treatment of bipolar depression: results of five double-blind, placebo-controlled clinical trials. Bipolar Disord 10 (2):323–333

Carlson BX, Ketter TA, Sun W et al (2012) Aripiprazole in combination with lamotrigine for the long-term treatment of patients with bipolar I disorder (manic or mixed): a randomized, multicenter, double-blind study (CN138-392). Bipolar Disord 14(1):41–53

Dilsaver SC, Chen YW, Swann AC et al (1997) Suicidality, panic disorder and psychosis in bipolar depression, depressive-mania and pure-mania. Psychiatry Res 73(1–2):47–56

Gao K, Ganocy SJ, Gajwani P et al (2008a) A review of sensitivity and tolerability of antipsychotics in patients with bipolar disorder or schizophrenia: focus on somnolence. J Clin Psychiatry 69(2):302–309

Gao K, Kemp DE, Ganocy SJ et al (2008b) Antipsychotic-induced extrapyramidal side effects in bipolar disorder and schizophrenia: a systematic review. J Clin Psychopharmacol 28 (2):203–209

Gao K, Kemp DE, Calabrese JR (2009) Pharmacological treatment of the maintenance phase of bipolar depression: focus on relapse prevention studies and the impact of design on generalizability. In: Parnham M, Bruinvels J (series eds) Milestones in drug therapy series, Zarate Jr CA, Manji HK (eds) Bipolar depression: molecular neurobiology, clinical diagnosis, and pharmacotherapy. Springer Science + Business Media, Birkhauser Verlag AG, Switzerland, pp 159–179

Gao K, Kemp DE, Fein E et al (2011) Number needed to treat to harm for discontinuation due to adverse events in the treatment of bipolar depression, major depressive disorder, and generalized anxiety disorder with atypical antipsychotics. J Clin Psychiatry 72(8):1063–1071

Gao K, Kemp DE, Wu RR et al (2015) Mood stabilizers. In: Tasman A, Liberman J, Kay J, First M, Riba M (eds) Psychiatry, 4th edn. Wiley, Chichester, UK, pp 2129–2153

Ghaemi SN, Pardo TB, Hsu DJ (2004) Strategies for preventing the recurrence of bipolar disorder. J Clin Psychiatry 65(Suppl 10):16–23

Goodwin F, Jamison K (2007) Manic-depressive illness: bipolar disorder and recurrent depression, 2nd edn. Oxford University Press, New York

Goodwin GM, Bowden CL, Calabrese JR et al (2004) A pooled analysis of 2 placebo-controlled 18-month trials of lamotrigine and lithium maintenance in bipolar I disorder. J Clin Psychiatry 65(3):432–441

Isometsa ET, Henriksson MM, Aro HM et al (1994) Suicide in bipolar disorder in Finland. Am J Psychiatry 151(7):1020–1024

Judd LL, Akiskal HS, Schettler PJ et al (2002) The long-term natural history of the weekly symptomatic status of bipolar I disorder. Arch Gen Psychiatry 59(6):530–537

Judd LL, Akiskal HS, Schettler PJ et al (2003) A prospective investigation of the natural history of the long-term weekly symptomatic status of bipolar II disorder. Arch Gen Psychiatry 60 (3):261–269

Kanba S, Kawasaki H, Ishigooka J et al (2014) A placebo-controlled, double-blind study of the efficacy and safety of aripiprazole for the treatment of acute manic or mixed episodes in Asian patients with bipolar I disorder (the AMAZE study). World J Biol Psychiatry 15(2):113–121

Keck PE Jr, Marcus R, Tourkodimitris S et al (2003) A placebo-controlled, double-blind study of the efficacy and safety of aripiprazole in patients with acute bipolar mania. Am J Psychiatry 160:1651–1658

Keck PE Jr, Calabrese JR, McQuade RD et al (2006) A randomized, double-blind, placebo-controlled 26-week trial of aripiprazole in recently manic patients with bipolar I disorder. J Clin Psychiatry 67(4):626–637

Keck PE Jr, Calabrese JR, McIntyre RS (2007) Aripiprazole monotherapy for maintenance therapy in bipolar I disorder: a 100-week, double-blind study versus placebo. J Clin Psychiatry 68 (10):1480–1491

Licht RW, Nielsen JN, Gram LF et al (2010) Lamotrigine versus lithium as maintenance treatment in bipolar I disorder: an open, randomized effectiveness study mimicking clinical practice. The 6th trial of the Danish University Antidepressant Group (DUAG-6). Bipolar Disord 12 (5):483–493

Macfadden W, Alphs L, Haskins JT et al (2009) A randomized, double-blind, placebo-controlled study of maintenance treatment with adjunctive risperidone long-acting therapy in patients with bipolar I disorder who relapse frequently. Bipolar Disord 11(8):827–839

MacQueen GM, Young LT, Robb JC et al (2000) Effect of number of episodes on wellbeing and functioning of patients with bipolar disorder. Acta Psychiatr Scand 101(5):374–381

Marcus R, Khan A, Rollin L et al (2011) Efficacy of aripiprazole adjunctive to lithium or valproate in the long-term treatment of patients with bipolar I disorder with an inadequate response to lithium or valproate monotherapy: a multicenter, double-blind, randomized study. Bipolar Disord 13(2):133–144

Muzina DJ, Gao K, Kemp DE et al (2011) Acute efficacy of divalproex sodium versus placebo in mood stabilizer-naive bipolar I or II depression: a double-blind, randomized, placebo-controlled trial. J Clin Psychiatry 72(6):813–819

Potkin SG, Keck PE Jr, Segal S (2005) Ziprasidone in acute bipolar mania: a 21-day randomized, double-blind, placebo-controlled replication trial. J Clin Psychopharmacol 25(4):301–310

Quiroz JA, Yatham LN, Palumbo JM et al (2010) Risperidone long-acting injectable monotherapy in the maintenance treatment of bipolar I disorder. Biol Psychiatry 68(2):156–162

Sachs G, Chengappa KN, Suppes T et al (2004) Quetiapine with lithium or divalproex for the treatment of bipolar mania: a randomized, double-blind, placebo-controlled study. Bipolar Disord 6(3):213–223

Sachs G, Sanchez R, Marcus R et al (2006) Aripiprazole in the treatment of acute manic or mixed episodes in patients with bipolar I disorder: A 3-week placebo-controlled study. J Psychopharmacol 20:536–546

Sachs GS, Ice KS, Chappell PB et al (2011) Efficacy and safety of adjunctive oral ziprasidone for acute treatment of depression in patients with bipolar I disorder: a randomized, double-blind, placebo-controlled trial. J Clin Psychiatry 72(10):1413–1422

Sachs GS, Vanderburg DG, Karayal ON et al (2012) Adjunctive oral ziprasidone in patients with acute mania treated with lithium or divalproex, part 1: results of a randomized, double-blind, placebo-controlled trial. J Clin Psychiatry 73(11):1412–1419

Shapiro DR, Quitkin FM, Fleiss JL (1989) Response to maintenance therapy in bipolar illness. Effect of index episode. Arch Gen Psychiatry 46(5):401–405

Suppes T, Vieta E, Liu S et al (2009) Maintenance treatment for patients with bipolar I disorder: results from a North American study of quetiapine in combination with lithium or divalproex (trial 127). Am J Psychiatry 166(4):476–488

Thase ME, Jonas A, Khan A et al (2008) Aripiprazole monotherapy in nonpsychotic bipolar I depression: results of 2 randomized, placebo-controlled studies. J Clin Psychopharmacol 28 (1):13–20

Tohen M, Jacobs TG, Grundy SL et al (2000) Efficacy of olanzapine in acute bipolar mania: a double-blind, placebo-controlled study. The Olanzapine HGGW Study Group. Arch Gen Psychiatry 57:841–849

Tohen M, Chengappa KN, Suppes T et al (2002) Efficacy of olanzapine in combination with valproate or lithium in the treatment of mania in patients partially nonresponsive to valproate or lithium monotherapy. Arch Gen Psychiatry 59:62–69

Tohen M, Vieta E, Calabrese J et al (2003) Efficacy of olanzapine and olanzapine-fluoxetine combination in the treatment of bipolar I depression. Arch Gen Psychiatry 60:1079–1088

Tohen M, Chengappa KN, Suppes T et al (2004) Relapse prevention in bipolar I disorder: 18-month comparison of olanzapine plus mood stabiliser v. mood stabiliser alone. Br J Psychiatry 184:337–345

Tohen M, Greil W, Calabrese JR et al (2005) Olanzapine versus lithium in the maintenance treatment of bipolar disorder: a 12-month, randomized, double-blind, controlled clinical trial. Am J Psychiatry 162(7):1281–1290

Tohen M, Calabrese JR, Sachs GS et al (2006) Randomized, placebo-controlled trial of olanzapine as maintenance therapy in patients with bipolar I disorder responding to acute treatment with olanzapine. Am J Psychiatry 163(2):247–256

Tohen M, Vieta E, Goodwin GM et al (2008) Olanzapine versus divalproex versus placebo in the treatment of mild to moderate mania: a randomized, 12-week, double-blind study. J Clin Psychiatry 69(11):1776–1789

Vieta E, Mullen J, Brecher M et al (2005) Quetiapine monotherapy for mania associated with bipolar disorder: combined analysis of two international, double-blind, randomised, placebo-controlled studies. Curr Med Res Opin 21:923–934

Vieta E, Cruz N, García-Campayo J et al (2008a) A double-blind, randomized, placebo-controlled prophylaxis trial of oxcarbazepine as adjunctive treatment to lithium in the long-term treatment of bipolar I and II disorder. Int J Neuropsychopharmacol 11(4):445–452

Vieta E, Suppes T, Eggens I et al (2008b) Efficacy and safety of quetiapine in combination with lithium or divalproex for maintenance of patients with bipolar I disorder (international trial 126). J Affect Disord 109(3):251–263

Vieta E, T'joen C, McQuade RD et al (2008c) Efficacy of adjunctive aripiprazole to either valproate or lithium in bipolar mania patients partially nonresponsive to valproate/lithium monotherapy: a placebo-controlled study. Am J Psychiatry 165(10):1316–1325

Vieta E, Nuamah IF, Lim P et al (2010) A randomized, placebo- and active-controlled study of paliperidone extended release for the treatment of acute manic and mixed episodes of bipolar I disorder. Bipolar Disord 12(3):230–243

Vieta E, Montgomery S, Sulaiman AH et al (2012) A randomized, double-blind, placebo-controlled trial to assess prevention of mood episodes with risperidone long-acting injectable in patients with bipolar I disorder. Eur Neuropsychopharmacol 22(11):825–835

Wang Z, Kemp DE, Chan PK (2011) Comparisons of the tolerability and sensitivity of quetiapine-XR in the acute treatment of schizophrenia, bipolar mania, bipolar depression, major depressive disorder, and generalized anxiety disorder. Int J Neuropsychopharmacol 14(1):131–142

Weisler RH, Nolen WA, Neijber A et al (2011) Continuation of quetiapine versus switching to placebo or lithium for maintenance treatment of bipolar I disorder (Trial 144: a randomized controlled study). J Clin Psychiatry 72(11):1452–1464

Woo YS, Bahk WM, Chung MY et al (2011) Aripiprazole plus divalproex for recently manic or mixed patients with bipolar I disorder: a 6-month, randomized, placebo-controlled, double-blind maintenance trial. Hum Psychopharmacol 26(8):543–553

Wu R, Kemp DE, Sajatovic M et al (2011) Communication of potential benefits and harm to patients and payers in psychiatry: a review and commentary. Clin Ther 33(12):B62–B76

Yatham LN, Paulsson B, Mullen J et al (2004) Quetiapine versus placebo in combination with lithium or divalproex for the treatment of bipolar mania. J Clin Psychopharmacol 24 (6):599–606

Young AH, McElroy SL, Bauer M et al (2010) A double-blind, placebo-controlled study of quetiapine and lithium monotherapy in adults in the acute phase of bipolar depression (EMBOLDEN I). J Clin Psychiatry 71(2):150–162

Young AH, McElroy SL, Olausson B et al (2014) A randomised, placebo-controlled 52-week trial of continued quetiapine treatment in recently depressed patients with bipolar I and bipolar II disorder. World J Biol Psychiatry 15(2):96–112

Chapter 11
Non-pharmacological Somatic Treatments for Bipolar Depression

Harold A. Sackeim

Abstract There has been an explosion of research interest in noninvasive and invasive forms of brain stimulation as treatments for bipolar depression and major depressive disorder (MDD). Electroconvulsive therapy (ECT) has the strongest evidence base. In the short term, ECT is more effective than any other intervention for MDD. MDD patients and those with bipolar disorder (BD) do not differ in response or remission rates, but BD patients respond more quickly. Magnetic Seizure Therapy (MST) and Focal Electrically Administered Seizure Therapy (FEAST) provide greater focality of stimulation than can be achieved with traditional ECT. Both techniques appear to have reduced cognitive effects, but equivalence in efficacy with traditional ECT is not established. Repetitive Transcranial Magnetic Therapy (rTMS) is an approved and widely used treatment for MDD. Efficacy appears stronger in the community than in the randomized controlled trials, perhaps due to concomitant use of pharmacotherapy. While commonly used in bipolar depression, as yet there is little information on efficacy specifically in this subgroup. Small studies have suggested that transcranial Direct Current Stimulation (tDCS) has antidepressant properties, including in bipolar depression. While promising, multisite randomized sham controlled trials are needed to test these claims. Vagus Nerve Stimulation (VNS) showed long-term antidepressant effects and good durability of benefit in treatment-resistant depression, including BD. Lack of insurance reimbursement has limited use in the USA for this indication. Initial open-label studies of Deep Brain Stimulation (DBS) at several targets were encouraging. However, two recent pivotal trials were terminated due to lack of an efficacy signal. These negative findings are leading to rethinking the role of DBS in the treatment of severe, treatment-resistant depression.

Keywords Electroconvulsive therapy (ECT) • Bipolar depression • Response rate • Magnetic seizure therapy (MST) • Focal electrically administered seizure therapy (FEAST) • Repetitive transcranial magnetic stimulation (rTMS) • Deep

H.A. Sackeim, PhD (✉)
Departments of Psychiatry and Radiology, College of Physicians and Surgeons of Columbia University, 2124 Moselem Springs Road, Fleetwood, PA 19522, USA
e-mail: has1@columbia.edu

© Springer International Publishing Switzerland 2016
C.A. Zarate Jr., H.K. Manji (eds.), *Bipolar Depression: Molecular Neurobiology, Clinical Diagnosis, and Pharmacotherapy*, Milestones in Drug Therapy,
DOI 10.1007/978-3-319-31689-5_11

brain stimulation (DBS) • Vagus nerve stimulation (VNS) • Transcranial direct current stimulation (tDCS)

11.1   Introduction

The pharmacological treatment of bipolar disorder (BD) has always presented key challenges. Results from a recent national study of major depressive disorder (MDD), the Sequenced Treatment Alternatives to Relieve Depression (STAR*D), have generally indicated that response and remission rates are disappointingly low and that large percentages of patients do not achieve substantial improvement if they have not benefited from two adequate treatment trials (Rush 2007). Furthermore, relapse is both more rapid and more likely in patients who prospectively manifest treatment resistance during sequential pharmacological trials. Similarly, the national study of bipolar depression, the Systematic Treatment Enhancement Program for Bipolar Disorder (STEP-BD), found disappointingly low rates of sustained recovery when paroxetine or bupropion were added to a mood-stabilizing agent, and these rates did not differ from the group receiving a mood-stabilizing agent and placebo (Sachs et al. 2007).

In addition to high rates of treatment resistance in bipolar depression, pharmacological management has been beset by two other major conundrums. There is considerable concern that exposure to antidepressant medications may induce or exacerbate symptoms of agitation in BD patients and, in some cases, result in a switch into a hypomanic or manic state. For example, 44 % of the first 500 patients to enter the STEP-BD study historically reported a switch to a hypomanic, manic, or mixed stated within 12 weeks of starting an antidepressant treatment (Truman et al. 2007). This was especially likely in patients with short duration of illness, exposure to multiple antidepressant trials, and a previous history of switching. An independent concern was raised during the era when tricyclic antidepressants (TCAs) and monoamine oxidase inhibitors (MAOIs) were the mainstay of antidepressant treatment. It was suggested that these agents, while often effective in the acute treatment of bipolar depression, could accelerate the progression of illness, resulting in shorter periods of euthymia and, in some cases, inducing rapid cycling (Wehr and Goodwin 1979). These concerns about the limitations of pharmacological treatment in bipolar depression are accentuated with respect to the management of bipolar mania, where there is a high rate of morbidity and mortality and especially great need for rapid and effective treatment.

Non-pharmacological somatic treatments have a long history in the care of patients with BD. Indeed, electroconvulsive therapy (ECT) is the biological intervention with the longest history of continuous use in psychiatry, and it remains the most effective acute treatment available for either MDD or BD (American Psychiatric Association 2001). This is a powerful statement that indicates that ECT is one of the few treatments with therapeutic properties in the acute treatment of either bipolar depression or mania and, even more remarkably, most likely the most effective acute treatment available for either condition. The essential limitations of ECT—adverse cognitive effects and high rates of relapse—are discussed below (Sackeim et al. 2001a, b; Prudic et al. 2004). New developments in this field created forms of ECT that dramatically reduce the frequency and severity of adverse cognitive effects. These include critical alterations in the administration of ECT, such as the use of ultrabrief electrical stimuli (Tor et al. 2015; Sackeim et al. 2008), and the development of new forms of convulsive therapy, particularly Magnetic Seizure Therapy (MST) (Sackeim 1994) and Focal Electrically Administered Seizure Therapy (FEAST) (Sackeim 2004).

ECT has developed to the point that the total exposure of the brain to an electrical stimulus over a complete course of treatment may be less than 1/10th of a second. The electrical stimulus that is applied is less than 1 amp at the scalp surface and markedly below that in neuronal tissue. Thus, a modest and remarkably transient electrical stimulus results in the most profound acute antidepressant and antimanic effects seen in BD. Since the intensity of the electrical stimulus is known not to result in neuronal injury, and since this stimulus is "ephemeral," having no "metabolites," residue, or other long-term physical existence in the brain, the therapeutic properties of ECT must result from the brain's response to being stimulated in this fashion. In essence, ECT is a paradigm for how endogenous neural processes can produce profound antidepressant and antimanic effects if triggered in an appropriate fashion.

This observation provides the essential rationale for a host of other brain stimulation technologies that do not rely on seizure induction as part of their therapeutic mechanism of action. ECT is a model where an intense single train of stimulation produces an ictal event that, in turn, results in a large set of neurochemical, neurophysiological, and neuroanatomic alterations, some of which are targeted at seizure suppression, some of which are intrinsic to the electrical stimulation (independent of whether a seizure occurs), and others that may be seizure induced but are not critical in seizure termination. In other words, it has become apparent in recent years that electrical stimulation of the brain, independent of whether seizures are produced, results in neurochemical release; the specifics of the magnitude and type of neurotransmitter and peptides involved depend on the intensity and patterns of stimulation. Beyond neurochemical alterations, electrical stimulation of the brain can enhance or block signal transmission and, perhaps in some cases, improve signal-to-noise ratios compromised by damage in distal regions. Consequently, the field of brain stimulation, currently in its early development, opens the possibility for focal control of neurochemical alterations, second messenger processes, and modulation of brain communication systems in ways that have never been achieved with pharmacological interventions. This chapter reviews both what is known about current brain stimulation technologies in the treatment of bipolar depression, as well as highlighting the potential for new developments.

11.1.1 Electroconvulsive Therapy

Meduna, acting under the view common at the time that there was an intrinsic antagonism between epilepsy and schizophrenia, introduced convulsive therapy. While others had tried blood transfusion across these illnesses, Meduna tested the bold concept that exogenously induced seizures might reduce symptoms of this disorder. Using camphor in oil as the induction method, he reported that a remarkable number of patients with a diagnosis of schizophrenia showed marked symptomatic improvement with this method (Meduna 1935). This assertion proved controversial, since the predominant view at the time in biological psychiatry was that the major forms of mental illness were due to congenital or degenerative conditions and could not be ameliorated, even palliatively, by any intervention. As a result of taking this position, Meduna lost his academic position. His method of chemical seizure induction was quickly replaced by the use of Metrazol, a gamma aminobutyric acid (GABA) antagonist, that more reliably resulted in seizures, and convulsive therapy was widely adopted worldwide.

In 1938, Cerletti and Bini in Rome demonstrated that electrical stimulation was a preferred method. It had the advantages of ensuring only one seizure occurred, whereas recirculation was always possible with chemical induction and, more critically, seizure induction was instantaneous following the electrical stimulus. This advantage was critical since chemical methods often involved a substantial delay, frequently resulting in full panic attacks prior to seizure onset, and subsequent refusal of treatment. The electrical stimulus itself was poorly conceived and basically varied only as a function of the amplitude of the sine wave voltage waveform output by the standard electrical grid with crude control over the duration of exposure. There was little consideration about whether this type of electrical signal was optimal for stimulating neural tissue. Subsequent developments during the 1950s introduced the use of muscle relaxants (first curare and then succinylcholine) to block convulsive motor manifestations. This innovation markedly reduced the rate of vertebral fractures, but required the introduction of general anesthesia. Whereas the application of the electrical stimulus invariably resulted in loss of consciousness, the pre-application of a muscle paralyzing agent, and the subsequent inability to breathe without assistance, necessitated the use of general anesthesia for psychological reasons.

Soon after the introduction of ECT, it was recognized that the intervention had greater success in the treatment of mood disorders than schizophrenia, at least in the short term. Of course, diagnosis at the time had questionable reliability, but the general consensus has been that mood disorders were under-recognized and schizophrenia over-diagnosed in the USA. Thus, the observation that mood disorders responded at remarkably high rates to ECT, and more so than in patients with schizophrenia, if anything, likely underestimated the true difference. Early on, Kalinowsky and others would claim that approximately 80–90 % of patients with depressive illness would achieve remission after receiving approximately six to 12 treatments with ECT (Kalinowsky and Hoch 1946).

This estimate, extending across MDD and BD depressive conditions, has not been realized in recent years. Regardless of treatment methods, remission rates with ECT are somewhat more modest (Prudic et al. 2004; Sackeim et al. 1993, 2000, 2009; Kellner et al. 2006). This shift is likely due to the fact that when ECT was introduced, there were few, if any, competitive treatments, and ECT was commonly used at the outset. Today, resistance to pharmacological treatments is the leading indication for the use of ECT. Several, but not all, studies have found that degree of medication resistance is predictive of ECT outcome and that, in general, patients who have not benefited from adequate psychopharmacology and/or who have long durations of their current episode of depression have somewhat inferior outcomes (Prudic et al. 1990; Sackeim et al. 2000; Heijnen et al. 2010). Thus, remission rates on the order of 60–70 % may be a more realistic estimate, especially if remission is defined as maintaining nearly complete symptomatic improvement for at least one week following ECT.

This extent of expected clinical improvement with ECT exceeds that of any other known antidepressant treatment (American Psychiatric Association 2001). Typically, in ECT research, the bar is set higher for what is defined as response or remission than in standard pharmacological trials, and yet the response and remission rates are higher, despite the concentration on patients with treatment resistance. For example, in the STAR*D study, remission rates among MDD patients who had not achieved adequate benefit after two pharmacological treatments were on the order of 10 % (Rush 2007). Such patients would be expected to remit at substantially higher rates if treated with ECT.

Part of the evidence base supporting the efficacy of ECT in MDD derives from randomized trials comparing ECT to antidepressant pharmacotherapy (Janicak et al. 1985; Folkerts et al. 1997). While such trials were not double-blind and had other limitations, ECT was consistently superior in efficacy. The first such trial was recently conducted in bipolar depression (Schoeyen et al. 2015). Schoeyen and colleagues randomized 73 patients with BD-I or BD-II depression to right unilateral ECT or algorithmic pharmacological treatment. Outcome was assessed by blinded raters after six weeks. All patients had failed two or more trials of antidepressants or mood stabilizers. The ECT group had greater symptomatic improvement and superior response rate (ECT = 74 % vs. pharmacotherapy = 35 %). However, there was no difference in remission rate, suggesting that ECT may have been terminated prior to full benefit.

There is no evidence that the distinction between MDD and bipolar depression has bearing on likelihood of achieving response or remission with ECT. Retrospective and prospective comparisons have generally indicated that both forms of depression respond or remit at approximately the same rate. However, there are two caveats to this generalization. First, patients with bipolar depression require fewer treatments to achieve response or remission than patients with MDD. This was first reported in samples treated in randomized protocols at the New York State Psychiatric Institute (NYSPI), with the observation that, on average, patients with bipolar depression required approximately 1.5 fewer treatments than MDD patients to meet the same outcome criteria (Daly et al. 2001). This observation was later replicated in a large naturalistic study of patients treated in community settings (Sackeim and Prudic 2005) and in Europe (Sienaert et al. 2009).

This observation reflects a large effect since the bulk of clinical gains with ECT are usually obtained within the first six treatments. Bipolar depressed patients appear to achieve this benefit more rapidly. The only factor known to have substantial impact on the speed of response with ECT is the extent to which electrical dosage exceeds seizure threshold, with higher dosage leading to more rapid improvement (Sackeim et al. 1993). However, multiple studies have failed to find a difference in initial seizure threshold in MDD and bipolar depression, and in the studies at NYSPI, dosage was always adjusted to a specific level relative to seizure threshold for all patients in a treatment condition. BD patients improved more rapidly regardless of whether they received right unilateral or bilateral ECT or of the particular dosage that was applied. This would suggest that the neurophysiological response to exogenous seizure induction differs in MDD and bipolar depression. For example, it has long been speculated that it is the endogenous anticonvulsant response of the brain in terminating the seizure that is critical to achieving antidepressant effects (Sackeim et al. 1983), while others have noted that ECT results in remarkably rapid onset of neuroplastic changes, including neurogenesis (Perera et al. 2007). Thus, a variety of avenues need exploration to account for the more rapid onset of benefit in bipolar depression.

The second area in which the efficacy of ECT for bipolar depression may be altered pertains to the subset of patients with psychotic or delusional depression. It has often been stated that psychotic features are overrepresented in patients with bipolar depression relative to MDD, although this is not firmly established. Regardless, most studies that have compared the efficacy of ECT in patients with and without psychotic features have found higher rates of response and remission in psychotic depression (American Psychiatric Association 2001). Until the advent of atypical antipsychotic medications, only a very small minority of patients with psychotic depression had received an adequate combined pharmacological trial prior to ECT, since the dosage of antipsychotic medication considered adequate was often intolerable, especially in the elderly, and especially when combined with the available antidepressant medications (Mulsant et al. 1997). Relatively low rates of established medication resistance continue to characterize patients with psychotic features, since ECT is also often considered due to clinical urgency, history of response, and patient preference.

Two principal issues distinguish the management of the patient with bipolar depression during ECT. The first pertains to concomitant pharmacological agents and the second to the emergence of hypomania or mania. In general, in the USA, it had long been recommended that all patients be withdrawn from psychotropic agents during ECT, with the exception of antipsychotics in patients with psychotic features (American Psychiatric Association 2001). There was little evidence that concomitant antidepressant medications enhanced clinical outcome and some concern that concomitant anxiolytics, especially benzodiazepines and perhaps anticonvulsants, interfered with the therapeutic process.

Recently, a large, multisite study randomized MDD patients and those with bipolar depression to concomitant treatment with placebo, nortriptyline, or venlafaxine during the course of ECT (Sackeim et al. 2009). There was significant enhancement of the therapeutic benefit in patients treated with nortriptyline or venlafaxine relative to placebo and some evidence that concurrent nortriptyline reduced the cognitive side effects of ECT. Over 20 % of the 316 participants in this study had bipolar depression, and there was no evidence that these effects differed with polarity. Thus, this recent evidence is leading to revision of the longstanding view that antidepressants should be stopped during the administration of ECT. For instance, in the intent-to-treat sample, the remission rates following ECT among MDD patients for those treated with nortriptyline or placebo were 61.2 % and 43.7 %, respectively. The comparable remission rates for bipolar depressed patients were 72.0 % and 59.3 %. This reflects a substantial enhancement of outcomes.

Research with schizophrenia has supported the safety and clinical utility of combining antipsychotic medications and ECT, with evidence for synergistic clinical effects (Sackeim 2003). However, there has long been concern that agents with anticonvulsant properties, especially benzodiazepines, may interfere with the seizure process and diminish efficacy. The evidence for diminished efficacy is entirely circumstantial, stemming mainly from naturalistic, retrospective studies. It is possible that the most agitated patients are the most likely to receive the highest doses of these agents, and these observations are confounded. Nonetheless, it is prudent to limit both benzodiazepine and anticonvulsant use during the ECT course. Since ECT has profound anticonvulsant properties, often leading to a decrease in anticonvulsant dosage in epilepsy patients, and since improvement is usually marked and rapid in psychic anxiety, these dosage limitations are usually well tolerated. Another problematic issue is exposure to lithium during the ECT course. It is well established that a minority of individuals will develop a severe organic brain syndrome when the two are combined, which diminishes rapidly once the lithium is stopped. For this reason, most expert groups recommend discontinuation of lithium during an acute ECT course or, at minimum, withholding doses the evening before a treatment (American Psychiatric Association 2001).

Major limitations in the use of ECT are its side effects, rates of relapse, and patient acceptability. There is always the concern that treatment of the patient in a mixed state or in bipolar depression will provoke a hypomanic or manic reaction. This certainly does happen with ECT. However, careful examination of the outcomes of hundreds of patients prospectively followed at NYSPI showed that such reactions occurred at remarkably small rates. It is not clear why this is so, but may reflect the antimanic properties of the treatment and/or its marked anticonvulsant effects. There is little consensus on how to manage emergent mania during ECT. Many practitioners will continue the treatment if symptoms are mild. Many would terminate the ongoing course of ECT, institute a new pharmacological regimen, and observe the patient if severe manic symptoms emerged.

Only in recent years have the adverse long-term effects of ECT on memory been documented. Both randomized and naturalistic studies have shown that methods of ECT administration substantially differ in their impact on the degree of retrograde amnesia observed six months following the treatment (Sackeim et al. 2007b, 2008; Sackeim 2014). Indeed, recent work has shown that the objective findings covary with patients' subjective reports of deficits (Berman et al. 2008; Brakemeier et al. 2011). It is established that ECT techniques alter the likelihood of long-term negative effects. For instance, the introduction of ultrabrief pulse stimulation, when coupled with right unilateral electrode placement, substantially reduces cognitive effects at all time points (Sackeim et al. 2008). There is also evidence that, at baseline, older patients with bipolar depression have greater cognitive impairment, especially memory deficits, than similarly aged MDD patients, presumably as a consequence of a history of more frequent episodes (Burt et al. 2000). However, there is no evidence that BD patients are more at risk than MDD patients with respect to ECT's cognitive effects.

ECT is one of the only psychiatric treatments that is typically discontinued once found effective. Relapse is common following ECT-induced remission, and modern prospective studies document that approximately 50 % of remitted patients relapse despite aggressive continuation therapy with pharmacological agents or ECT, with medication resistance a strong predictor (Kellner et al. 2006; Sackeim et al. 1990, 2001a; Jelovac et al. 2013). However, STAR*D and other recent studies have reported similarly high rates of relapse despite continuation of the same pharmacological regimen that produced response or remission, with medication resistance again predicting more rapid and frequent relapse (Rush 2007). Durability of benefit appears to be a significant and general problem in the management of depression. Although sample sizes have generally been small, there is no evidence that relapse risk following ECT differs in MDD and bipolar depression.

11.1.2 Magnetic Seizure Therapy

It has been established that the current paths of the ECT stimulus and the dosing within those paths have profound effects on the efficacy and side effects of the treatment (Sackeim et al. 1987, 1993, 2000). Yet, with traditional ECT, the high impedance of the skull and other anatomic reasons limit the capacity to restrict current paths. A treatment method that offered superior control over anatomic distribution of current and greater precision in intracerebral dosing (current densities) would provide a major advance. Sackeim proposed that use of a time-varying train of magnetic pulses might achieve these goals, terming the intervention MST (Sackeim 1994). Compared to traditional ECT, the transparency of the scalp and skull to the magnetic field allows for greater anatomic precision, and the fact that dosage is primarily determined by distance from the coil limits deep stimulation and allows for greater dosing precision (Deng et al. 2015).

The future of this modality is uncertain. Preliminary studies have generally shown a relatively low level of cognitive side effects but uncertain efficacy (Lisanby et al. 2001, 2003). The major limitation in MST development, making its future uncertain, is largely engineering issues. It has been difficult to develop

MST systems sufficiently powerful to elicit seizures from regions in frontal cortex using coils that maximize focality of stimulation. This limitation is especially problematic since the extent that dosage is substantially above seizure threshold can be a critical determinant of efficacy. MST is also largely limited to seizure initiation in superficial cortex.

11.1.3 Focal Electrically Applied Seizure Therapy (FEAST)

FEAST is another new intervention that also offers the possibility of greater anatomic precision in site of seizure initiation. Sackeim reasoned that by using unidirectional current flow, which would establish a consistent anode and cathode, as well as altering the geometry and positioning of electrodes, one could achieve greater precision in the anatomic distribution of currents paths (Sackeim 2004). The basic principles underlying FEAST have been validated in research with nonhuman primates (Cycowicz et al. 2008) and in small open clinical investigations (Nahas et al. 2013). Relative to ECT, FEAST appears to have fewer cognitive side effects, but equivalence in efficacy is not yet established.

11.1.4 Repetitive Transcranial Magnetic Stimulation (rTMS)

One can induce current in neural tissue by exposing the tissue to a time-varying magnetic field. With a magnetic coil placed on the surface of the head, anatomic resolution and distribution are determined mainly by coil geometry, and detectable current densities can generally reach 2 cm deep. There are a large number of open and blinded studies that raised the possibility that repetitive stimulation at high frequency (>5 Hz) over the left dorsolateral prefrontal cortex (DLPFC) has anti-depressant effects, and a smaller set of studies suggested that slow stimulation (≤1 Hz) over the right DLPFC has similar effects. Several meta-analyses have concluded that randomized sham-controlled trials have shown consistent antide-pressant effects (Burt et al. 2002). A large industry-sponsored multisite trial reported generally positive findings (O'Reardon et al. 2007), and superior antide-pressant effects with left DLPFC high frequency relative to an active sham condi-tion were also observed in the NIH-supported multisite trial (George et al. 2010). These findings led to FDA approval of rTMS specifically for treatment-resistant MDD.

There is considerable skepticism in the field regarding the clinical utility of rTMS in the treatment of depression. While it was incontrovertible that active rTMS exerted greater antidepressant properties than sham interventions, it was uncertain whether the magnitude of benefit was clinically significant given the relatively low remission rates (Sackeim 2000). However, a recent multisite open study of patients receiving rTMS in the community reports an impressively high rate of short-term remission (Carpenter et al. 2012). It is speculated that rTMS coupled with antidepressant pharmacology may be particularly potent. Surprisingly, follow-up of this same sample suggested strong durability of benefit when combined with rapid reintroduction of rTMS with emergent symptoms (Dunner et al. 2014). Thus, rTMS has become frequently used. Although originally approved only for MDD, use in bipolar depression is common. Some practitioners claim that rTMS can at times fundamentally alter the course of BD. However, as yet there is no evidence that patients with bipolar depression differ in response to rTMS from patients with MDD.

11.1.5 Transcranial Direct Current Stimulation (tDCS) and Related Technologies

There has been an explosion in recent years in research using a variety of methods to stimulate the brain with low intensity noninvasive current. The most studied technique, transcranial Direct Current Stimulation (tDCS), involves passage of a low amperage (e.g., 1 mA) direct current between anode and cathode electrodes placed on the scalp. It is believed that exposure to the direct current alters the firing rate of neuronal populations due to a change in neuronal membrane electrical potential. A variety of other techniques involve alterations in the electrical signal, such as use of alternating current or pulsed current.

Scores of studies have reported enhancement or decrement in human cognitive abilities after exposure to tDCS, although there is dispute regarding the reliability of these effects (Horvath et al. 2015). A small literature has examined potential antidepressant effects of these techniques, with most work concentrating on tDCS. While meta-analysis indicates that tDCS has antidepressant effects greater than sham, the findings derived from small samples were heterogeneous (Shiozawa et al. 2014). More rigorous, large-scale studies are needed to determine whether tDCS (or related techniques) deserve a clinical role. There is initial evidence of efficacy in bipolar depression (Brunoni et al. 2011), and a randomized controlled trial in bipolar depression is underway (Pereira Junior Bde et al. 2015).

11.1.6 Vagus Nerve Stimulation (VNS)

VNS is a treatment approved by the FDA and labeled specifically for treatment-resistant depression, both MDD and BD. Eighty percent of the fibers in the vagus nerve are afferent to brain, and basic research has shown that repetitive electrical stimulation of the vagus nerve can have widespread effects on brain physiology and neurochemistry. It became established that VNS had anticonvulsant properties and was approved for the treatment of epilepsy in 1997. An initial pilot study in

60 patients suggested that VNS had clinically significant long-term effects in patients with marked medication resistance (Sackeim et al. 2001b). A subsequent randomized, sham-controlled, multisite study failed to detect a difference between active and sham VNS after a 10-week treatment period (Rush et al. 2005). However, as in the pilot study, a substantial number of patients were improved after a year. Of special note, it also seems that VNS has remarkable durability of benefit (Sackeim et al. 2007a). A surprisingly large percentage of patients who showed clinical benefit after starting VNS maintained the benefit for periods of up to two years. Thus, this intervention may take a considerable time to show antidepressant effects and has a high capacity to maintain benefit if achieved. As yet, there is no evidence that MDD and BD patients differ in response to VNS.

Despite FDA approval, the absence of controlled data establishing the claims of late onset of action and strong durability of benefit has limited access to VNS in the USA, due to the reluctance of insurers to reimburse for the procedure. Recent developments in this field include the development of noninvasive techniques to stimulate the vagus nerve using peripheral electrical or magnetic stimulation.

11.1.7 Deep Brain Stimulation (DBS)

Stimulation through electrodes indwelling in specified locations in the brain offers unique opportunities to modulate specific pathways for therapeutic benefit. DBS is an FDA-approved treatment for dystonia, essential tremor, and tremor in Parkinson's disease. Therapeutic effects in Parkinson's disease may be marked and evident from first onset of stimulation and highly dependent on the contact placement within a small neural structure and stimulation parameters. In Parkinson's disease, long-term follow-up (five years) indicates that retention of benefit is remarkably high, especially in the context of a degenerative, medication-resistant disorder. In Parkinson's disease, DBS either in the subthalamic nucleus or the globus pallidus is effective, while this is not true for dystonia. Thus, there may be multiple entry points to modulate a network for therapeutic purposes, and these networks differ anatomically among the movement disorders (Hardesty and Sackeim 2007).

DBS in mood disorders is an experimental procedure with a small knowledge base. The morbidity/mortality risk of DBS is significant due to the invasive procedure. Therefore, DBS in mood disorder patients is only conducted in a research context with patients with markedly resistant and severe MDD. The initial experience was limited to open-label, pilot studies with small sample sizes that suggested impressive clinical effects. The targets for stimulation have been the subgenual anterior cingulate in the work led by Mayberg (Mayberg et al. 2005), the anterior limb of the internal capsule (Greenberg et al. 2006), and the nucleus accumbens and the medial forebrain bundle (Schlaepfer et al. 2008).

Mood and movement disorders may differ in how rapidly treatment paradigms are developed. First, knowledge of specific circuitry is less advanced in the case of mood disorders. Second, the nuclei targeted within the striatum are relatively small in the case of movement disorders, and yet specific location within a nucleus is critical to outcome. In the case of MDD, the structures most often implicated as targets for modulation are large gray matter areas like the anterior cingulate or right orbital frontal cortex. However, the DBS signal does not broadcast well over wide regions of tightly packed gray matter. Thus, it is difficult to modulate over broad areas and, consequently, the work by Mayberg involved stimulating the white matter under the anterior cingulate, thus hoping to modulate activity within the cingulate itself (Mayberg et al. 2005). Similarly, the group stimulating in the internal capsule are also stimulating white matter tracts that may act at distant structures. Initial observations with this target suggested that therapeutic effects in MDD might be contingent on use of high intensity stimulation.

The initial experience with DBS in resistant MDD was largely positive. With the sample sizes small, the trials unblinded and uncontrolled, and many other caveats, the groups focusing on the anterior cingulate, internal capsule, accumbens, and medial forebrain bundle have been encouraged by the clinical outcomes observed, including effectiveness in MDD and bipolar depression and strong durability of benefit (Holtzheimer et al. 2012). However, two randomized controlled pivotal trials were conducted by industry to establish the safety and efficacy of DBS to the subgenual cingulate target (BROADEN trial) or to the ventral capsule/ventral striatum (Dougherty et al. 2015). Both trials were prematurely stopped due to lack of an efficacy signal. The negative findings have resulted in rethinking the role of DBS in the treatment of severe, treatment-resistant depression and in considerable discussion about the source of differences with the results of the original, open-label investigations.

References

American Psychiatric Association (2001) The practice of ECT: recommendations for treatment, training and privileging, 2nd edn. American Psychiatric Press, Washington, DC

Berman RM, Prudic J, Brakemeier EL, Olfson M, Sackeim HA (2008) Subjective evaluation of the therapeutic and cognitive effects of electroconvulsive therapy. Brain Stimul 1(1):16–26

Brakemeier EL, Berman R, Prudic J, Zwillenberg K, Sackeim HA (2011) Self-evaluation of the cognitive effects of electroconvulsive therapy. J ECT 27(1):59–66

Brunoni AR, Ferrucci R, Bortolomasi M, Vergari M, Tadini L, Boggio PS, Giacopuzzi M, Barbieri S, Priori A (2011) Transcranial direct current stimulation (tDCS) in unipolar vs. bipolar depressive disorder. Prog Neuropsychopharmacol Biol Psychiatry 35(1):96–101

Burt T, Prudic J, Peyser S, Clark J, Sackeim HA (2000) Learning and memory in bipolar and unipolar major depression: effects of aging. Neuropsychiatry Neuropsychol Behav Neurol 13 (4):246–253

Burt T, Lisanby SH, Sackeim HA (2002) Neuropsychiatric applications of transcranial magnetic stimulation: a meta analysis. Int J Neuropsychopharmacol 5(1):73–103

Carpenter LL, Janicak PG, Aaronson ST, Boyadjis T, Brock DG, Cook IA, Dunner DL, Lanocha K, Solvason HB, Demitrack MA (2012) Transcranial magnetic stimulation (TMS) for major depression: a multisite, naturalistic, observational study of acute treatment outcomes in clinical practice. Depress Anxiety 29(7):587–596

Cycowicz YM, Luber B, Spellman T, Lisanby SH (2008) Differential neurophysiological effects of magnetic seizure therapy (MST) and electroconvulsive shock (ECS) in non-human primates. Clin EEG Neurosci 39(3):144–149

Daly JJ, Prudic J, Devanand DP, Nobler MS, Lisanby SH, Peyser S, Roose SP, Sackeim HA (2001) ECT in bipolar and unipolar depression: differences in speed of response. Bipolar Disord 3 (2):95–104

Deng ZD, Lisanby SH, Peterchev AV (2015) Effect of anatomical variability on electric field characteristics of electroconvulsive therapy and magnetic seizure therapy: a parametric modeling study. IEEE Trans Neural Syst Rehabil Eng 23(1):22–31

Dougherty DD, Rezai AR, Carpenter LL, Howland RH, Bhati MT, O'Reardon JP, Eskandar EN, Baltuch GH, Machado AD, Kondziolka D, Cusin C, Evans KC, Price LH, Jacobs K, Pandya M, Denko T, Tyrka AR, Brelje T, Deckersbach T, Kubu C, Malone DA Jr (2015) A randomized sham-controlled trial of deep brain stimulation of the ventral capsule/ventral striatum for chronic treatment-resistant depression. Biol Psychiatry 78(4):240–248

Dunner DL, Aaronson ST, Sackeim HA, Janicak PG, Carpenter LL, Boyadjis T, Brock DG, Bonneh-Barkay D, Cook IA, Lanocha K, Solvason HB, Demitrack MA (2014) A multisite, naturalistic, observational study of transcranial magnetic stimulation for patients with pharmacoresistant major depressive disorder: durability of benefit over a 1-year follow-up period. J Clin Psychiatry 75(12):1394–1401

Folkerts HW, Michael N, Tolle R, Schonauer K, Mucke S, Schulze-Monking H (1997) Electroconvulsive therapy vs. paroxetine in treatment-resistant depression – a randomized study. Acta Psychiatr Scand 96(5):334–342

George MS, Lisanby SH, Avery D, McDonald WM, Durkalski V, Pavlicova M, Anderson B, Nahas Z, Bulow P, Zarkowski P, Holtzheimer PE III, Schwartz T, Sackeim HA (2010) Daily left prefrontal transcranial magnetic stimulation therapy for major depressive disorder: a sham-controlled randomized trial. Arch Gen Psychiatry 67(5):507–516

Greenberg BD, Malone DA, Friehs GM, Rezai AR, Kubu CS, Malloy PF, Salloway SP, Okun MS, Goodman WK, Rasmussen SA (2006) Three-year outcomes in deep brain stimulation for highly resistant obsessive-compulsive disorder. Neuropsychopharmacology 31 (11):2384–2393

Hardesty DE, Sackeim HA (2007) Deep brain stimulation in movement and psychiatric disorders. Biol Psychiatry 61(7):831–835

Heijnen WT, Birkenhager TK, Wierdsma AI, van den Broek WW (2010) Antidepressant pharmacotherapy failure and response to subsequent electroconvulsive therapy: a meta-analysis. J Clin Psychopharmacol 30(5):616–619

Holtzheimer PE, Kelley ME, Gross RE, Filkowski MM, Garlow SJ, Barrocas A, Wint D, Craighead MC, Kozarsky J, Chismar R, Moreines JL, Mewes K, Posse PR, Gutman DA, Mayberg HS (2012) Subcallosal cingulate deep brain stimulation for treatment-resistant unipolar and bipolar depression. Arch Gen Psychiatry 69(2):150–158

Horvath JC, Forte JD, Carter O (2015) Quantitative review finds no evidence of cognitive effects in healthy populations from single-session transcranial Direct Current Stimulation (tDCS). Brain Stimul 8(3):535–550

Janicak P, Davis J, Gibbons R, Ericksen S, Chang S, Gallagher P (1985) Efficacy of ECT: a meta-analysis. Am J Psychiatry 142(3):297–302

Jelovac A, Kolshus E, McLoughlin DM (2013) Relapse following successful electroconvulsive therapy for major depression: a meta-analysis. Neuropsychopharmacology 38(12):2467–2474

Kalinowsky LB, Hoch PH (1946) Shock treatments and other somatic procedures in psychiatry. Grune & Stratton, New York

Kellner CH, Knapp RG, Petrides G, Rummans TA, Husain MM, Rasmussen K, Mueller M, Bernstein HJ, O'Connor K, Smith G, Biggs M, Bailine SH, Malur C, Yim E, McClintock S, Sampson S, Fink M (2006) Continuation electroconvulsive therapy vs pharmacotherapy for relapse prevention in major depression: a multisite study from the Consortium for Research in Electroconvulsive Therapy (CORE). Arch Gen Psychiatry 63(12):1337–1344

Lisanby SH, Schlaepfer TE, Fisch HU, Sackeim HA (2001) Magnetic seizure therapy of major depression. Arch Gen Psychiatry 58(3):303–305

Lisanby SH, Luber B, Schlaepfer TE, Sackeim HA (2003) Safety and feasibility of magnetic seizure therapy (MST) in major depression: randomized within-subject comparison with electroconvulsive therapy. Neuropsychopharmacology 28(10):1852–1865

Mayberg HS, Lozano AM, Voon V, McNeely HE, Seminowicz D, Hamani C, Schwalb JM, Kennedy SH (2005) Deep brain stimulation for treatment-resistant depression. Neuron 45 (5):651–660

Meduna LJ (1935) Versuche über die biologische Beeinflussung des Abaufes der Schizophrenie: I. Campher und Cardiozolkrämpfe. Z Neurol Psychiatry 152:235–262

Mulsant BH, Haskett RF, Prudic J, Thase ME, Malone KM, Mann JJ, Pettinati HM, Sackeim HA (1997) Low use of neuroleptic drugs in the treatment of psychotic major depression. Am J Psychiatry 154(4):559–561

Nahas Z, Short B, Burns C, Archer M, Schmidt M, Prudic J, Nobler MS, Devanand DP, Fitzsimons L, Lisanby SH, Payne N, Perera T, George MS, Sackeim HA (2013) A feasibility study of a new method for electrically producing seizures in man: focal electrically administered seizure therapy [FEAST]. Brain Stimul 6(3):403–408

O'Reardon JP, Solvason HB, Janicak PG, Sampson S, Isenberg KE, Nahas Z, McDonald WM, Avery D, Fitzgerald PB, Loo C, Demitrack MA, George MS, Sackeim HA (2007) Efficacy and safety of transcranial magnetic stimulation in the acute treatment of major depression: a multisite randomized controlled trial. Biol Psychiatry 62(11):1208–1216

Pereira Junior Bde S, Tortella G, Lafer B, Nunes P, Bensenor IM, Lotufo PA, Machado-Vieira R, Brunoni AR (2015) The bipolar depression electrical treatment trial (BETTER): design, rationale, and objectives of a randomized, sham-controlled trial and data from the pilot study phase. Neural Plast 2015:684025

Perera TD, Coplan JD, Lisanby SH, Lipira CM, Arif M, Carpio C, Spitzer G, Santarelli L, Scharf B, Hen R, Rosoklija G, Sackeim HA, Dwork AJ (2007) Antidepressant-induced neurogenesis in the hippocampus of adult nonhuman primates. J Neurosci 27(18):4894–4901

Prudic J, Sackeim HA, Devanand DP (1990) Medication resistance and clinical response to electroconvulsive therapy. Psychiatry Res 31(3):287–296

Prudic J, Olfson M, Marcus SC, Fuller RB, Sackeim HA (2004) Effectiveness of electroconvulsive therapy in community settings. Biol Psychiatry 55(3):301–312

Rush AJ (2007) STAR*D: what have we learned? Am J Psychiatry 164(2):201–204

Rush AJ, Marangell LB, Sackeim HA, George MS, Brannan SK, Davis SM, Howland R, Kling MA, Rittberg BR, Burke WJ, Rapaport MH, Zajecka J, Nierenberg AA, Husain MM, Ginsberg D, Cooke RG (2005) Vagus nerve stimulation for treatment-resistant depression: a randomized, controlled acute phase trial. Biol Psychiatry 58(5):347–354

Sachs GS, Nierenberg AA, Calabrese JR, Marangell LB, Wisniewski SR, Gyulai L, Friedman ES, Bowden CL, Fossey MD, Ostacher MJ, Ketter TA, Patel J, Hauser P, Rapport D, Martinez JM, Allen MH, Miklowitz DJ, Otto MW, Dennehy EB, Thase ME (2007) Effectiveness of adjunctive antidepressant treatment for bipolar depression. N Engl J Med 356(17):1711–1722

Sackeim HA (1994) Magnetic stimulation therapy and ECT. Convulsive Ther 10:255–258

Sackeim HA (2000) Repetitive transcranial magnetic stimulation: What are the next steps? Biol Psychiatry 48(10):959–961

Sackeim HA (2003) Electroconvulsive therapy and schizophrenia. In: Hirsch SR, Weinberger D (eds) Schizophrenia, 2nd edn. Balckwell, Oxford, pp 517–551

Sackeim HA (2004) The convulsant and anticonvulsant properties of electroconvulsive therapy: towards a focal form of brain stimulation. Clin Neurosci Rev 4:39–57

Sackeim HA (2014) Autobiographical memory and electroconvulsive therapy: Do not throw out the baby. J ECT 30(3):177–186

Sackeim HA, Prudic J (2005) Length of the ECT course in bipolar and unipolar depression. J ECT 21(3):195–197

Sackeim HA, Decina P, Prohovnik I, Malitz S, Resor SR (1983) Anticonvulsant and antidepressant properties of electroconvulsive therapy: a proposed mechanism of action. Biol Psychiatry 18 (11):1301–1310

Sackeim HA, Decina P, Kanzler M, Kerr B, Malitz S (1987) Effects of electrode placement on the efficacy of titrated, low-dose ECT. Am J Psychiatry 144(11):1449–1455

Sackeim HA, Prudic J, Devanand DP, Decina P, Kerr B, Malitz S (1990) The impact of medication resistance and continuation pharmacotherapy on relapse following response to electroconvulsive therapy in major depression. J Clin Psychopharmacol 10(2):96–104

Sackeim HA, Prudic J, Devanand DP, Kiersky JE, Fitzsimons L, Moody BJ, McElhiney MC, Coleman EA, Settembrino JM (1993) Effects of stimulus intensity and electrode placement on the efficacy and cognitive effects of electroconvulsive therapy. N Engl J Med 328(12):839–846

Sackeim HA, Prudic J, Devanand DP, Nobler MS, Lisanby SH, Peyser S, Fitzsimons L, Moody BJ, Clark J (2000) A prospective, randomized, double-blind comparison of bilateral and right unilateral electroconvulsive therapy at different stimulus intensities. Arch Gen Psychiatry 57 (5):425–434

Sackeim HA, Haskett RF, Mulsant BH, Thase ME, Mann JJ, Pettinati HM, Greenberg RM, Crowe RR, Cooper TB, Prudic J (2001a) Continuation pharmacotherapy in the prevention of relapse following electroconvulsive therapy: a randomized controlled trial. JAMA 285(10):1299–1307

Sackeim HA, Rush AJ, George MS, Marangell LB, Husain MM, Nahas Z, Johnson CR, Seidman S, Giller C, Haines S, Simpson RK Jr, Goodman RR (2001b) Vagus nerve stimulation (VNS) for treatment-resistant depression: efficacy, side effects, and predictors of outcome. Neuropsychopharmacology 25(5):713–728

Sackeim HA, Brannan SK, John Rush A, George MS, Marangell LB, Allen J (2007a) Durability of antidepressant response to vagus nerve stimulation (VNSTM). Int J Neuropsychopharmacol 10:817–826

Sackeim HA, Prudic J, Fuller RB, Keilp J, Lavori PW, Olfson M (2007b) The cognitive effects of electroconvulsive therapy in community settings. Neuropsychopharmacology 32:244–254

Sackeim HA, Prudic J, Nobler MS, Fitzsimons L, Lisanby SH, Payne N, Berman RM, Brakemeier EL, Perera T, Devanand DP (2008) Effects of pulse width and electrode placement on the efficacy and cognitive effects of electroconvulsive therapy. Brain Stimul 1(2):71–83

Sackeim HA, Dillingham EM, Prudic J, Cooper T, McCall WV, Rosenquist P, Isenberg K, Garcia K, Mulsant BH, Haskett RF (2009) Effect of concomitant pharmacotherapy on electroconvulsive therapy outcomes: short-term efficacy and adverse effects. Arch Gen Psychiatry 66(7):729–737

Schlaepfer TE, Cohen MX, Frick C, Kosel M, Brodesser D, Axmacher N, Joe AY, Kreft M, Lenartz D, Sturm V (2008) Deep brain stimulation to reward circuitry alleviates anhedonia in refractory major depression. Neuropsychopharmacology 33(2):368–377

Schoeyen HK, Kessler U, Andreassen OA, Auestad BH, Bergsholm P, Malt UF, Morken G, Oedegaard KJ, Vaaler A (2015) Treatment-resistant bipolar depression: a randomized controlled trial of electroconvulsive therapy versus algorithm-based pharmacological treatment. Am J Psychiatry 172(1):41–51

Shiozawa P, Fregni F, Bensenor IM, Lotufo PA, Berlim MT, Daskalakis JZ, Cordeiro Q, Brunoni AR (2014) Transcranial direct current stimulation for major depression: an updated systematic review and meta-analysis. Int J Neuropsychopharmacol 17(9):1443–1452

Sienaert P, Vansteelandt K, Demyttenaere K, Peuskens J (2009) Ultra-brief pulse ECT in bipolar and unipolar depressive disorder: differences in speed of response. Bipolar Disord 11 (4):418–424

Tor P-C, Bautovich A, Wang M-J, Martin D, Harvey SB, Loo C (2015) A systematic review and meta-analysis of brief versus ultrabrief right unilateral electroconvulsive therapy for depression. J Clin Psychiatry 76(9):e1092–e1098

Truman CJ, Goldberg JF, Ghaemi SN, Baldassano CF, Wisniewski SR, Dennehy EB, Thase ME, Sachs GS (2007) Self-reported history of manic/hypomanic switch associated with antidepressant use: data from the Systematic Treatment Enhancement Program for Bipolar Disorder (STEP-BD). J Clin Psychiatry 68(10):1472–1479

Wehr TA, Goodwin FK (1979) Rapid cycling in manic-depressives induced by tricyclic antidepressants. Arch Gen Psychiatry 36(5):555–559

Chapter 12
Potential Novel Treatments in Bipolar Depression

Rodrigo Machado-Vieira, Ioline D. Henter, Husseini K. Manji, and Carlos A. Zarate Jr.

Abstract Currently available therapeutics for bipolar disorder (BD)—and bipolar depression in particular—are scarce and often ineffective. This is particularly troubling because the long-term course of bipolar depression comprises recurrent depressive episodes and persistent residual symptoms. Glutamate and its cognate receptors have consistently been implicated in the pathophysiology of mood disorders, as well as in the development of novel therapeutics for these disorders. Since the rapid and robust antidepressant effects of the glutamatergic modulator ketamine were first observed, similar agents have been studied in both major depressive disorder (MDD) and BD. This chapter reviews the clinical and preclinical evidence supporting the use of novel glutamate receptor modulators for the treatment of bipolar depression. We also discuss other promising, non-glutamatergic targets for potential rapid antidepressant effects in mood disorders, including the cholinergic system, the melatonergic system, the glucocorticoid system, the arachidonic acid (AA) cascade, and oxidative stress and bioenergetics. The chapter discusses several specific agents, including *N*-acetyl cysteine (NAC), scopolamine, biperiden, agomelatine, riluzole, ketamine, memantine, creatine, metyrapone, ketoconazole, mifepristone, and celecoxib. Non-pharmacological somatic treatments are not reviewed.

Keywords Antidepressant • Bipolar disorder • Cholinergic • Depression • Dopamine • Glutamate • Melatonin • Oxidative stress • Treatment

R. Machado-Vieira, MD, PhD (✉) • I.D. Henter
Experimental Therapeutics and Pathophysiology Branch, 10 Center Dr, Room 7-5341, Bethesda, MD 20892, USA
e-mail: machadovieirar@mail.nih.gov

H.K. Manji
Janssen Pharmaceuticals of Johnson and Johnson, Janssen Research and Development, 1125 Trenton-Harbourton Road, Titusville, NJ 08560, USA

C.A. Zarate Jr.
Experimental Therapeutics and Pathophysiology Branch, National Institute of Mental Health, National Institutes of Health, Room 7-5342, Bethesda, MD 20892, USA

© Springer International Publishing Switzerland 2016
C.A. Zarate Jr., H.K. Manji (eds.), *Bipolar Depression: Molecular Neurobiology, Clinical Diagnosis, and Pharmacotherapy*, Milestones in Drug Therapy,
DOI 10.1007/978-3-319-31689-5_12

12.1 Introduction

Bipolar disorder (BD) is a chronic, severe, and heterogeneous disorder character-ized by frequent episode relapses and considerable subsyndromal symptoms (Judd et al. 2002). In contrast to the manic phase of the illness where a fairly large variety of effective treatments are available—antiepileptic agents in particular—effective therapeutics are scarce in bipolar depression. A large, 26-week study funded by the National Institute of Mental Health (NIMH) found no benefit to antidepressant use for patients with BD-I or BD-II depression (Sachs et al. 2007). Furthermore, when currently available therapeutics for depression—both for major depressive disorder (MDD) and BD—do work, they are often poorly tolerated or associated with a delayed onset of action of several weeks; this latency period significantly increases risk of suicide and self-harm and is a key public health issue (Machado-Vieira et al. 2009b).

Surprisingly, to date, no agent has been developed specifically to treat BD. Thus, there is a vital need to evaluate the efficacy and safety of novel treatments for bipolar depression. Recent clinical evidence—much of which will be reviewed in this chapter—suggests that rapid antidepressant effects are indeed achievable in humans. This lends additional urgency to developing new treatments for bipolar depression that target alternate neurobiological systems, especially for patient sub-groups that do not respond to currently available pharmacological agents.

This chapter reviews new compounds and targets with antidepressant efficacy in bipolar depression or those that show a signal in preclinical models or preliminary studies. In addition, we highlight some data for targets drawn largely from MDD studies, but that we consider particularly relevant for future drug development in BD. Promising drug targets include the glutamatergic system, the cholinergic system, the melatonergic system, the glucocorticoid system, the arachidonic acid (AA) cascade, and oxidative stress and bioenergetics.

12.2 The Glutamatergic System

Glutamate is the most abundant excitatory neurotransmitter in the mammalian brain and activates different ionotropic and metabotropic receptors pre- and postsynap-tically. Glutamate controls neurotransmission across excitatory synapses and is directly involved in several brain functions such as synaptic plasticity, learning, mood regulation, and memory (Machado-Vieira et al. 2009a, 2012). Altered glutamatergic regulation is also directly involved in the altered neuroplasticity and cellular resilience described in individuals with BD. Thus, diverse glutamate-modulating agents have been tested in "proof-of-concept" studies in subjects with bipolar depression (Iadarola et al. 2015).

Glutamate receptor subtypes include the ionotropic (N-methyl-D-aspartate (NMDA) and α-amino-3-hydroxy-5-methyl-4-isoxazolepropionic acid (AMPA)) and metabotropic glutamate receptors (mGluRs); both types have a wide range of effects and downstream targets. The novel experimental agents currently being explored have targeted both receptor subtypes as well as glia (Machado-Vieira et al. 2012).

12.2.1 Ionotropic Glutamate Receptors

12.2.1.1 NMDA Receptors

NMDA receptors are heteromeric complexes whose subunits form three different subtypes: GluN1, GluN2, and GluN3. Eight GluN1 subunits, four GluN2 subunits (A, B, C, and D), and two GluN3 subunits (A and B) exist.

Preclinical evidence strongly suggests that NMDA receptors are involved in the pathophysiology of mood disorders and the mechanism of action of antidepressants (Skolnick 2002). Because chronic treatment with various classes of antidepressant agents affects NMDA receptor function, it has been hypothesized that the NMDA receptor could even represent a convergent mechanistic target for the antidepressant action of conventional antidepressants and mood stabilizers as well as novel experimental therapeutics (Skolnick 1999). In preclinical models of depression, NMDA receptor antagonists have been shown to exert broad antidepressant-like effects (Machado-Vieira et al. 2009b, 2012). Clinical evidence for the utility of NMDA antagonists is described below.

12.2.1.2 AMPA Receptors

The ionotropic AMPA receptors play a key role in memory and learning and mediate the fast component of excitatory neurotransmission. Positive modulators of AMPA receptors, also known as AMPAkines, are allosteric effectors of the receptors that have been investigated in recent years because of their potential use as a treatment for various diseases. AMPAkines are involved in activity-dependent regulation of synaptic strength and behavioral plasticity and have been found to decrease receptor desensitization or deactivation rates (Bleakman et al. 2007; Sanacora et al. 2008). Chronic treatment with traditional antidepressants was found to elevate the expression of hippocampal AMPA receptors (Sanacora et al. 2008). In preclinical studies, the rapid antidepressant effects of the NMDA receptor antagonist ketamine (discussed below) appeared to involve AMPA receptor activation (Maeng et al. 2008).

In preclinical models of depression, AMPA modulators had broad antidepressant-like effects (Du et al. 2007); for instance, ampalex induced antidepressant-like effects in the first few days of treatment (Knapp et al. 2002). Similarly, the competitive AMPA receptor antagonist NS1209 had rapid and consistent anticonvulsant effect compared to diazepam in preclinical and clinical studies and was well-tolerated (Pitkanen et al. 2007; Rogawski 2006).

Because preclinical studies found that AMPA agonists and positive allosteric modulators have antidepressant-like effects (Bleakman et al. 2007; O'Neill and Witkin 2007), several AMPA agonists and positive allosteric modulators were developed to treat MDD. These include ORG-26576, an AMPA receptor positive allosteric modulator; the AMPA agonist farampator (CX-691/ORG 2448); older AMPAkines such as levetiracetam; and more potent compounds such as coluracetam (BCI-540). However, the results of these initial clinical studies were disappointing, and the agents were not subsequently explored in the treatment of BD. As a result, they will not be discussed further in this chapter.

12.2.2   Metabotropic Glutamate Receptors (mGluRs)

mGluRs are G protein-coupled receptors that modulate glutamate levels. These receptors are located pre- and postsynaptically at glutamatergic synapses, and their activation lowers glutamate release and excitotoxic damage (Machado-Vieira et al. 2009a). Eight different classes of mGluRs have been identified and subcategorized into three subtypes (Kim et al. 2008): the Group 1 metabotropic receptors—mGluR1 and mGluR5—are generally stimulatory and located post-synaptically; the Group 2 metabotropic receptors—mGluR2 and mGluR3—as well as the Group 3 metabotropic receptors—mGluR4-8—share major sequence homology (~70 %). They generally inhibit glutamatergic neurotransmission and target mTOR signaling (Conn and Pin 1997; Dwyer et al. 2012; Koike et al. 2011). These Group 2 and 3 receptors are mainly located presynaptically at glutamatergic synapses, and their activation lowers glutamate release and blocks glutamate-induced excitotoxicity (Machado-Vieira et al. 2009a). Several mGluR2/3 antagonists and negative allosteric modulators have shown antidepressant-like efficacy in preclinical models of depression (reviewed in Chaki et al. (2013)), and these effects appear to be associated with enhanced presynaptic glutamate release (Dwyer et al. 2012; Karasawa et al. 2005). The mGluR5s are also physically and physiologically interconnected with AMPA and NMDA receptors and regulate synaptic plasticity (Pilc et al. 2013). A number of mGluR5 antagonists were found to have antidepressant-like effects in behavioral models (Chaki et al. 2013). The clinical evidence is reviewed in Sect. 12.2.5.

12.2.3 Nonselective/Noncompetitive NMDA Receptor Antagonists

12.2.3.1 Ketamine

Ketamine is a glutamatergic modulator that releases presynaptic glutamate after disinhibiting gamma-aminobutyric acid (GABA)-ergic inputs (Moghaddam et al. 1997). Ketamine has demonstrated significant antidepressant effects in diverse preclinical models (Maeng et al. 2008; Aguado et al. 1994; Mickley et al. 1998; Silvestre et al. 1997; Garcia et al. 2008).

Single-Dose Ketamine Studies

An initial pilot study showed that individuals with treatment-resistant MDD improved within 72 hours after ketamine infusion (Berman et al. 2000). In a subsequent placebo-controlled, double-blind, crossover study conducted in subjects with treatment-resistant MDD, a single ketamine infusion (0.5 mg/kg for 40 min) had rapid (two hours) antidepressant effects (Zarate et al. 2006); more than 70 % of patients responded to ketamine at 24 hours post-infusion, and 35 % maintained a relative sustained response one week post-infusion. These antidepressant effects were replicated in another single-blind, non-counter-balanced design study of 10 patients with treatment-resistant depression (Valentine et al. 2011). A more recent study used the short-acting benzodiazepine midazolam as a psychoactive placebo in order to mimic ketamine's sedative and anxiolytic effects (Murrough et al. 2013a). Consistent with the aforementioned MDD studies, the investigators found that at 24 hours post-infusion, subanesthetic-dose ketamine infusion (0.5 mg/kg × 40 min) was more effective than placebo (response rates were 64 % and 28 %, respectively, in subjects randomized to ketamine and midazolam).

Building on this work, researchers investigated the possible rapid antidepressant effects of ketamine in patients with treatment-resistant bipolar depression; these placebo-controlled studies used ketamine adjunctively with lithium or valproate monotherapy. In the first study, 18 subjects with treatment-resistant bipolar depression maintained on therapeutic levels of mood stabilizers received a single subanesthetic dose of ketamine. A rapid (within 40 minutes) and relatively sustained (up to three days) antidepressant response was observed; maximal antidepressant response occurred two days after infusion (Diazgranados et al. 2010a). A confirmatory study using an identical design ($n = 15$) obtained similar results (Zarate et al. 2012). A recent, larger, single open-label study of ketamine used adjunctively with mood stabilizers in patients with bipolar depression ($n = 42$) confirmed these previous findings (Permoda-Osip et al. 2014). Notably, ketamine appears safe for use in individuals with bipolar depression and has very low risk (similar to placebo) of inducing hypo/mania (Niciu et al. 2013).

It should be noted that the APA Council of Research Task Force on Novel Biomarkers and Treatments concluded in their meta-analysis of ketamine studies (seven trials encompassing 147 participants who received ketamine) that this agent produced a rapid, yet transient, antidepressant effect accompanied by brief psychotomimetic and dissociative effects; odds ratios for response and transient remission of symptoms at 24 hours were 9.87 (4.37–22.29) and 14.47 (2.67–78.49), respectively (Newport et al. 2015).

Repeated-Dose Ketamine Studies

The rapid antidepressant effect size of ketamine is large in patients with mood disorders; however, these effects are also transient in most patients. As a result, repeated dosing strategies may offer more sustained antidepressant benefits. Only one repeated-dose study has been conducted in patients with bipolar depression. In that study, 28 medicated treatment-resistant patients with either MDD ($n = 22$) or bipolar depression ($n = 6$) received a standard dose (0.5 mg/kg × 40 min) of ketamine weekly or bi-weekly over three weeks for a total of three or six infusions, respectively (Diamond et al. 2014). Of the eight patients, 29 % achieved response; no cognitive impairment was noted over time. One BD patient experienced rapid cycling (hypomanic switches) after three infusions. Three other patients had mood instability over the course of infusions with worsening mood symptoms and increased suicidal ideation.

In MDD, a preliminary study investigated the safety, efficacy, and tolerability of repeated-dose ketamine in treatment-resistant depression (aan het Rot et al. 2010). Ten unmedicated patients with treatment-resistant depression were given six open-label subanesthetic dose (0.5 mg/kg × 40 min) ketamine infusions over a 12-day period. Antidepressant effects and a mild, transient side effect profile were observed. In another study of 24 medication-free patients with treatment-resistant depression (including the initial 10 in the study described above (aan het Rot et al. 2010)), the investigators found an antidepressant response rate of 70.8 % in the patients who had received six infusions over a 12-day period (Murrough et al. 2013b). These patients were then followed naturalistically over the next 83 days (which allowed for traditional antidepressant treatment); mean time to relapse was 18 days, but about 30 % of responders maintained antidepressant response until the end of the naturalistic observation.

Another study of repeated subanesthetic-dose (0.3 mg/kg over 100 min) ketamine examined the effects of an open-label infusion in 10 patients with treatment-resistant depression. Patients received infusions twice per week for two weeks until they had either received four total infusions or their symptoms remitted (Rasmussen et al. 2013). By the end of the study, six subjects had received the maximum number of doses (four), and three and five patients were responders and remitters, respectively; there were two nonresponders. When the patients were monitored for four additional weeks, 50 % of the responders achieved remission and another two patients retained their initial symptom remission.

Recently, Singh and colleagues showed that serial intravenous ketamine (two or three times per week over a four-week period) induced a good overall response compared to placebo in 67 patients with treatment-resistant MDD (Singh et al. 2014). Interestingly, no difference was found when comparing response between those who received infusions two or three times a week. No systematic repeat dose studies have been conducted with ketamine in bipolar depression.

Finally, a case report described a 44-year-old patient with treatment-resistant MDD who received more than 40 ketamine infusions over several months. An initial improvement of depressive symptoms was observed with no cognitive deficits; however, improvement waned over time, suggesting tachyphylaxis (Blier et al. 2012).

Ketamine's Anti-suicidal Effects

In addition to its antidepressant effects, ketamine also has rapid-onset anti-suicidal properties. In one study, measures of suicidal ideation decreased within 40 minutes of a single open-label ketamine infusion (0.5 mg/kg) in 33 patients with treatment-resistant MDD; the effect was maintained for up to four hours post-infusion (DiazGranados et al. 2010b). Other studies observed that ketamine infusion reduced both explicit suicidal ideation and implicit suicidal thinking in patients with treatment-resistant MDD (Price et al. 2014). One study that evaluated 14 acutely suicidal patients in a psychiatric emergency room found that open-label ketamine (0.2 mg/kg intravenous ketamine, administered over one to two minutes) had rapid anti-suicidal and antidepressant efficacy (Larkin and Beautrais 2011).

With regard to bipolar depression in particular, a recent study of 108 patients with treatment-resistant MDD or bipolar depression who received a single, standard-dose infusion of ketamine (0.5 mg/kg over 40 min) noted rapid improvements in suicidal ideation compared to placebo (Ballard et al. 2014).

12.2.3.2 Memantine

Memantine is a low-trapping, noncompetitive NMDA antagonist with antidepressant-like effects in preclinical models (Rogoz et al. 2002). In humans, a three-year, mirror-image, naturalistic study that added memantine to the treatment of 30 treatment-resistant patients with BD found that this agent substantially improved course of illness by preventing or ameliorating both depressive and manic episodes (Serra et al. 2015). In addition, an eight-week, placebo-controlled, randomized investigation of individuals with bipolar depression who had previously not responded to lamotrigine found that escalating-dose memantine was superior to placebo at four weeks, although efficacy was not maintained at end point (eight weeks) (Anand et al. 2012). A case series of two patients with bipolar depression concomitantly treated with mood stabilizers found that add-on memantine (10–20 mg/day) improved depressive symptoms and cognitive performance (Teng and Demetrio 2006).

In MDD, a case report found that treatment with memantine subsequent to repeated-dose ketamine had antidepressant effects; although this patient was prescribed seven psychotropic drugs, remission was sustained for 13 weeks (Kollmar et al. 2008). However, a small ($n = 31$), double-blind, placebo-controlled trial using memantine as an augmentation strategy in patients with MDD found that it had no antidepressant effects (Smith et al. 2013).

12.2.3.3 Dextromethorphan

The cough suppressant dextromethorphan—a nonselective, noncompetitive NMDA receptor antagonist—is a derivative of morphine with sedative and dissociative properties. Dextromethorphan had antidepressant-like effects in animal models, similar to those observed with both conventional and rapid-acting antidepressants (Nguyen and Matsumoto 2015). In BD, dextromethorphan was studied in a placebo-controlled, randomized trial as add-on therapy to valproic acid; no significant group differences were seen between groups (as assessed by mean Young Mania Rating Scale (YMRS) scores and Hamilton Rating Scale for Depression (HAM-D) scores) (Lee et al. 2012). A retrospective chart review of BD-II or BD not otherwise specified (BD-NOS) patients ($n = 22$) found that adding 20 mg dextromethorphan and 10 mg quinidine (a cytochrome 2D6 inhibitor) once or twice daily to their current medications significantly improved Clinical Global Impression (CGI) scale scores (Kelly and Lieberman 2014). This dextromethorphan–quinidine combination—Nuedexta—is now being studied for treatment-resistant depression (NCT01882829). This agent is presently approved for the treatment of pseudobulbar affect, and a case report found that Nuedexta had antidepressant effects in a depressed patient with emotional lability (Messias and Everett 2012).

12.2.4 Glutamatergic Modulators that Act Indirectly to Alter Glutamate Release

12.2.4.1 Riluzole

Riluzole, which has blood–brain penetrant glutamatergic and neuroprotective effects, is FDA-approved for amyotrophic lateral sclerosis. Riluzole increases glutamate reuptake and the synthesis of neurotrophic factors (Mizuta et al. 2001; Frizzo et al. 2004). In preclinical models, pretreatment with 10 mg/kg riluzole (but not 3 mg/kg) moderately reduced amphetamine-induced hyperlocomotion (Lourenco Da Silva et al. 2003) in a preclinical model of mania. It also reversed or blocked depressive phenotype-induced models of depression (Banasr et al. 2010).

In initial open-label clinical trials, riluzole exerted antidepressant effects and was well-tolerated. In patients with bipolar depression, riluzole (100–200 mg/day for six weeks) was evaluated in 14 subjects as add-on therapy to lithium (Zarate et al. 2005); there was a significant treatment effect, and no switch into hypo/mania was observed. In patients with MDD, the 13 patients who completed the study (68 %) had all improved at week six (Zarate et al. 2004). Another study of patients with MDD found that riluzole (50 mg/twice daily) led to a 36 % decrease in HAM-D scores after one week of treatment (Sanacora et al. 2007). In contrast, a recent four-week, double-blind, placebo-controlled study of riluzole (100–200 mg/day) in 42 subjects with treatment-resistant MDD who had initially received a single ketamine infusion found that riluzole had no antidepressant effects compared to placebo (Ibrahim et al. 2012). Another study found a similar lack of efficacy when riluzole was used as an augmentation strategy (Mathew et al. 2010), suggesting that additional controlled trials are needed to clarify its potential efficacy. In addition, controlled studies are needed in BD to confirm the results of these earlier studies.

12.2.5 mGluR Antagonists

Presynaptic mGluR2 agonists seem to work by reducing excessive glutamate release. In contrast, mGluR2/3 antagonists appear to enhance synaptic glutamate levels, thereby boosting extracellular monoamine levels as well as AMPA receptor transmission and firing rates. In rodent models of depression (reviewed in Chaki et al. (2013)), several mGluR2/3 antagonists, as well as the negative allosteric modulator RO4995819, were found to have antidepressant-like effects. The tolerability and safety of mGluR2/3 modulators have also been studied in healthy volunteers (NCT01547703 and NCT01546051). While animal studies suggest that this class of glutamate modulators might be clinically useful for treating bipolar depression, mGluR2/3 modulators have yet to show convincing results in clinical trials for mood disorders. For instance, despite positive preliminary findings, the mGluR2 negative allosteric modulator RG1578 was tested clinically and final results were disappointing (Dale et al. 2015).

The mGluR5s are expressed pre- and postsynaptically and are involved in AMPA receptor internalization, which is key to modulating synaptic plasticity (Pilc et al. 2013). mGluR5s are also physiologically and physically interconnected with NMDA receptors. A variety of preclinical studies noted that mGluR5 antagonists had antidepressant-like effects (Chaki et al. 2013), and several have been tested in clinical trials for treatment-resistant MDD, including AZD2066 and the mGluR5 negative allosteric modulator basimglurant (RO4917523) (Quiroz et al. 2014). Basimglurant showed mostly negative results in two clinical trials (NCT00809562, NCT01437657). Specifically, a large, nine-week, placebo-controlled, double-blind study of 333 individuals with treatment-resistant MDD found that basimglurant (0.5 mg or 1.5/day adjunctive to ongoing treatment with selective serotonin reuptake inhibitors (SSRIs) or serotonin–noradrenaline reuptake inhibitors (SNRIs)) was no more effective than placebo, though significant antidepressant effects were observed in a secondary analysis (Quiroz et al. 2014); greater improvements were noted with the higher dose of basimglurant (1.5 mg).

In addition, a phase IIa, multicenter, randomized, double-blind, double-dummy, active- and placebo-controlled, parallel group study to assess the effectiveness and safety of AZD2066 found that six weeks of treatment (12–18 mg/day) with this agent was no more effective in patients with MDD than either placebo or the SNRI duloxetine (NCT01145755). Studies have also evaluated the potent and selective mGluR5 antagonist fenobam, but these were discontinued because of psychostimulant effects (Palucha and Pilc 2007).

In addition, mGluR7 knockout mice displayed an antidepressant phenotype in the forced swim and tail suspension tests (Cryan et al. 2003). Group 3 GluR6 knockout mice displayed altered motor activity when exposed to amphetamine, with elevated risk-taking and aggressive behavior, which was reversed by lithium (Shaltiel et al. 2008). Although the preclinical data suggest that this class of glutamate modulators might display clinical utility for both manic and depressive episodes, to date no Group 3 agonist has been clinically evaluated in the treatment of BD.

12.3   The Cholinergic System

Over 40 years ago, Janowsky introduced the adrenergic–cholinergic balance hypothesis of BD; depression was postulated to be associated with high cholinergic compared to adrenergic activity, while mania was proposed to be associated with hypocholinergia and increased adrenergic signaling (Janowsky et al. 1972).

Cholinergic receptors are divided into the metabotropic muscarinic (M_1 to M_5) and ionotropic nicotinic (diverse α and β) acetylcholine receptors. In animal models, rodents bred for enhanced sensitivity of muscarinic receptors showed a depressive-like phenotype characterized by the presence of lethargy, anhedonia, and behavioral despair (Overstreet 1993). Elevated muscarinic sensitivity has been associated with depressive-like behaviors that were reversed with the use of antimuscarinic agents (Betin et al. 1982; Gao and Jacobson 2013). In addition, neuroplasticity targets such as mTORC1, AMPA, and cyclic adenosine monophosphate response element-binding protein (CREB) are thought to underlie the rapid antidepressant effects of cholinergic agents (e.g., scopolamine, reviewed below) in preclinical studies (Voleti et al. 2013; Hasselmann 2014) and might represent potential therapeutic targets for new drug development in bipolar depression.

The cholinergic system appears to be dysregulated in BD; indeed, one study linked 19 cholinergic genes and BD (Shi et al. 2007). In addition, a positron emission tomography (PET) imaging study found reduced M_2 receptor binding in the anterior cingulate cortex of BD subjects (Cannon et al. 2006), and lower M_2 and

M_3 receptor binding was observed in the frontal cortex of subjects with BD in a postmortem study (Gibbons et al. 2009). Cholinergic hyperactivity has also been linked to worsening depressive symptoms in subjects with MDD. Furthermore, increased sensitivity to cholinergic activity (as assessed via pupillary and neuroendocrine response) was observed in individuals with depression (Dilsaver 1986), and this was found to be blunted during manic episodes (Sokolski and DeMet 2000). It should also be noted that the mood stabilizers lithium and valproate normalized pupillary response in conjunction with their anti-manic efficacy (Sokolski and DeMet 2000). Finally, a single-photon emission computed tomography (SPECT) study found decreased β_2 nicotinic receptor availability in individuals with bipolar depression compared to both euthymic and control subjects (Hannestad et al. 2013).

Anticholinergic drugs have been tested for several decades in mood disorders research. Acetylcholinesterase inhibitors, which increase central acetylcholine tone, have consistently been found to induce symptoms of depression in humans (van Enkhuizen et al. 2015). Controlled pilot studies with physostigmine, an acetylcholinesterase inhibitor, demonstrated that single or multiple injections of this agent rapidly but transiently decreased manic symptoms (Khouzam and Kissmeyer 1996). Similar studies with donepezil, a cholinesterase inhibitor, in manic episodes obtained mixed results (Eden Evins et al. 2006). One early study found that the anticholinergic agent biperiden had antidepressant effects in severely depressed inpatients ($n = 10$) (Kasper et al. 1981).

A number of randomized, double-blind, placebo-controlled studies were conducted with intravenous doses of the anticholinergic agent scopolamine as add-on or monotherapy in subjects with either MDD or bipolar depression (Drevets et al. 2013; Jaffe et al. 2013; Khajavi et al. 2012), and antidepressant effects were typically observed in the first week of treatment. Scopolamine is a muscarinic receptor antagonist that acts at all muscarinic receptor subtypes (M_1–M_5) with comparatively equivalent potency. A crossover, double-blind, placebo-controlled pilot study of patients with either MDD ($n = 9$) or bipolar depression ($n = 9$) found that scopolamine exerted rapid and robust antidepressant and anxiolytic effects, particularly at doses of 4 µg/kg (compared to 2 or 3 µg/kg) (Furey and Drevets 2006). The investigators found that scopolamine's antidepressant effects persisted for two weeks and, furthermore, repeated dosing provided extended response and additional benefits. No significant increase in manic symptoms was noted.

Interestingly, scopolamine's antidepressant actions also seem to involve glutamatergic NMDA receptor antagonism; for instance, scopolamine-induced mAChR inhibition was shown to limit the mRNA expression of NMDA receptor subunits (NR1A and NR2A) (Liu et al. 2004). These findings support the presence of a hypercholinergic state in bipolar depression. It should be noted that while scopolamine seems to be a promising target for drug development, few studies with anticholinergic drugs have been performed; thus, limited information exists with regard to dose-response curves, pharmacodynamics, safety, and treatment duration. Furthermore, this agent is associated with several limitations, including anticholinergic side effects and risk of psychosis at higher doses (Khajavi et al. 2012); both may limit its broad clinical use.

12.4   The Melatonergic System

The neurohormone melatonin regulates sleep-related and circadian-related responses. It also has immunomodulatory, antioxidant, neuroprotective, and other chronobiological effects (Cardinali et al. 2011; Pacchierotti et al. 2001) and seems to directly target synaptic plasticity by increasing hippocampal long-term potentiation (Etain et al. 2012).

The two melatonin receptors (MT_1 and MT_2) are high affinity G_i/G_0 protein-coupled receptors with elevated expression in the brain (MT_1 in the pituitary gland and SCN and MT_2 in the retina). During both depressive and manic episodes, individuals with BD display circadian rhythm abnormalities and changes in sleep patterns. These include sleep–wake irregularities (Harvey 2008), circadian preference for evening (Ahn et al. 2008), and abnormal actimetric parameters (Millar et al. 2004). Supersensitivity to melatonin suppression by light has also been demonstrated in individuals with BD, monozygotic twins discordant for BD, and the non-affected offspring of BD probands; however, a similar study evaluating euthymic BD subjects found no such association (Hallam et al. 2005c). Individuals with BD also appear to display variations in the melatonin biosynthesis pathway; for example, BD subjects displayed altered melatonin suppression in response to light (Nathan et al. 1999; Nurnberger et al. 2000), which suggests a genetic trait marker in BD (Hallam et al. 2005c, 2006). A polymorphism in *GPR50* (H9, melatonin-related receptor) was also linked to elevated risk of BD, but this finding was not subsequently replicated (Alaerts et al. 2006; Thomson et al. 2005).

No controlled studies of exogenous melatonin have been published in BD. However, the mood stabilizers lithium and valproate lowered melatonin light sensitivity in healthy volunteers (Hallam et al. 2005a, b), with mixed results in BD subjects (Bersani and Garavani 2000; Leibenluft et al. 1997). Interestingly, adjunctive use of melatonin agonists to treat sleep disorders improved the metabolic profile of BD patients (Geoffrey et al. 2015; Romo-Nava et al. 2014).

As regards specific therapeutics, adjunctive use of ramelteon—a selective MT_1 and MT_2 receptor agonist—effectively maintained mood stabilization in euthymic BD subjects who were also experiencing sleep disturbances (8 mg/day, 23 weeks double-blind, $n = 83$) (Norris et al. 2013). The group taking ramelteon had half the relapse rate of those receiving placebo (Norris et al. 2013).

Agomelatine is a melatonin MT_1 and MT_2 receptor agonist and a 5-HT(2C) receptor antagonist (Bourin and Prica 2009) that was found to induce significant antidepressant-like effects in preclinical models of depression (Bertaina-Anglade et al. 2006; Millan et al. 2005; Papp et al. 2003). In bipolar depression, a six-week trial of agomelatine (25 mg/day, $n = 21$) observed an 81 % response rate at study endpoint; notably, 47 % of patients responded during the first week of treatment (Calabrese et al. 2007) and there were no dropouts secondary to adverse events. In three large, multicenter, controlled clinical trials in individuals with MDD, agomelatine was well-tolerated and more effective than placebo (Kennedy and Emsley 2006; Loo et al. 2002; Montgomery and Kasper 2007). In a 24-week, randomized, double-blind, placebo-controlled trial in individuals with MDD, agomelatine was more effective than placebo. Furthermore, a significantly lower relapse rate was observed with agomelatine ($n = 133$) than placebo ($n = 174$) during a six-month extension phase (Goodwin et al. 2009). Finally, a recent meta-analysis of 20 trials involving 7460 subjects found that agomelatine was more effective than placebo—and as effective as other antidepressants—in significantly reducing depressive symptoms (Taylor et al. 2014). Further studies are required to confirm these promising preliminary findings.

12.5 The Glucocorticoid System

Dysfunction of the hypothalamic-pituitary adrenal (HPA) axis—the major "stress pathway"—has been associated with BD. Specifically, hypercortisolemia has been implicated in the etiopathogenesis of both depressive symptoms and the neurocognitive deficits observed in BD. Thus, attempts to normalize the effects of cortisol have been the focus of recent research. Interestingly, glucocorticoids acutely increase glutamate release in the prefrontal cortex and amygdala (Musazzi et al. 2013) and also limit hippocampal brain derived neurotrophic factor (BDNF) expression, which may possibly explain the deleterious effects on cellular resilience associated with the use of chronic corticosteroids. Several antiglucocorticoid agents have been studied in the treatment of mood disorders, including corticosteroid receptor antagonists (mifepristone), cortisol synthesis inhibitors (aminoglutethimide, ketoconazole, and metyrapone), dehydroepiandrosterone (DHEA), and pregnenolone (reviewed in Quiroz et al. (2004)). Only a handful of these compounds have been tested in bipolar depression.

Mifepristone (RU-486), a synthetic, nonselective glucocorticoid receptor antagonist was reported to have antidepressant and antipsychotic effects in inpatients with psychotic depression who had HAM-D21 scores of 18 or greater and who received 600 or 1200 mg/day RU-486; significant decreases in Brief Psychiatric Rating Scale (BPRS) and HAM-D21 scores were observed (Belanoff et al. 2002). In another study of drug-free subjects with psychotic depression, seven days of treatment with mifepristone followed by treatment as usual was both effective and well tolerated (DeBattista et al. 2006). Nevertheless, at least one small study found that RU-486 had no antidepressant efficacy in psychotic depression (Carroll and Rubin 2008). In individuals with treatment-resistant bipolar depression, a six-week pilot study found that mifepristone (600 mg/day) improved depressive symptoms and cognition over placebo (Young et al. 2004). Notably, cognitive improvement was inversely associated with basal cortisol levels, supporting the presence of an anti-glucocorticoid effect. Another placebo-controlled, randomized, double-blind trial of 60 patients with bipolar depression evaluated 600 mg/day of mifepristone for one week as an adjunctive treatment. Mifepristone treatment was associated with a time-limited increase in cortisol awakening response and with sustained improvement in spatial working memory (the primary outcome measure).

These effects were evident seven weeks after cessation of treatment, with no improvement in depressive symptoms (Watson et al. 2012). Taken together, the evidence suggests that RU-486 may have predominantly short-term benefits during acute depressive episodes. It should be noted, however, that long-term treatment could be associated with significant side effects, and these have yet to be properly assessed (Rothschild 2003; Grunberg et al. 2006).

Pregnenolone is an endogenous steroid hormone that is the precursor of progestogens, mineralocorticoids, glucocorticoids, androgens, and estrogens. It is also biologically active in its own right, acting as a neurosteroid. In a recent, randomized, double-blind, 12-week, placebo-controlled trial of pregnenolone for bipolar depression ($n = 80$), depression remission rates were higher in the pregnenolone group (61 %) than in the placebo group (37 %) as assessed by the Inventory of Depressive Symptomatology Self-Report, but not by the HAM-D (Brown et al. 2014).

Glucocorticoid synthesis inhibitors, which include ketoconazole and metyrapone, are thought to potentiate the efficacy of antidepressants. In both clinical and preclinical studies, these agents have been found to have antidepressant effects (Quiroz et al. 2004). These compounds lower cortisol levels and reduce tissue-specific gluconeogenesis and fatty acid metabolism. However, this effect may decrease feedback to the HPA, which may then upregulate cortisol after long-term treatment. In a controlled, randomized, double-blind trial in MDD patients, add-on metyrapone therapy was superior to placebo and accelerated the onset of antidepressant action (Jahn et al. 2004). A large, multicenter, randomized, placebo-controlled study evaluating metyrapone augmentation in treatment-resistant depression was recently completed, but results are not yet available (NCT01375920).

Add-on treatment with ketoconazole (up to 800 mg/day) was investigated in six patients with treatment-resistant BD (Brown et al. 2001). In addition to decreasing cortisol levels, ketoconazole also significantly improved depressive symptoms in three patients who received at least 400 mg/day, with no induction of manic symptoms. Another study found that ketoconazole improved depression ratings in subjects with hypercortisolemia compared to placebo; interestingly, this effect was not observed in nonhypercortisolemic patients (Wolkowitz et al. 1999a). One study analyzed results from five trials conducted with ketoconazole in nonpsychotic depression (either BD or MDD) and found a significant difference in favor of treatment (summarized in Gallagher et al. (2008)). It should be noted, however, that the relative risk for drug interactions associated with this agent may limit its chronic use in mood disorders.

DHEA has also been investigated as a possible antidepressant. A randomized, controlled trial of 22 subjects with MDD found that DHEA (up to 90 mg/day for six weeks) had significant antidepressant effects (Wolkowitz et al. 1999b). These initial findings were confirmed by another six-week trial of DHEA as monotherapy that found significant improvement in depression rating scale scores compared to placebo (Schmidt et al. 2005).

In the search for effective novel antidepressants, trials have also examined neuropeptides such as neurokinin 1 (NK1), vasopressin, orexin antagonists, and corticotropin releasing factor (CRF). Much of this work has been conducted in MDD trials. Despite promising preclinical evidence, particularly for CRF-1 antagonists, clinical results have been mixed and largely disappointing (Dale et al. 2015).

12.6   Arachidonic Acid (AA) Cascade and Inflammation Pathway

The AA signaling pathway has been implicated in the pathophysiology and therapeutics of mood disorders. This pathway modulates several brain second messenger pathways associated with the release of AA and cyclooxygenase (COX)-mediated generation of eicosanoid metabolites such as prostaglandins and thromboxanes. Chronic administration of mood stabilizers in preclinical models limited the production of prostaglandin E(2) and turnover in brain phospholipids, as well as blocked the expression of AA cascade enzymes, including cytosolic phospholipase A(2), COX-2, and/or acyl-CoA synthetase (Rapoport et al. 2009).

A six-week, placebo-controlled, double-blind study of the COX-2 inhibitor celecoxib in bipolar depression (400 mg/day adjunctive to mood stabilizers) found that this agent demonstrated superior antidepressant effects only during the first week of treatment (Nery et al. 2008). A placebo-controlled study of individuals with MDD found that celecoxib (400 mg/day as add-on therapy to fluoxetine, $n = 37$) significantly decreased depression scores compared to placebo (Akhondzadeh et al. 2009). Another six-week study of individuals with MDD found that celecoxib (400 mg/day adjunctive to reboxetine) was more effective than placebo (Muller et al. 2006). Finally, an eight-week study of first-episode women with MDD also found that celecoxib (200 mg/day as add-on therapy to sertraline, $n = 30$) led to a greater antidepressant response than placebo (Majd et al. 2015). Nevertheless, it should be noted that the ability of celecoxib to enter the blood–brain barrier remains uncertain. In addition, selective COX-2 inhibitors have been associated with higher risk of adverse cardiovascular effects, which may limit their long-term use (Velentgas et al. 2006).

Interestingly, a 12-week, placebo-controlled, randomized study of 60 patients with MDD found that, compared to placebo, the tumor necrosis factor (TNF) antagonist infliximab (5 mg/kg infusion) had no antidepressant effects at baseline, at two weeks, or at four weeks (Raison et al. 2013). However, post hoc analyses found that baseline concentrations of TNF and its soluble receptors were significantly higher in infliximab-treated responders than nonresponders ($p < .05$). In addition, infliximab-treated responders had significantly greater decreases in high-sensitivity C-reactive protein (hs-CRP) from baseline to week 12 compared with placebo-treated responders ($p < .01$). Dropouts and adverse events were limited and did not differ between groups.

Overall, the AA cascade may represent a target for specific subtypes of depression with an as yet unknown and unique biosignature. Future studies using new pharmacological interventions targeting this system may clarify its therapeutic relevance for BD. However, further studies are required given that these findings have not been replicated, which may limit evaluations of potential utility for this class of agents.

12.7 Oxidative Stress and Bioenergetics

12.7.1 N-Acetyl Cysteine (NAC)

Levels of glutathione, the most abundant antioxidant protein in the brain, are altered in individuals with BD (Andreazza et al. 2007; Kuloglu et al. 2002). N-acetylcysteine (NAC) is a precursor of glutathione and increases its levels. NAC enhances glial cystine uptake and increases glial cystine levels via a cystine-glutamate antiporter that induces glutamate release into the extracellular space.

Several studies have investigated NAC for the treatment of both acute episodes and maintenance in BD. The first randomized, double-blind, placebo-controlled investigation in BD evaluating NAC ($n = 75$; 1 g/twice daily added on to treatment) found that this agent was more effective than placebo after eight weeks. NAC improved measures of depression, quality of life, and functionality (Berk et al. 2008); however, results did fluctuate over time and antidepressant efficacy did not separate from placebo at the post-discontinuation visit (after 24 weeks). Another study of 14 individuals with BD-II found that NAC improved both depressive and manic symptoms more consistently than placebo (Magalhaes et al. 2011). Finally, a large, open-label, eight-week trial of 149 patients with bipolar depression showed that adjunctive use of NAC significantly improved depression rating scale scores, overall functioning, and quality of life (Berk et al. 2011). In MDD, a large, 12-week, controlled, randomized, add-on trial ($n = 252$) found that patients receiving NAC showed antidepressant improvements similar to the placebo group (Berk et al. 2014). However, at week 12, scores on the Longitudinal Interval Follow-Up Evaluation-Range of Impaired Functioning Tool (LIFE-RIFT) differed from placebo. Taken together, the evidence suggests that NAC might improve depressive symptoms during mood episodes, but studies evaluating the frequency of cycling or mood stability are required. It should also be noted that NAC has been tested in other psychiatric disorders such as schizophrenia and obsessive compulsive disorder (OCD), suggesting it may have nonspecific therapeutic effects.

12.7.2   Creatine

Creatine is a nonessential dietary element and precursor of phosphocreatine (PCr). It plays a key role in brain energy homeostasis. PCr is converted to creatine when energy is required, and this increases intracellular concentrations of adenosine triphosphate (ATP) (Sauter and Rudin 1993). Hippocampal changes in brain creatine kinase, which phosphorylates creatine, were demonstrated in preclinical models of BD and in postmortem BD studies (Streck et al. 2008; Segal et al. 2007). Lower hippocampal and prefrontal cortex creatine kinase mRNA expression has also been described in postmortem BD (MacDonald et al. 2006). Interestingly, ketamine and imipramine were both found to increase cerebellar and prefrontal cortex creatine kinase activity in rats (Assis et al. 2009).

Because it alters brain high-energy phosphate metabolism, creatine supplementation may be a novel approach for treating BD. Adjunctive treatment with 3–5 mg/day of creatine monohydrate in patients with either MDD or bipolar depression significantly improved depressive symptoms (Roitman et al. 2007); however, two patients with bipolar depression experienced a transient switch to hypomania. Another study found that creatine improved depressive symptoms in comorbid depression and fibromyalgia (Amital et al. 2006).

12.7.3   Cytidine

Cytidine is a pyrimidine component of RNA that regulates dysfunctional neuronal-glial glutamate cycling and affects mitochondrial function, cerebral phospholipid metabolism, and catecholamine synthesis. This agent was tested in a 12-week, placebo-controlled investigation of 35 subjects with bipolar depression who received cytidine as add-on therapy to valproate. Depressive symptoms remitted more rapidly for those receiving cytidine (within the first week). This effect was mediated by a decrease in cerebral glutamate and/or glutamine levels assessed via proton magnetic resonance spectroscopy (MRS) at baseline and at two, four, and 12 weeks after oral cytidine administration (Yoon et al. 2009). The investigators posited that cytidine supplementation might be effective in bipolar depression because of its ability to directly modulate brain glutamate/glutamine levels.

Overall, further studies are required given that these findings have not been replicated, which may limit evaluations of potential utility for this class of agents.

12.8 Final Remarks

A variety of new agents with therapeutic potential for bipolar depression have recently been tested in animal models as well as in clinical proof-of-concept trials. As reviewed above, promising targets for the development of new, improved therapeutics for acute depressive episodes in BD include: (1) the glutamatergic system, (2) the cholinergic system, (3) the melatonergic system, (4) the glucocorticoid system, (5) the AA cascade, and (6) oxidative stress and bioenergetics. None of these new pharmacological approaches are FDA-approved for BD. Though a good number of placebo-controlled studies have been carried out—particularly with glutamatergic agents—to date, most of the clinical evidence has come from case reports, case series, or early proof-of-concept studies, most with relatively small samples. Nevertheless, these findings may help guide future directions in drug development for BD. Some agents for which less evidence presently exists—such as creatine or cytidine—will require further double-blind, controlled studies to assess their potential utility in the treatment of mood disorders.

As this chapter has highlighted, the most promising novel targets for achieving rapid antidepressant effects appear to be the ionotropic glutamate receptors. Moving forward towards the goal of personalized medicine will require health professionals to preemptively identify potential responders. Towards this goal, the search for the unique biosignatures of rapid-acting antidepressants—those whose validity has been tested in larger samples—will require evaluation of drug kinetics, ability to pass through the blood–brain barrier, and information regarding brain distribution. Future trials using enriched samples would thus likely be more efficient, given that participants would be more likely to receive the potentially most effective agent available based on their biomarker signatures. Relatedly, an improved understanding of the neurobiological underpinnings of BD will be key to developing system-targeted approaches that act more rapidly, whose effects last longer, and whose overall efficacy compared to standard agents is improved. Nevertheless, the novel therapeutics reviewed in this chapter may prove clinically useful in treating bipolar depression, particularly treatment-resistant cases.

Acknowledgments Funding for this work was supported by the Intramural Research Program at the National Institute of Mental Health and National Institutes of Health (IRP-NIMH-NIH; ZIA MH002927). A patent for the use of ketamine in depression has been awarded that lists Drs. Zarate and Manji among the inventors. Dr. Zarate has assigned his rights on the patent to the US government, but will share a percentage of any royalties that may be received by the government. Dr. Manji has also assigned his rights on the patent to the US government, and has waived his personal rights to this patent and will not receive any direct financial benefit therefrom. Dr. Manji is a full-time employee of Janssen Research & Development, LLC. The remaining authors have no conflicts of interest to disclose, financial or otherwise.

References

aan het Rot M, Collins KA, Murrough JW, Perez AM, Reich DL, Charney DS, Mathew SJ (2010) Safety and efficacy of repeated-dose intravenous ketamine for treatment-resistant depression. Biol Psychiatry 67(2):139–145

Aguado L, San Antonio A, Perez L, del Valle R, Gomez J (1994) Effects of the NMDA receptor antagonist ketamine on flavor memory: conditioned aversion, latent inhibition, and habituation of neophobia. Behav Neural Biol 61(3):271–281

Ahn YM, Chang J, Joo YH, Kim SC, Lee KY (2008) Chronotype distribution in bipolar I disorder and schizophrenia in a Korean sample. Bipolar Disord 10:271–275

Akhondzadeh S, Jafari S, Raisi F, Nasehi AA, Ghoreishi A, Salehi B, Mohebbi-Rasa S, Raznahan M, Kamalipour A (2009) Clinical trial of adjunctive celecoxib treatment in patients with major depression: a double blind and placebo controlled trial. Depress Anxiety 26:607–611

Alaerts M, Venken T, Lenaerts AS, De Zutter S, Norrback KF, Adolfsson R, Del-Favero J (2006) Lack of association of an insertion/deletion polymorphism in the G protein-coupled receptor 50 with bipolar disorder in a Northern Swedish population. Psychiatr Genet 16(6):235–236

Amital D, Vishne T, Rubinow A, Levine J (2006) Observed effects of creatine monohydrate in a patient with depression and fibromyalgia. Am J Psychiatry 163:1840–1841

Anand A, Gunn AD, Barkay G, Karne HS, Nurnberger JI, Mathew SJ, Ghosh S (2012) Early antidepressant effect of memantine during augmentation of lamotrigine inadequate response in bipolar depression: a double-blind, randomized, placebo-controlled trial. Bipolar Disord 14 (1):64–70

Andreazza AC, Cassini C, Rosa AR, Leite MC, de Almeida LM, Nardin P, Cunha AB, Cereser KM, Santin A, Gottfried C, Salvador M, Kapczinski F, Goncalves CA (2007) Serum S100B and antioxidant enzymes in bipolar patients. J Psychiatr Res 41(6):523–529

Assis LC, Rezin GT, Comim CM, Valvassori SS, Jeremias IC, Zugno AI, Quevedo J, Streck EL (2009) Effect of acute administration of ketamine and imipramine on creatine kinase activity in the brain of rats. Rev Bras Psiquiatr 31:247–252

Ballard ED, Ionescu DF, Vande Voort JL, Niciu MJ, Richards EM, Luckenbaugh DA, Brutsche NE, Ameli R, Furey ML, Zarate CA Jr (2014) Improvement in suicidal ideation after ketamine infusion: relationship to reductions in depression and anxiety. J Psychiatr Res 58:161–166

Banasr M, Chowdhury GM, Terwilliger R, Newton SS, Duman RS, Behar KL, Sanacora G (2010) Glial pathology in an animal model of depression: reversal of stress-induced cellular, metabolic and behavioral deficits by the glutamate-modulating drug riluzole. Mol Psychiatry 15:501–511

Belanoff JK, Rothschild AJ, Cassidy F, DeBattista C, Baulieu EE, Schold C, Schatzberg AF (2002) An open label trial of C-1073 (mifepristone) for psychotic major depression. Biol Psychiatry 52(5):386–392

Berk M, Copolov DL, Dean O, Lu K, Jeavons S, Schapkaitz I, Anderson-Hunt M, Bush AI (2008) N-acetyl cysteine for depressive symptoms in bipolar disorder--a double-blind randomized placebo-controlled trial. Biol Psychiatry 64(6):468–475

Berk M, Dean O, Cotton SM, Gama CS, Kapczinski F, Fernandes BS, Kohlmann K, Jeavons S, Hewitt K, Allwang C, Cobb H, Bush AI, Schapkaitz I, Dodd S, Malhi GS (2011) The efficacy of N-acetylcysteine as an adjunctive treatment in bipolar depression: An open label trial. J Affect Disord 135(1–3):389–394

Berk M, Dean OM, Cotton SM, Jeavons S, Tanious M, Kohlmann K, Hewitt K (2014) The efficacy of adjunctive N-acetylcysteine in major depressive disorder: a double-blind, randomized, placebo-controlled trial. J Clin Psychiatry 75:628–636

Berman RM, Cappiello A, Anand A, Oren DA, Heninger GR, Charney DS, Krystal JH (2000) Antidepressant effects of ketamine in depressed patients. Biol Psychiatry 47(4):351–354

Bersani G, Garavani A (2000) Melatonin add-on in manic patients with treatment resistant insomnia. Biol Psychiatry 24:185–191

Bertaina-Anglade V, la Rochelle CD, Boyer PA, Mocaer E (2006) Antidepressant-like effects of agomelatine (S 20098) in the learned helplessness model. Behav Pharmacol 17(8):703–713

Betin C, DeFeudis FV, Blavet N, Clostre F (1982) Further characterization of the behavioral despair test in mice: positive effects of convulsants. Physiol Behav 28:307–311

Bleakman D, Alt A, Witkin JM (2007) AMPA receptors in the therapeutic management of depression. CNS Neurol Disord Drug Targets 6:117–126

Blier P, Zigman D, Blier J (2012) On the safety and benefits of repeated intravenous injections of ketamine for depression. Biol Psychiatry 72(4):e11–e12

Bourin M, Prica C (2009) Melatonin receptor agonist agomelatine: a new drug for treating unipolar depression. Curr Pharm Des 15:1675–1682

Brown ES, Bobadilla L, Rush AJ (2001) Ketoconazole in bipolar patients with depressive symptoms: a case series and literature review. Bipolar Disord 3:23–29

Brown ES, Park J, Marx CE, Hynan LS, Gardner C, Davila D, Nakamura A, Sunderajan P, Lo A, Holmes T (2014) A randomized, double-blind, placebo-controlled trial of pregnenolone for bipolar depression. Neuropsychopharmacology 39:2867–2873

Calabrese JR, Guelfi JD, Perdrizet-Chevalier C (2007) Agomelatine adjunctive therapy for acute bipolar depression: preliminary open data. Bipolar Disord 9:628–635

Cannon DM, Carson RE, Nugent AC, Eckelman WC, Kiesewetter DO, Williams J, Rollis D, Drevets M, Gandhi S, Solorio G, Drevets WC (2006) Reduced muscarinic type 2 receptor binding in subjects with bipolar disorder. Arch Gen Psychiatry 63:741–747

Cardinali DP, Cano P, Jimenez-Ortega V, Esquifino AI (2011) Melatonin and the metabolic syndrome: physiopathologic and therapeutical implications. Neuroendocrinology 93:133–142

Carroll BJ, Rubin RT (2008) Mifepristone in psychotic depression? Biol Psychiatry 63(1):e1; author reply e3

Chaki S, Ago Y, Palucha-Paniewiera A, Matrisciano F, Pilc A (2013) mGlu2/3 and mGlu5 receptors: potential targets for novel antidepressants. Neuropharmacology 66:40–52

Conn PJ, Pin JP (1997) Pharmacology and functions of metabotropic glutamate receptors. Annu Rev Pharmacol Toxicol 37:205–237

Cryan JF, Kelly PH, Neijt HC, Sansig G, Flor PJ, van Der Putten H (2003) Antidepressant and anxiolytic-like effects in mice lacking the group III metabotropic glutamate receptor mGluR7. Eur J Neurosci 17(11):2409–2417

Dale E, Bang-Andersen B, Sanchez C (2015) Emerging mechanisms and treatments for depression beyond SSRIs and SNRIs. Biochem Pharmacol 95:81–97

DeBattista C, Belanoff J, Glass S, Khan A, Horne RL, Blasey C, Carpenter LL, Alva G (2006) Mifepristone versus placebo in the treatment of psychosis in patients with psychotic major depression. Biol Psychiatry 60(12):1343–1349

Diamond PR, Farmery AD, Atkinson S, Haldar J, Williams N, Cowen PJ, Geddes JR, McShane R (2014) Ketamine infusions for treatment resistant depression: a series of 28 patients treated weekly or twice weekly in an ECT clinic. J Psychopharmacol 28(6):536–544

Diazgranados N, Ibrahim L, Brutsche N, Newberg A, Kronstein P, Khalife S, Kammerer WA, Quezado Z, Luckenbaugh DA, Salvadore G, Machado-Vieira R, Manji HK, Zarate CA (2010a) A randomized add-on trial of an N-methyl-D-aspartate antagonist in treatment-resistant bipolar depression. Arch Gen Psychiatry 67:793–802

DiazGranados N, Ibrahim LA, Brutsche NE, Ameli R, Henter ID, Luckenbaugh DA, Machado-Vieira R, Zarate CA Jr (2010b) Rapid resolution of suicidal ideation after a single infusion of an N-methyl-D-aspartate antagonist in patients with treatment-resistant major depressive disorder. J Clin Psychiatry 71(12):1605–1611

Dilsaver SC (1986) Pathophysiology of "cholinoceptor supersensitivity" in affective disorders. Biol Psychiatry 21:813–829

Drevets WC, Zarate CAJ, Furey ML (2013) Antidepressant effects of the muscarinic cholinergic receptor antagonist scopolamine: a review. Biol Psychiatry 73:1156–1163

Du J, Suzuki K, Wei Y, Wang Y, Blumenthal R, Chen Z, Falke C, Zarate CA Jr, Manji HK (2007) The anticonvulsants lamotrigine, riluzole, and valproate differentially regulate AMPA receptor membrane localization: relationship to clinical effects in mood disorders. Neuropsychopharmacology 32:793–802

Dwyer JM, Lepack AE, Duman RS (2012) mTOR activation is required for the antidepressant effects of mGluR(2)/(3) blockade. Int J Neuropsychopharmacol 15:429–434

Eden Evins A, Demopulos C, Nierenberg A, Culhane MA, Eisner L, Sachs GS (2006) A double-blind, placebo-controlled trial of adjunctive donepezil in treatment-resistant mania. Bipolar Disord 8:75–80

Etain B, Dumaine A, Bellivier F, Pagan C, Francelle L, Goubran-Botros H, Moreno S, Deshommes J, Moustafa K, Le Dudal K, Mathieu F, Henry C, Kahn JP, Launay JM, Muhleisen TW, Cichon S, Bougeron T, Leboyer M, Jamain S (2012) Genetic and functional abnormalities of the melatonin biosynthesis pathway in patients with bipolar disorder. Hum Mol Genet 21:4030–4037

Frizzo ME, Dall'Onder LP, Dalcin KB, Souza DO (2004) Riluzole enhances glutamate uptake in rat astrocyte cultures. Cell Mol Neurobiol 24(1):123–128

Furey ML, Drevets WC (2006) Antidepressant efficacy of the antimuscarinic drug scopolamine: a randomized, placebo-controlled clinical trial. Arch Gen Psychiatry 63:1121–1129

Gallagher P, Malik N, Newham J, Young AH, Ferrier IN, Mackin P (2008) Antiglucocorticoid treatments for mood disorders. Cochrane Database Syst Rev Jan 23(1):CD005168

Gao ZG, Jacobson KA (2013) Allosteric modulation and functional selectivity of G protein-coupled receptors. Drug Discov Today Technol 10:e237–e243

Garcia LS, Comim CM, Valvassori SS, Reus GZ, Barbosa LM, Andreazza AC, Stertz L, Fries GR, Gavioli EC, Kapczinski F, Quevedo J (2008) Acute administration of ketamine induces antidepressant-like effects in the forced swimming test and increases BDNF levels in the rat hippocampus. Prog Neuropsychopharmacol Biol Psychiatry 32(1):140–144

Geoffrey PA, Etain B, Franchi JA, Bellivier F, Ritter P (2015) Melatonin and melatonin agonists as adjunctive treatments in bipolar disorders. Curr Pharm Des 21:3352–3358

Gibbons AS, Scarr E, McLean C, Sundram S, Dean B (2009) Decreased muscarinic receptor binding in the frontal cortex of bipolar disorder and major depressive disorder subjects. J Affect Disord 116:184–191

Goodwin GM, Emsley R, Rembry S, Rouillon F, Agomelatine Study Group (2009) Agomelatine prevents relapse in patients with major depressive disorder without evidence of a discontinuation syndrome: a 24-week randomized, double-blind, placebo-controlled trial. J Clin Psychiatry 70:1128–1137

Grunberg SM, Weiss MH, Russell CA, Spitz IM, Ahmadi J, Sadun A, Sitruk-Ware R (2006) Long-term administration of mifepristone (RU486): clinical tolerance during extended treatment of meningioma. Cancer Invest 24(8):727–733

Hallam KT, Olver JS, Horgan JE, McGrath C, Norman TR (2005a) Low doses of lithium carbonate reduce melatonin light sensitivity in healthy volunteers. Int J Neuropsychopharmacol 8:255–259

Hallam KT, Olver JS, Norman TR (2005b) Effect of sodium valproate on nocturnal melatonin sensitivity to light in healthy volunteers. Neuropsychopharmacology 30:1400–1404

Hallam KT, Olver JS, Norman TR (2005c) Melatonin sensitivity to light in monozygotic twins discordant for bipolar I disorder. Aust N Z J Psychiatry 39(10):947

Hallam KT, Olver JS, Chambers V, Begg DP, McGrath C, Norman TR (2006) The heritability of melatonin secretion and sensitivity to bright nocturnal light in twins. Psychoneuroendocrinology 31:867–875

Hannestad JO, Cosgrove KP, Dellagioia NF, Perkins E, Bois F, Bhagwagar Z, Seibyl JP, McClure-Begley TD, Picciotto MR, Esterlis I (2013) Changes in the cholinergic system between bipolar depression and euthymia as measured with [I]5IA single photon emission computed tomography. Biol Psychiatry 74:768–776

Harvey AG (2008) Sleep and circadian rhythms in bipolar disorder: seeking synchrony, harmony, and regulation. Am J Psychiatry 165(7):820–829

Hasselmann H (2014) Scopolamine and depression: a role for muscarinic antagonism. CNS Neurol Disord Drug Targets 13:673–683

Iadarola ND, Niciu MJ, Richards EM, Vande Voort JL, Ballard ED, Lundin NB, Nugent AC, Machado-Vieira R, Zarate CAJ (2015) Ketamine and other N-methyl-D-aspartate receptor antagonists in the treatment of depression: a perspective review. Ther Adv Chronic Dis 6:97–114

Ibrahim L, Diazgranados N, Franco-Chaves J, Brutsche N, Henter ID, Kronstein P, Moaddel R, Wainer I, Luckenbaugh DA, Manji HK, Zarate CA (2012) Course of improvement in depressive symptoms to a single intravenous infusion of ketamine vs add-on riluzole: results from a 4-week, double-blind, placebo-controlled study. Neuropsychopharmacology 37:1526–1533

Jaffe RJ, Novakovic V, Peselow ED (2013) Scopolamine as an antidepressant: a systematic review. Clin Neuropharmacol 36:24–26

Jahn H, Schick M, Kiefer F, Kellner M, Yassouridis A, Wiedemann K (2004) Metyrapone as additive treatment in major depression: a double-blind and placebo-controlled trial. Arch Gen Psychiatry 61(12):1235–1244

Janowsky DS, el-Yousef MK, Davis JM, Sekerke HJ (1972) A cholinergic-adrenergic hypothesis of mania and depression. Lancet 2:632–635

Judd LL, Akiskal HS, Schettler PJ, Endicott J, Maser J, Solomon DA, Leon AC, Rice JA, Keller MB (2002) The long-term natural history of the weekly symptomatic status of bipolar I disorder. Arch Gen Psychiatry 59:530–537

Karasawa J, Shimazaki T, Kawashima N, Chaki S (2005) AMPA receptor stimulation mediates the antidepressant-like effect of a group II metabotropic glutamate receptor antagonist. Brain Res 1042:92–98

Kasper S, Moises HW, Beckmann H (1981) The anticholinergic biperiden in depressive disorders. Pharmacopsychiatria 14:195–198

Kelly TF, Lieberman DZ (2014) The utility of the combination of dextromethorphan and quinidine in the treatment of bipolar II and bipolar NOS. J Affect Disord 167:333–335

Kennedy SH, Emsley R (2006) Placebo-controlled trial of agomelatine in the treatment of major depressive disorder. Eur Neuropsychopharmacol 16(2):93–100

Khajavi D, Farokhnia M, Modabbernia A, Ashrafi M, Abbasi SH, Tabrizi M, Akhondzadeh S (2012) Oral scopolamine augmentation in moderate to severe major depressive disorder: a randomized, double-blind, placebo-controlled study. J Clin Psychiatry 73:1428–1433

Khouzam HR, Kissmeyer PM (1996) Physostigmine temporarily and dramatically reversing acute mania. Gen Hosp Psychiatry 18:203–204

Kim CH, Lee J, Lee JY, Roche KW (2008) Metabotropic glutamate receptors: phosphorylation and receptor signaling. J Neurosci Res 86(1):1–10

Knapp RJ, Goldenberg R, Shuck C, Cecil A, Watkins J, Miller C, Crites G, Malatynska E (2002) Antidepressant activity of memory-enhancing drugs in the reduction of submissive behavior model. Eur J Pharmacol 440(1):27–35

Koike H, Iijima M, Chaki S (2011) Involvement of AMPA receptor in both the rapid and sustained antidepressant-like effects of ketamine in animal models of depression. Behav Brain Res 224:107–111

Kollmar R, Markovic K, Thurauf N, Schmitt H, Kornhuber J (2008) Ketamine followed by memantine for the treatment of major depression. Aust N Z J Psychiatry 42(2):170

Kuloglu M, Ustundag B, Atmaca M, Canatan H, Tezcan AE, Cinkilinc N (2002) Lipid peroxidation and antioxidant enzyme levels in patients with schizophrenia and bipolar disorder. Cell Biochem Funct 20(2):171–175

Larkin GL, Beautrais AL (2011) A preliminary naturalistic study of low-dose ketamine for depression and suicide ideation in the emergency department. Int J Neuropsychopharmacol 14:1–5

Lee SY, Chen SL, Chang YH, Chen SH, Chu CH, Huang SY, Tzeng NS, Wang CL, Lee IH, Yeh TL, Yang YK, Lu RB (2012) The DRD2/ANKK1 gene is associated with response to add-on dextromethorphan treatment in bipolar disorder. J Affect Disord 138:295–300

Leibenluft E, Feldman-Naim S, Turner EH, Wehr TA, Rosenthal NE (1997) Effects of exogenous melatonin administration and withdrawal in five patients with rapid-cycling bipolar disorder. J Clin Psychiatry 58:383–388

Liu HF, Zhou WH, Xie XH, Cao JL, Gu J, Yang GD (2004) Muscarinic receptors modulate the mRNA expression of NMDA receptors in brainstem and the release of glutamate in periaqueductal grey during morphine withdrawal in rats. Sheng Li Xue Bao 56:95–100

Loo H, Hale A, D'Haenen H (2002) Determination of the dose of agomelatine, a melatoninergic agonist and selective 5-HT(2C) antagonist, in the treatment of major depressive disorder: a placebo-controlled dose range study. Int Clin Psychopharmacol 17(5):239–247

Lourenco Da Silva A, Hoffmann A, Dietrich MO, Dall'Igna OP, Souza DO, Lara DR (2003) Effect of riluzole on MK-801 and amphetamine-induced hyperlocomotion. Neuropsychobiology 48 (1):27–30

MacDonald ML, Naydenov A, Chu M, Matzilevich D, Konradi C (2006) Decrease in creatine kinase messenger RNA expression in the hippocampus and dorsolateral prefrontal cortex in bipolar disorder. Bipolar Disord 8(3):255–264

Machado-Vieira R, Manji HK, Zarate CA (2009a) The role of the tripartite glutamatergic synapse in the pathophysiology and therapeutics of mood disorders. Neuroscientist 15(5):525–539

Machado-Vieira R, Salvadore G, Ibrahim LA, Diaz-Granados N, Zarate CA Jr (2009b) Targeting glutamatergic signaling for the development of novel therapeutics for mood disorders. Curr Pharm Des 15(14):1595–1611

Machado-Vieira R, Ibrahim L, Henter ID, Zarate CA Jr (2012) Novel glutamatergic agents for major depressive disorder and bipolar disorder. Pharmacol Biochem Behav 100(4):678–687

Maeng S, Zarate CA Jr, Du J, Schloesser RJ, McCammon J, Chen G, Manji HK (2008) Cellular mechanisms underlying the antidepressant effects of ketamine: role of alpha-amino-3-hydroxy-5-methylisoxazole-4-propionic acid receptors. Biol Psychiatry 63(4):349–352

Magalhaes PV, Dean OM, Bush AI, Copolov DL, Malhi GS, Kohlmann K, Jeavons S, Schapkaitz I, Anderson-Hung M, Berk M (2011) N-acetyl cysteine add-on treatment for bipolar II disorder: a subgroup analysis of a randomized placebo-controlled trial. J Affect Disord 129:317–320

Majd M, Hashemian F, Hosseini SM, Vahdat Shariatpanahi M, Sharifi A (2015) A randomized double-blind, placebo-controlled trial of celecoxib augmentation of sertraline in treatment of drug-naive depressed women: a pilot study. Iran J Pharm Res 14:891–899

Mathew SJ, Murrough JW, aan het Rot M, Collins KA, Reich DL, Charney DS (2010) Riluzole for relapse prevention following intravenous ketamine in treatment-resistant depression: a pilot randomized, placebo-controlled continuation trial. Int J Neuropsychopharmacol 13(1):71–82

Messias E, Everett B (2012) Dextromethorphan and quinidine combination in emotional lability associated with depression: a case report. Prim Care Companion CNS Disord 14: PCC.12l01400

Mickley GA, Schaldach MA, Snyder KJ, Balogh SA, Len T, Neimanis K, Goulis P, Hug J, Sauchak K, Remmers-Roeber DR, Walker C, Yamamoto BK (1998) Ketamine blocks a conditioned taste aversion (CTA) in neonatal rats. Physiol Behav 64(3):381–390

Millan MJ, Brocco M, Gobert A, Dekeyne A (2005) Anxiolytic properties of agomelatine, an antidepressant with melatoninergic and serotonergic properties: role of 5-HT2C receptor blockade. Psychopharmacology (Berl) 177(4):448–458

Millar A, Espie CA, Scott J (2004) The sleep of remitted bipolar outpatients: a controlled naturalistic study using actigraphy. J Affect Disord 80:145–153

Mizuta I, Ohta M, Ohta K, Nishimura M, Mizuta E, Kuno S (2001) Riluzole stimulates nerve growth factor, brain-derived neurotrophic factor and glial cell line-derived neurotrophic factor synthesis in cultured mouse astrocytes. Neurosci Lett 310(2–3):117–120

Moghaddam B, Adams B, Verma A, Daly D (1997) Activation of glutamatergic neurotransmission by ketamine: a novel step in the pathway from NMDA receptor blockade to dopaminergic and cognitive disruptions associated with the prefrontal cortex. J Neurosci 17:2921–2927

Montgomery SA, Kasper S (2007) Severe depression and antidepressants: focus on a pooled analysis of placebo-controlled studies on agomelatine. Int Clin Psychopharmacol 22 (5):283–291

Muller N, Schwarz MJ, Dehning S, Douhe A, Cerovecki A, Goldstein-Muller B, Spellmann I, Hetzel G, Maino K, Kleindienst N, Moller HJ, Arolt V, Riedel M (2006) The cyclooxygenase-2 inhibitor celecoxib has therapeutic effects in major depression: results of a double-blind, randomized, placebo controlled, add-on pilot study to reboxetine. Mol Psychiatry 11 (7):680–684

Murrough JW, Iosifescu DV, Chang LC, Al Jurdi RK, Green CE, Perez AM, Iqbal S, Pillemer S, Foulkes A, Shah A, Charney DS, Mathew SJ (2013a) Antidepressant efficacy of ketamine in treatment-resistant major depression: a two-site randomized controlled trial. Am J Psychiatry 170(10):1134–1142

Murrough JW, Perez AM, Pillemer S, Stern J, Parides MK, aan het Rot M, Collins KA, Mathew SJ, Charney DS, Iosifescu DV (2013b) Rapid and longer-term antidepressant effects of repeated ketamine infusions in treatment-resistant major depression. Biol Psychiatry 74(4):250–256

Musazzi L, Treccani G, Mallei A, Popoli M (2013) The action of antidepressants on the glutamate system: regulation of glutamate release and glutamate receptors. Biol Psychiatry 73:1180–1188

Nathan PJ, Burrows GD, Norman TR (1999) Melatonin sensitivity to dim white light in affective disorders. Neuropsychopharmacology 21:408–413

Nery FG, Monkul ES, Hatch JP, Fonseca M, Zunta-Soares GB, Frey BN, Bowden CL, Soares JC (2008) Celecoxib as an adjunct in the treatment of depressive or mixed episodes of bipolar disorder: a double-blind, randomized, placebo-controlled study. Hum Psychopharmacol 23 (2):87–94

Newport DJ, Carpenter LL, McDonald WM, Potash JB, Tohen M, Nemeroff CB, APA Council of Research Task Force on Novel Biomarkers and Treatments (2015) Ketamine and other NMDA antagonists: early clinical trials and possible mechanisms in depression. Am J Psychiatry 172:950–966

Nguyen L, Matsumoto RR (2015) Involvement of AMPA receptors in the antidepressant-like effects of dextromethorphan in mice. Behav Brain Res 295:26–34

Niciu MJ, Luckenbaugh D, Ionescu DF, Mathews D, Richards EM, Zarate CA (2013) Subanesthetic dose ketamine does not induce an affective switch in three independent samples of treatment-resistant major depression. Biol Psychiatry 74:e23–e24

Norris ER, Burke K, Correll JR, Zemanek KJ, Lerman J, Primelo RA, Kaufmann MW (2013) A double-blind, randomized, placebo-controlled trial of adjunctive ramelteon for the treatment of insomnia and mood stability in patients with euthymic bipolar disorder. J Affect Disord 144:141–147

Nurnberger JI Jr, Adkins S, Lahiri DK, Mayeda A, Hu K, Lewy A, Miller A, Bowman ES, Miller MJ, Rau L, Smiley C, Davis-Singh D (2000) Melatonin suppression by light in euthymic bipolar and unipolar patients. Arch Gen Psychiatry 57(6):572–579

O'Neill MJ, Witkin JM (2007) AMPA receptor potentiators: application for depression and Parkinson's disease. Curr Drug Targets 8:603–620

Overstreet DH (1993) The Flinders sensitive line rats: a genetic animal model of depression. Neurosci Biobehav Rev 17:51–68

Pacchierotti C, Iapichino S, Bossini L, Pieraccini F, Castrogiovanni P (2001) Melatonin in psychiatric disorders: a review on the melatonin involvement in psychiatry. Front Neuroendocrinol 22:18–32

Palucha A, Pilc A (2007) Metabotropic glutamate receptor ligands as possible anxiolytic and antidepressant drugs. Pharmacol Ther 115:116–147

Papp M, Gruca P, Boyer PA, Mocaer E (2003) Effect of agomelatine in the chronic mild stress model of depression in the rat. Neuropsychopharmacology 28(4):694–703

Permoda-Osip A, Kisielewski J, Bartkowska-Sniatkowska A, Rybakowski JK (2014) Single ketamine infusion and neurocognitive performance in bipolar depression. Pharmacopsychiatry 48(2):78–79

Pilc A, Wieronska JM, Skolnick P (2013) Glutamate-based antidepressants: preclinical psychopharmacology. Biol Psychiatry 73:1125–1132

Pitkanen A, Mathiesen C, Ronn LC, Moller A, Nissinen J (2007) Effect of novel AMPA antagonist, NS1209, on status epilepticus. An experimental study in rat. Epilepsy Res 74 (1):45–54

Price RB, Iosifescu DV, Murrough JW, Chang LC, Al Jurdi RK, Iqbal SZ, Soleimani L, Charney DS, Foulkes AL, Mathew SJ (2014) Effects of ketamine on explicit and implicit suicidal cognition: a randomized controlled trial in treatment-resistant depression. Depress Anxiety 31 (4):335–343

Quiroz JA, Singh J, Gould TD, Denicoff KD, Zarate CA Jr, Manji HK (2004) Emerging experimental therapeutics for bipolar disorder: clues from the molecular pathophysiology. Mol Psychiatry 9:756–776

Quiroz J, Tamburri P, Deptula D, Banken L, Beyer U, Fontoura P, Santarelli L (2014) The efficacy and safety of basimglurant as adjunctive therapy in major depression: a randomized, double-blind, placebo controlled study. Neuropsychopharmacology 39:S376–S377

Raison CL, Rutherford RE, Woolwine BJ, Shuo C, Schettler P, Drake DF, Haroon E, Miller AH (2013) A randomized controlled trial of the tumor necrosis factor antagonist infliximab for treatment-resistant depression: the role of baseline inflammatory biomarkers. JAMA Psychiatry 70(1):31–41

Rapoport SI, Basselin M, Kim HW, Rao JS (2009) Bipolar disorder and mechanisms of action of mood stabilizers. Brain Res Rev 61:185–209

Rasmussen KG, Lineberry TW, Galardy CW, Kung S, Lapid MI, Palmer BA, Ritter MJ, Schak KM, Sola CL, Hanson AJ, Frye MA (2013) Serial infusions of low-dose ketamine for major depression. J Psychopharmacol 27(5):444–450

Rogawski MA (2006) Diverse mechanisms of antiepileptic drugs in the development pipeline. Epilepsy Res 69(3):273–294

Rogoz Z, Skuza G, Maj J, Danysz W (2002) Synergistic effect of uncompetitive NMDA receptor antagonists and antidepressant drugs in the forced swimming test in rats. Neuropharmacology 42:1024–1030

Roitman S, Green T, Osher Y, Karni N, Levine J (2007) Creatine monohydrate in resistant depression: a preliminary study. Bipolar Disord 9(7):754–758

Romo-Nava F, Alvarez-Icaza Gonzalez D, Fresan-Orellana A, Saracco Alvarez R, Becerra-Palars-C, Moreno J, Ontiveros Uribe MP, Berlanga C, Heinze G, Buijs RM (2014) Melatonin attenuates antipsychotic metabolic effects: an eight-week randomized, double-blind, parallel-group, placebo-controlled clinical trial. Bipolar Disord 16:410–421

Rothschild AJ (2003) Challenges in the treatment of depression with psychotic features. Biol Psychiatry 53(8):680–690

Sachs GS, Nierenberg AA, Calabrese JR, Marangell LB, Wisniewski SR, Gyulai L, Friedman ES, Bowden CL, Fossey MD, Ostacher MJ, Ketter TA, Patel J, Hauser P, Rapport D (2007) Effectiveness of adjunctive antidepressant treatment for bipolar depression. N Engl J Med 356:1711–1722

Sanacora G, Kendell SF, Levin Y, Simen AA, Fenton LR, Coric V, Krystal JH (2007) Preliminary evidence of riluzole efficacy in antidepressant-treated patients with residual depressive symptoms. Biol Psychiatry 61(6):822–825

Sanacora G, Zarate CA Jr, Krystal JH, Manji HK (2008) Targeting the glutamatergic system to develop novel, improved therapeutics for mood disorders. Nat Rev Drug Discov 7:426–437

Sauter A, Rudin M (1993) Determination of creatine kinase kinetic parameters in rat brain by NMR magnetization transfer. Correlation with brain function. J Biol Chem 268:13166–13171

Schmidt PJ, Daly RC, Bloch M, Smith MJ, Danaceau MA, St. Clair LS, Murphy JH, Haq N, Rubinow DR (2005) Dehydroepiandrosterone monotherapy in midlife-onset major and minor depression. Arch Gen Psychiatry 62:154–162

Segal M, Avital A, Drobot M, Lukanin A, Derevenski A, Sandbank S, Weizman A (2007) CK levels in unmedicated bipolar patients. Eur Neuropsychopharmacol 17(12):763–767

Serra G, Koukopoulos A, De Chiara L, Koukopoulos AE, Tondo L, Girardi P, Baldessarini RJ, Serra G (2015) Three-year, naturalistic, mirror-image assessment of adding memantine to the treatment of 30 treatment-resistant patients with bipolar disorder. J Clin Psychiatry 76:e91–e97

Shaltiel G, Maeng S, Malkesman O, Pearson B, Schloesser RJ, Tragon T, Rogawski M, Gasior M, Luckenbaugh D, Chen G, Manji HK (2008) Evidence for the involvement of the kainate receptor subunit GluR6 (GRIK2) in mediating behavioral displays related to behavioral symptoms of mania. Mol Psychiatry 13(9):858–872

Shi J, Hattori E, Zou H, Badner JA, Christian SL, Gershon ES, Liu C (2007) No evidence for association between 19 cholinergic genes and bipolar disorder. Am J Med Genet B Neuropsychiatr Genet 144:715–723

Silvestre JS, Nadal R, Pallares M, Ferre N (1997) Acute effects of ketamine in the holeboard, the elevated-plus maze, and the social interaction test in Wistar rats. Depress Anxiety 5(1):29–33

Singh J, Fedgchin M, Daly E, De Boer P, Cooper K, Lim P, Pinter C, Murrough J, Sanacora G, Shelton R, Kurian B, Winokur A, Fava M, Manji HK, Drevets W, van Nueten L (2014) A double-blind, randomized, placebo-controlled, parallel group, dose frequency study of intravenous ketamine in patients with treatment-resistant depression. In: Poster Session 1, Janssen Research and Development, LLC, June 17 2014

Skolnick P (1999) Antidepressants for the new millennium. Eur J Pharmacol 375(1–3):31–40

Skolnick P (2002) Modulation of glutamate receptors: strategies for the development of novel antidepressants. Amino Acids 23(1–3):153–159

Smith EG, Deligiannidis KM, Ulbricht CM, Landolin CS, Patel JK, Rothschild AJ (2013) Antidepressant augmentation using the N-methyl-D-aspartate antagonist memantine: a randomized, double-blind, placebo-controlled trial. J Clin Psychiatry 74:966–973

Sokolski KN, DeMet EM (2000) Cholinergic sensitivity predicts severity of mania. Psychiatry Res 95:195–200

Streck EL, Amboni G, Scaini G, Di-Pietro PB, Rezin GT, Valvassori SS, Luz G, Kapczinski F, Quevedo J (2008) Brain creatine kinase activity in an animal model of mania. Life Sci 82 (7–8):424–429

Taylor D, Sparshatt A, Varma S, Olofinjana O (2014) Antidepressant efficacy of agomelatine: meta-analysis of published and unpublished studies. BMJ 348:g1888

Teng CT, Demetrio FN (2006) Memantine may acutely improve cognition and have a mood stabilizing effect in treatment-resistant bipolar disorder. Rev Bras Psiquiatr 28:252–254

Thomson PA, Wray NR, Thomson AM, Dunbar DR, Grassie MA, Condie A, Walker MT, Smith DJ, Pulford DJ, Muir W, Blackwood DH, Porteous DJ (2005) Sex-specific association between bipolar affective disorder in women and GPR50, an X-linked orphan G protein-coupled receptor. Mol Psychiatry 10(5):470–478

Valentine GW, Mason GF, Gomez R, Fasula M, Watzl J, Pittman B, Krystal JH, Sanacora G (2011) The antidepressant effect of ketamine is not associated with changes in occipital amino acid neurotransmitter content as measured by [(1)H]-MRS. Psychiatry Res 191(2):122–127

van Enkhuizen J, Janowsky DS, Olivier B, Minassian A, Perry W, Young JW, Geyer MA (2015) The catecholaminergic-cholinergic balance hypothesis of bipolar disorder revisited. Eur J Pharmacol 753:114–126

Velentgas P, West W, Cannuscio CC, Watson DJ, Walker AM (2006) Cardiovascular risk of selective cyclooxygenase-2 inhibitors and other non-aspirin non-steroidal anti-inflammatory medications. Pharmacoepidemiol Drug Saf 15(9):641–652

Voleti B, Navarria A, Liu RJ, Banasr M, Li N, Terwilliger R, Sanacora G, Eid T, Aghajanian G, Duman RS (2013) Scopolamine rapidly increases mammalian target of rapamycin complex

1 signaling, synaptogenesis, and antidepressant behavioral responses. Biol Psychiatry 74:742–749

Watson S, Gallagher P, Porter RJ, Smith MS, Herron LJ, Bulmer S, North-East Mood Disorders Clinical Research Group, Young AH, Ferrier IN (2012) A randomized trial to examine the effect of mifepristone on neuropsychological performance and mood in patients with bipolar depression. Biol Psychiatry 72:943–949

Wolkowitz OM, Reus VI, Chan T, Manfredi F, Raum W, Johnson R, Canick J (1999a) Antiglucocorticoid treatment of depression: double-blind ketoconazole. Biol Psychiatry 45:1070–1074

Wolkowitz OM, Reus VI, Keebler A, Nelson N, Friedland M, Brizendine L, Roberts E (1999b) Double-blind treatment of major depression with dehydroepiandrosterone. Am J Psychiatry 156:646–649

Yoon SJ, Lyoo IK, Haws C, Kim TS, Cohen BM, Renshaw PF (2009) Decreased glutamate/glutamine levels may mediate cytidine's efficacy in treating bipolar depression: a longitudinal proton magnetic resonance spectroscopy study. Neuropsychopharmacology 34(7):1810–1818

Young AH, Gallagher P, Watson S, Del-Estal D, Owen BM, Ferrier IN (2004) Improvements in neurocognitive function and mood following adjunctive treatment with mifepristone (RU-486) in bipolar disorder. Neuropsychopharmacology 29:1538–1545

Zarate CA, Payne JL, Quiroz J, Sporn J, Denicoff KK, Luckenbaugh D, Charney DS, Manji HK (2004) An open-label trial of riluzole in patients with treatment-resistant major depression. Am J Psychiatry 161(1):171–174

Zarate CA, Quiroz JA, Singh JB, Denicoff KD, De Jesus G, Luckenbaugh DA, Charney DS, Manji HK (2005) An open-label trial of the glutamate-modulating agent riluzole in combination with lithium for the treatment of bipolar depression. Biol Psychiatry 57(4):430–432

Zarate CA Jr, Singh JB, Carlson PJ, Brutsche NE, Ameli R, Luckenbaugh DA, Charney DS, Manji HK (2006) A randomized trial of an N-methyl-D-aspartate antagonist in treatment-resistant major depression. Arch Gen Psychiatry 63(8):856–864

Zarate CA, Brutsche N, Ibrahim L, Franco-Chaves J, Diazgranados N, Cravchick A, Selter J, Marquardt C, Liberty V, Luckenbaugh DA (2012) Replication of ketamine's antidepressant efficacy in bipolar depression: a randomized controlled add-on trial. Biol Psychiatry 71:939–946

Part IV
Treatment of Bipolar Disorder in Special Populations

Chapter 13
Treatment of Bipolar Disorder in Special Populations

John L. Beyer and K. Ranga R. Krishnan

Abstract Though often considered a young person's disease because its mean age of onset is in late adolescence, bipolar disorder (BD) frequently demonstrates recurrent episodes and morbidities that continue throughout the lifetime of the patient and into old age. In fact, presentation of mood episodes in older age is quite common and requires clinicians to understand the unique challenges that the interaction of aging and mental illness present. Further, it has been well established that the depressive polarity of episodes increases in frequency over the life span with a decline in manic and mixed episode presentations. Finally, while new onset of illness is relatively rare in later ages, it is not unknown, and the variability of age of onset may allow for a better understanding of the disease process. This chapter reviews our current understanding of BD in the elderly and highlights the implications of age in understanding the heuristic causes of the disease, challenges to treatment, and the limitations of our knowledge for clinical care.

Keywords Bipolar disorder • Elderly • Age of onset • Treatment

13.1 Introduction

The diagnosis of bipolar disorder (BD) is based on presentation of symptoms and course rather than etiological criteria (American Psychiatric Association 2013). Therefore, there is a large heterogeneity of presentations in patients that may affect both course and treatment of the disorder. Several aspects of presentation provide guidance in treatment decisions. For example, the current phase of illness or the severity of symptoms at presentation for treatment has been recognized by the American Psychiatric Association (2002) and multiple other expert consensus treatment guidelines (Grunze et al. 2009, 2010; Goodwin et al. 2009; Yatham

J.L. Beyer, MD (✉)
Duke University Medical Center, Box 3519 DUMC, Durham, NC 27710, USA
e-mail: john.beyer@duke.edu

K.R.R. Krishnan, MD
Dean Rush Medical College, Rush University Medical Center, Chicago, IL, USA

© Springer International Publishing Switzerland 2016
C.A. Zarate Jr., H.K. Manji (eds.), *Bipolar Depression: Molecular Neurobiology, Clinical Diagnosis, and Pharmacotherapy*, Milestones in Drug Therapy,
DOI 10.1007/978-3-319-31689-5_13

et al. 2013) as being meaningful in clinical decision-making. Patients may be treated differently if they are present in a manic, depressive, or mixed state or have psychotic symptoms.

Other groupings within the bipolar spectrum may also have heuristic meaning for understanding the etiology of BD or for modifying clinical decision-making. For example, efforts are currently underway to determine if structural or functional neuroimaging findings, genetic variability, or other biological markers may inform treatment decisions.

Similarly, clinical factors may serve as markers for modifying clinical treatment. One clinical characteristic that is a modifying risk factor is the effect of age and age-of-onset in BD patients. Here, we focus on the impact of older age and aging in late-life BD. Questions addressed include whether patients with BD early in life "burn out" with age; whether there is a difference in BD symptoms through the life cycle; whether there are etiological and phenomenological differences if the disease begins early in life versus later in life; and whether aging affects treatment response.

This chapter is divided into two parts. The first part will discuss the current literature on aging in BD with an emphasis on late-life and late-onset disease. The second part will discuss treatment in late-life BD, noting especially the current literature on bipolar depression in the elderly.

13.2   Bipolar Disorder in Late Life

13.2.1   Prevalence

Though BD has been a recognized mental illness since the mid-1800s, the prevalence of BD in the geriatric population remains unclear. It is known that between six and eight percent of psychiatric admissions are for geriatric BD (Depp and Jeste 2004; Ettner and Hermann 1998); however, the Epidemiologic Catchment Area (ECA) study failed to capture any active manic elderly cases during their community survey of psychiatric disorders in the United States (Weissman et al. 1988). Instead, using a statistical weighted analysis, the authors reported a one-year prevalence range of elderly with BD between 0.0 and 0.5 % (with a cross-site mean of 0.1 %). This was markedly lower than the prevalence of BD reported among young (18–44 years; 1.4 %) and middle-aged (45–64 years; 0.4 %) adults. This range, though, was consistent with three other community-based studies that included assessments of the prevalence of BD in the elderly. Unutzer and colleagues (1998) reviewed a large HMO database and found a prevalence rate of 0.25 % (compared with 0.46 % in adults aged 40–64). Klap and colleagues (2003) reported on the HealthCare for Communities (HCC) Household Telephone Survey of 9585 households and found a prevalence rate of 0.08 % (compared with 1.17 % for adults aged 30–64). Finally, Hirschfeld and colleagues (2003) reported results of a screening questionnaire (Mood Disorder Questionnaire) sent to 125,000 individuals (85,258 responders). They found the screen rate for adults 65 and older was 0.5 % (compared with 3.4 % in adults younger than 65).

Interestingly, each of these surveys suggested that the prevalence of BD declines with age or in aging cohorts. This has led some researchers to suggest that bipolar episodes decrease with age (Angst et al. 1973). Winokur (1975) was the first to propose the concept that manic patients may "burn out" after a finite number of episodes. In the Iowa 500 study, Winokur and colleagues followed 109 patients admitted for mania up to 20 years. The authors observed that bipolar episodes occurred in "bursts," and then became quiescent. However in a prospective study, Angst and Preisig (1995) followed 209 BD patients over a period of 40 years (median age 68). They found that manic episodes did not decrease with age, and many patients continued to have episodes into their seventh decade.

Overall, the decline in the prevalence of BD with age noted by the community surveys is similar to that seen in prevalence rates of other mental illnesses (such as depression and schizophrenia) and may actually represent a cohort effect or an increased mortality rate noted in patients with mental illness.

In general, the development of BD in late life can be divided into four patterns: (1) those who had early-onset of BD and have reached old age; (2) those who were previously diagnosed with major depressive disorder (MDD) but had a switch to mania in late life; (3) those whose bipolar symptoms have never been recognized or were misdiagnosed; and (4) those who have never had an affective illness but develop mania in late life (possibly due to a specific medical or neurologic event or for reasons unknown). It is not known how common each presentation may be, though the most frequent experience is a patient who developed BD earlier in life and is now seeking treatment (Sajatovic et al. 2005c). However, based on findings by Hirschfeld and Vornik (2004), it is not uncommon for the diagnosis of BD to have been missed previously.

13.2.2   Age of Onset

Though BD is a life-long illness, the literature tends to focus on the disease in younger individuals. Indeed, the mean age of onset is just under age 20 (Weissman et al. 1996). However, some researchers have found heuristic evidence in dividing BD into early- and late-onset subtypes. Most surveys have found that the onset of BD tends to be unimodal, with a declining incidence in first-onset mania after the age of 40. A few studies, however, have noted two peaks: the first in the early/mid 20s, and a second peak (much smaller) closer to middle age (Petterson 1977; Angst 1978; Goodwin and Jamison 1984; Kessing 2006). This bimodal distribution is more prominent in women, with the second peak occurring around the time of menopause (Sibisi 1990; Petterson 1977; Zis et al. 1979; Angst 1978). A few studies have identified a second peak occurring in males in the eighth (Spicer et al. 1973) or ninth (Sibisi 1990) decade.

More recently, Bellivier and colleagues (2001) conducted an admixture analysis of age of onset and identified three distinct subgroups of early, intermediate, and late-onset, peaking at 17, 27, and 46 years, respectively. This finding of three distinct subgroups has been subsequently replicated in other independent samples, identifying very similar mean ages of onset (Bellivier et al. 2003; Manchia et al. 2008; Hamshere et al. 2009). Thus, there is support for distinguishing subgroups of BD patients based on age of onset, but does this classification have any meaningful difference?

It should be noted that in reviewing the literature, there actually are no clear definitions of late onset. The Diagnostic and Statistical Manual of Mental Disorders, Fifth Edition (DSM-5) (American Psychiatric Association 2013) does not make a distinction, and various studies have used ages as young as 30 or as old as 60 to mark late onset (Loranger and Levine 1978; Eagles and Whalley 1985; Ghadirian et al. 1986; James 1977; Taylor and Abrams 1973; Hopkinson 1964; Sajatovic et al. 2005c; Kessing 2006; Chu et al. 2010; Oostervink et al. 2009, 2015). However, even with this variability, a few findings have stood out.

For example, several studies (Mendlewicz et al. 1972; Taylor and Abrams 1973; James 1977; Baron et al. 1981; Stenstedt 1952; Post et al. 2016; Hopkinson 1964; Snowdon 1991; Chu et al. 2010) found that patients with early onset of illness have more family members with affective disorders compared to those with a later onset of illness. Researchers have postulated that early-onset patients have a higher genetic loading than those who develop the disease later in life. Countering this argument are several other studies that found no differences in family mental illness between the two groups (Depp and Jeste 2004; Hays et al. 1998; Tohen et al. 1994; Broadhead and Jacoby 1990; Glasser and Rabins 1984; Carlson et al. 1977), though this could be obscured by the consistent finding that all patients (but especially the younger groups) had high numbers of affectively ill relatives (as many as four to 22 % in the late-onset groups) (Mendlewicz et al. 1972; Taylor and Abrams 1973; James 1977; Stenstedt 1952; Post et al. 2016).

While family inheritance data is especially indicative in the younger onset groups, the relationship of late-onset illness to neurological abnormalities is much more consistent across the studies. Though the nature of a neurological illness varied, of the five studies assessing this, three (Tohen et al 1994; Wylie et al. 1999; Almeida and Fenner 2002) showed significantly higher rates in the late-onset patients, while the other two (Broadhead and Jacoby 1990; Hays et al. 1998) showed trends toward increased levels. Further, Tamashiro and colleagues (2008) noted that BD patients with late-onset illness (illness onset after age 60) had a greater prevalence of white matter hyperintensities in the deep parietal and basal ganglia regions and more severe white matter hyperintensities in the deep frontal, parietal, and putamen regions. Subramaniam and colleagues (2007), in a cross-sectional survey of elderly BD patients, found that the late-onset group (illness onset after age 60) had a higher stroke risk score even though cognitive function and physical health were no different from the early-onset group. These data suggest that a neurological insult (especially cerebrovascular disease)—either known or silent—may induce BD, especially late-onset. The kind and location of neurological insult has yet to be determined (see below). Alternatively, it is possible that neurological developmental abnormalities that eventually cause neurological illness may also be associated with late-onset BD.

In regard to clinical symptomatology and course of treatment, age of onset may suggest only limited differences. In the European Mania In Bipolar Longitudinal Evaluation of Medications (EMBLEM) study, a two-year, prospective, observational study that evaluated treatment and outcome in 3459 BD patients (475 of whom were >60 years of age), the researchers divided the older group based on their age of onset (before or after age 50). They noted no differences in baseline severity scores between the early-onset group and the late-onset group, but more patients in the late-onset group recovered, and they recovered faster (Oostervink et al. 2009, 2015). Similarly, Carlson and colleagues (1977) noted that age of onset (adolescent-onset vs onset after age 45) as an independent variable did not predict either the course or prognosis of BD in their sample, though early-onset BD patients were less likely to experience complete episode remission during the following 24 months than late-onset BD patients (Carlson et al. 2000). Depp and colleagues (2004) also found that a later age of onset had few significant clinical differences from earlier onset, except that later onset of BD predicted lower intensity of psychiatric pathology. Biffin and colleagues (2009) reported findings from the Bipolar Comprehensive Outcomes Study in Australia and noted that the earliest-onset group had more depression, suicidal ideation, binge drinking, and poorer quality of life than the later-onset groups. Kessing (2006) reviewed data from the Denmark nationwide register and found that patients who were older at their first psychiatric hospitalization (>50 years) presented with fewer psychotic manic episodes but more severe depressive episodes with psychosis than younger patients. Finally, Chu and colleagues (2010) evaluated 61 older adults with BD and noted no significant differences on demographic or clinical variables, except for a slightly higher percentage of days spent depressed for the early-onset group.

Thus, the overall clinical usefulness of age of onset is limited, but it may prove important in genetic epidemiologic studies in order to reduce underlying genetic heterogeneity (Leboyer et al. 2005) or in neuroimaging to understand mood regulation circuitry.

13.2.3 Mortality and Comorbidity

It is well known that individuals with BD suffer a disproportionate amount of morbidity and die earlier than the general population (Sajatovic et al. 2013). Standardized mortality ratios in BD are 2.5 for men and 2.7 for women, with cardiovascular disorders, suicide, and cancer the most frequent causes of premature mortality (Osby et al. 2001; Laursen et al. 2007). For patients with BD who have survived to older age, mortality rates continue to be high (Dhingra and Rabins 1991). Further, Shulman and colleagues (1992) found that the mortality rate of elderly hospitalized BD patients was significantly higher than that of elderly hospitalized MDD patients over a 10–15 year follow-up (50 % versus 20 %). They suggested that late-life mania was either a more severe form of affective illness than MDD, with a poorer prognosis, or was associated with increased medical and neurological comorbidities.

Indeed, elderly BD patients average three to four chronic medical conditions (Lala and Sajatovic 2012), the most common being cardiovascular and neurological in origin (Depp and Jeste 2004; Beyer et al. 2005; McIntyre et al. 2007). Approximately two-thirds of elderly BD patients have hypertension and a third have diabetes (Kemp et al. 2010). Neurological diseases are especially prevalent. Shulman and colleagues (1992) compared 50 geriatric patients hospitalized for mania to 50 age-matched patients hospitalized for MDD. They found that the rates of neurological illness in manic patients were significantly higher (36 % versus 8 %), supporting the hypothesis that neurological disease is a risk factor for the development of mania in late life. A review of the stroke literature (Starkstein and Robinson 1989) demonstrated that strokes occurring in the right hemisphere (especially the limbic region) are more likely to be associated with manic symptoms than left hemispheric strokes (see also Starkstein et al. 1987, 1990, 1991).

A review of the literature reveals multiple case reports and case series that generally support a tentative association between mania and vascular risk factors and also between mania and cerebrovascular disease (Cassidy and Carroll 2002; Subramaniam et al. 2007; Wijeratne and Malhi 2007). In the limited neuroimaging literature, the few studies that have reported findings in elderly BD patients repeatedly note increased white matter hyperintense lesions compared with healthy controls (Altshuler et al. 1995; Beyer et al. 2004), a finding also noted to a lesser extent in children and adolescents with BD (Beyer et al. 2009). It is thought that these hyperintense lesions, which are not uncommon findings in aging brains, represent areas of ischemia, possibly a consequence of having more atherosclerotic risk factors. This association has been termed "vascular mania" (Steffens and Krishnan 1998). Proposed diagnostic criteria have defined a late age at onset (50 years +) subtype of mania, with associated neuroimaging and neuropsychological changes that are not specific to this age group.

Given these findings, one would also suspect that decline in cognitive functioning and dementia would be more prevalent in late-life BD. It should be realized that for all ages, having BD is associated with cognitive dysfunction (Sajatovic et al. 2013), especially in the areas of attention, working memory, executive function, verbal memory, and processing speed. While these changes are not "neurodegenerative," cognitive dysfunction in BD is thought to be a "neuroprogressive" process that includes neurodevelopment aspects, medical comorbidities, lifestyle causes, compounded by the aging process (Gildengers et al. 2012). Further, many medications used for the treatment of BD have been associated with cognitive blunting/ impairment, often worsening the aging and underlying changes of BD. Thus, cognitive problems are not uncommonly seen in late-life BD, and comorbid dementia is often seen as well, ranging from 3 % to 25 % depending on the population being assessed (Broadhead and Jacoby 1990; Stone 1989; Ponce et al. 1999; Himmelhoch et al. 1980; Sajatovic et al. 2006).

13.3 Treatment

13.3.1 Pharmacological Interventions

Treatment of BD at any age is a challenge. It is a complex disease with varying intensities of mood and behavioral alterations set in a variable cycle of frequent relapses and residual symptoms. Further, as noted above, the disorder has a high number of medical and psychiatric comorbidities that demand an individualistic treatment focused on the whole person. Finally, the high incidence of poor insight and resulting poor adherence to medications (or poorly tolerated medications) has made the disease especially challenging to control. In the last few years, there have been several publications of structured guidelines or algorithms for treatment of acute BD based on systematic reviews of the literature or expert opinions (Grunze et al. 2009, 2010; Goodwin et al. 2009; Yatham et al. 2013). These guidelines have been constructed to help clinicians navigate the complexity of pharmacotherapy in BD; however, recent studies have found that clinical practice frequently differs from guideline recommendations (Lim et al. 2001; Perlis 2005; Sachs 2003).

For the geriatric BD patient, there are four additional complications. First, the aging body may affect pharmacologic tolerance or sensitivity. Multiple pharmacological considerations, such as changes in the absorption, distribution, and elimination of medications must be understood when prescribing medications in this population (Van Gerpen et al. 1999). While we will refer to specific examples below, a fuller review can be found in Catterson and colleagues (1997). Second, aging is associated with an increasing number of medical problems (see Beyer et al. 2005) and associated medication use. A recent review of BD treatment in geriatric patients found that the average number of total medications prescribed to a patient was 8.0 ± 4.6 (range 1–24) (Beyer et al. 2008). The presence of medical problems may limit treatment options or cause problems secondary to the treatment. Further, with the increased number of medications used, there is an increased risk of problematic medication interactions. The higher number of associated medical problems may be associated with the higher mortality rate found in BD patients compared with similarly aged non-psychiatrically ill and MDD groups (Shulman et al. 1992; Dhingra and Rabins 1991). As noted previously, the incidence of a new onset mania in late life is relatively uncommon; therefore, every patient should be evaluated for potential medical illnesses that cause manic symptoms. This evaluation would include a thorough neurological examination. Also, since geriatric patients are usually taking multiple medications, these must be reviewed for a temporal association with the illness presentation. Laboratory tests should include basic health panels (complete blood count and blood chemistries) as well as a thyroid panel. Consideration should also be given to conducting a neuroimaging test such as an MRI or CT scan. This would be especially important if the new presentation includes psychosis. Third, older adults frequently have age-related psychosocial problems that potentially complicate treatment (such as loss of ability to drive or limited social support) (Sajatovic 2002; Beyer et al. 2003).

Finally, an additional challenge for optimal treatment of the geriatric BD patient is the limited data available about treatment response of older adults to BD medications, even common treatments currently approved by the FDA (Young et al. 2004). While mixed-aged studies have included some geriatric subjects, there have been no controlled prospective studies of acute or long-term management of mania in the geriatric patient population. And of the mixed-aged studies that did include older subjects, only a few have examined the effects of age within their study population (see Mirchandani and Young 1993; Young et al. 2004), thus making informed, evidence-based treatment even more difficult.

13.3.2 Treatment Efficacy

13.3.2.1 Lithium

Until the turn of the century, lithium was the most commonly prescribed medication for treatment of BD in the elderly (Oshima and Higuchi 1999; Umapathy et al. 2000; Shulman et al. 2003), despite the fact that no placebo-controlled, double-blind clinical trials had been conducted in geriatric patients. In 2004, Young and colleagues reviewed studies that reported on the use of lithium in which more than 10 elderly BD subjects were enrolled (van der Velde 1970; Himmelhoch et al. 1980; Schaffer and Garvey 1984; Chen et al. 1999). They found that 66 % of elderly manic patients improved overall, but certain groups of elderly BD patients did more poorly than others. Patients with dementia and drug abuse were found to be especially resistant to treatment, in part due to the increased difficulty in tolerating lithium. This may have contributed to the large variation in reported lithium concentrations among studies (0.3 mEq/l to 2.0 mEq/l).

The recommended lithium level for elderly BD patients has been debated. Case series (Roose et al. 1979; Prien et al. 1972) have suggested that elderly patients in acute mania may respond to lower lithium levels (0.5–0.8 mEq/l) than what is recommended for younger adults, while other reports have not found a difference (Young et al. 1992; DeBattista and Schatzberg 2006). A large retrospective study (Paton et al. 2010) in the United Kingdom (UK) found that for maintenance treatment, therapeutic threshold was comparable between old and young patients. Older patients with a lithium level under 0.4 mEq/l were more likely to relapse than those with a higher level. Chen and colleagues (1999) noted that patients who were able to achieve a serum lithium concentration ≥ 0.8 were much more improved at discharge than those who did not obtain or could not tolerate this level.

In the UK study, Paton and colleagues (2010) reported that older patients required lower doses of lithium than younger patients to achieve a therapeutic lithium level. This was also reported in a secondary analysis of the Systematic Treatment Enhancement Program for Bipolar Disorder (STEP-BD) (Al Jurdi et al. 2008). The average lithium dose in older adults was 689 mg/day, compared with 1006 mg/day in the younger patients.

While serum levels of lithium are important in guiding dosage, the correlation between serum and brain levels of lithium appear to diminish, if not disappear, in older adults (Forester et al. 2009). It is speculated that this change is due to the age-related decline in the integrity of the blood–brain barrier or changes in sodium–lithium countertransport that result in a higher ratio of brain/serum concentration than in younger patients.

Possibly the most helpful study guiding the use of lithium in older adults with BD is the recent NIMH-sponsored Geriatric Enhancement Treatment in Bipolar Disorders (GERI-BD) clinical trial, which sought to establish the efficacy and tolerability of lithium and valproate in this fragile population. Preliminary results reported on 224 BD subjects aged 60 years and older who presented in a manic, hypomanic, or mixed episode. All subjects were randomized to double-blind treatment with either lithium or valproate at a targeted level of 0.80–0.99 mEq/l or 80–99 mcg/ml, respectively, over a nine-week period. Most subjects were able to tolerate the medications and achieve the targeted plasma concentrations (lithium: 57 %; valproate: 56 %) at week nine, though at week three achievement of targeted plasma levels was relatively much lower for both groups (lithium: 35 %; valproate: 33 %). The study found that both groups had a good response to treatment; however, the effect of lithium was significantly larger, especially in the more severely manic subjects. There were no significant differences in side effects, though the lithium group did experience more tremor.

This concern about side effects has complicated dosing recommendations because older patients, especially those over the age of 70, are more likely to have adverse effects with lithium, and these occur even at "therapeutic" levels (Tueth et al. 1998; McDonald 2000). Commonly reported adverse effects of lithium in the elderly include cognitive impairment, ataxia, urinary frequency, weight gain, edema, tremor, worsening of psoriasis or arthritis, or disruption of normal thyroid activity. Thus, in practical clinical application, lithium dosing and "adequate" serum levels in the elderly are primarily determined by the patient's medical status and frailty (Young et al. 2004; Sajatovic et al. 2005a).

Guidelines for use of lithium in the elderly recommend starting at half the normally recommended dosage in younger patients because aging significantly affects lithium pharmacokinetics. Although absorption is generally unchanged, the renal clearance of lithium and the distribution volume are decreased while the elimination half life is increased (Sproule et al. 2000; Foster 1992; Shulman et al. 1987). Thus, the risk of toxicity increases in the elderly. Further, the age-related decline in renal function is compounded by the deleterious effect lithium has on the kidneys. In a cross-sectional study comparing glomerular filtration rates of 61 patients treated with lithium for about 16 years, with that of 53 patients who only receive electroconvulsive therapy (ECT), 34.4 % of the lithium-treated patients had stage 3 chronic renal disease, compared with 15.1 % of the ECT group. When patients older than 70 were evaluated separately, about 70 % of the lithium-treated patients had stage 3 chronic renal disease, compared with 36.4 % of the ECT group (Tredget et al. 2010).

Finally, medical comorbidities that may increase risk of lithium toxicity (such as dehydration, heart failure, hyponatremia, etc.) and medications commonly prescribed to the elderly (such as thiazide diuretics, nonsteroidal anti-inflammatory agents, and angiotensin-converting enzyme inhibitors) may affect lithium levels. Prior to starting lithium, a preliminary medical workup should include laboratory assessment of renal function, electrolytes, thyroid function tests, fasting blood glucose, and ECG (McDonald 2000). These should also be rechecked every few months. McDonald (2000) has also suggested that slow-release forms of lithium may be better tolerated by elderly patients. Because of all these challenges, approximately one-fifth of geriatric patients have experienced lithium toxicity (Foster 1992).

13.3.2.2 Anticonvulsants

Valproate

Approved by the FDA for the treatment of bipolar mania in 1993 (and originally approved for use as an anticonvulsant), valproate is currently the most frequently prescribed medication for the treatment of BD among the elderly (Shulman et al. 2003; Beyer et al. 2008). This may in part be due to its reported efficacy for patients with non-classic manic symptoms and prominent depressive symptoms (Evans et al. 1995; McDonald 2000). This increased use is even more remarkable considering that, similar to lithium, there are no prospective trials comparing valproate with placebo in the elderly. Rather, with the exception of the lithium/valproate comparison trial in elderly manic subjects (GERI-BD; Young et al. 2010), only retrospective and open-label studies in the geriatric population have been published.

Young and colleagues reviewed the five published studies of valproate that have included more than 10 elderly manic subjects (Chen et al. 1999; Niedermier and Nasrallah 1998; Noaguil et al. 1998; Kando et al. 1996; Puryear et al. 1995). They found that 59 % of the combined sample met the various improvement criteria, though again the dose concentrations varied widely (25–120 mcg/ml). As noted previously, the NIMH-sponsored GERI-BD trial attempted to establish efficacy and tolerability data for the use of lithium and valproate in the elderly. Valproate was as well tolerated as lithium, and both medications were equally effective (though lithium appeared to be more effective in more severely manic patients).

In the general population, recommended blood levels for valproate are 50–120 mcg/ml (Bowden and Singh 2005), though Chen and colleagues (1999) noted that higher, compared with lower, concentrations (65–90 mcg/ml) were associated with more improvement in elderly manic patients. It should be noted that the blood level measurement should be used only as a guide in treatment. As patients age, the elimination half-life of valproate may be prolonged and the free fraction of plasma valproate increased. Thus, the total valproate level (which is the most common laboratory test for valproate concentration) may underreport the amount of valproate clinically available. The clinical significance of this is unknown (Young et al. 2004; Sajatovic et al. 2005a). Further, common medications may also influence the level of valproate. For instance, aspirin can increase valproate free fraction while phenytoin and carbamazepine may decrease valproate levels. Valproate itself may influence the pharmacokinetics of other medications. It inhibits the metabolism of lamotrigine (thus requiring lower doses) and may also increase the unbound fraction of warfarin (thus requiring careful monitoring of coagulation times) (Panjehshahin et al. 1991).

Prior to initiating valproate therapy, medical workup should include liver enzymes, complete blood count (with platelets), and an ECG. Starting doses for elderly patients are 125–250 mg per day with a gradual titration every two to five days of 125–250 mg depending on the medical condition/frailty of the patient. Extended release preparations of valproate appear to be well tolerated by the elderly, but it should be noted that correctly drawn trough blood levels may need to be collected 24–36 hours after the last dose (Reed and Dutta 2006).

The most common side effects associated with valproate are nausea, somnolence, and weight gain, while less common side effects of special concern in older adults are the possibility of hair thinning, thrombocytopenia, hepatotoxicity, and pancreatitis (though the latter two are less likely to occur with age) (Bowden et al. 2002; Fenn et al. 2006). It should also be noted that valproate is available in sprinkle and liquid formulations for patients who may have difficulty swallowing. In addition, Regenold and Prasad (2001) reported on the intravenous use of valproate in three geriatric patients with good success.

Carbamazepine

Carbamazepine has been approved for the treatment of bipolar mania since 1996, while the extended release form was approved in 2005. Despite this, there is very limited information on the use of either carbamazepine preparation in elderly BD patients. The literature is currently limited only to case reports and the inclusion of some elderly patients in larger studies. Okuma and colleagues (1990) noted that seven elderly manic patients were included in the larger sample of 50 treated with carbamazepine in a double-blind study that showed good efficacy. Some researchers have suggested that in contrast to lithium, carbamazepine may best be utilized as a preferred agent in secondary mania (Evans et al. 1995; Sajatovic 2002).

Possible adverse effects associated with carbamazepine include sedation, ataxia, nystagmus/blurred vision, leukopenia, hyponatremia (secondary to SIADH), and agranulocytosis. Severe and sometimes life-threatening skin reactions have been noted to be a rare side effect. These include toxic epidermal necrolysis and Stevens–Johnson syndrome. The FDA (2007) recently recommended that patients of Asian ancestry have a genetic blood test to identify an inherited variant of the gene $HLA = B*1502$ (found almost exclusively in people of Asian ancestry) before starting therapy. Those patients testing positive should not be treated with carbamazepine.

Prior to beginning carbamazepine, medical workup should include assessment of liver enzymes, electrolytes, complete blood count, and ECG. In the elderly, carba- mazepine doses should be initiated at 100 mg either once or twice daily and gradually increased every three to five days to 400–800 mg/day (McDonald 2000). As in the younger population, targeted serum levels are between 6 and 12 mcg/l. Since carbamazepine can induce its own metabolism, dose increases may need to be adjusted in the first one to two months.

Carbamazepine is metabolized in the liver by cytochrome P450 enzyme 3A4/5. Studies in patients with epilepsy found that carbamazepine clearance decreased in an age-dependent manner, presumably due to a reduction in CYP 3A4/5 metabo- lism (Battino et al. 2003). The implication is that elderly patients may require lower doses to achieve similar levels of drug as younger patients. Carbamazepine can also alter the pharmacokinetics of other medications, including oral hormones, calcium channel blockers, cimetidine, terfenadine, and erythromycin (Sajatovic 2002).

Lamotrigine

Lamotrigine is another anticonvulsant medication that has more recently been found to be effective in the treatment of BD. Though lamotrigine has not demon- strated efficacy in the treatment of acute mania or depression in BD, it was approved by the FDA in 2003 for use in the maintenance phase. Sajatovic and colleagues (2005b) conducted a secondary analysis of two placebo-controlled, double-blind, clinical trials for maintenance therapy that had included 98 subjects over the age of 55. Focusing on this "older" group, they found that older patients on lamotrigine who had been stabilized from either an acute episode of mania or depression demonstrated a significant delay until the recurrence of another mood episode. Response was consistent with that seen in younger patients. When the results were further evaluated, the authors found that lamotrigine was significantly more effective than lithium or placebo in increasing the time-to-intervention for depressive recurrences; however lithium was more effective in increasing the time- to-intervention for manic recurrences. The mean daily dose of lamotrigine in this older group was 243 mg/day, and the mean daily dose of lithium was 736 mg/day. Sajatovic and colleagues (2011) then conducted a prospective open-label augmen- tation trial of 57 elderly subjects with bipolar depression who had been treatment resistant to current medications. Over the 13-week trial, 57 % of the subjects achieved remission while 65 % achieved treatment response (average daily dose was 114 mg). Overall, the authors found that lamotrigine was well tolerated in both studies by the older BD patients, and no increased incidence of rash was noted (Sajatovic et al. 2005b, 2007, 2011).

Lamotrigine is metabolized in the liver and eliminated through the hepatic glucuronide conjugation. Aging may decrease hepatic glucuronidation but the effect does not appear to significantly change lamotrigine dosing (Posner et al. 1991; Hussein and Posner 1997). The dose of lamotrigine should be halved when administered with valproate since valproate inhibits the metabolism of lamotrigine (Calabrese et al. 2002).

The most common adverse effects are headache and nausea, though serious skin rashes (Stevens–Johnson Syndrome) have also been reported to be associated with lamotrigine. With studies suggesting that lamotrigine may be better tolerated than lithium (Sajatovic et al. 2005b) or carbamazepine (Aldenkamp et al. 2003) in elderly patients, some researchers have suggested that lamotrigine will have an increasingly important role in late-life treatment of BD.

13.3.2.3 Antipsychotic Agents

Antipsychotic medications have been used empirically for the treatment of acute bipolar mania for many years, either as monotherapy or adjunctive treatment. However, use of conventional antipsychotics has always been problematic in the elderly because of anticholinergic effects, higher risks of extrapyramidal symptoms, and tardive dyskinesia (Sajatovic et al. 2005a). In the past two decades, the "atypical" antipsychotic agents have largely supplanted the use of conventional antipsychotics as first-line antipsychotic treatment in geriatric patients (Jeste et al. 1999). However, in the treatment of late-life BD, published controlled clinical trials of atypical antipsychotics are lacking. Most of the current practice recommendations are based on extrapolated data from mixed population trials or studies conducted in elderly populations of patients with schizophrenia or dementia. Information is especially crucial since a black box warning (FDA 2005) was added to each of the atypical antipsychotic agents indicating that clinical trials of atypical antipsychotics for the treatment of elderly patients with dementia-related psychosis had an increased risk of death compared to placebo. Presumably, the fatalities were related to increased cerebrovascular or cardiovascular incidents, problems that have been noted to be particularly relevant to patients with late-life BD.

The other major concern with use of atypical antipsychotics is increased risk of metabolic abnormalities, such as obesity, diabetes, and dyslipidemia (ADA 2004). While the elderly may have less weight gain associated with atypical antipsychotic use (Meyer 2002), each of these medical conditions are frequently observed in elderly BD patients. When using these medications, it is recommended to monitor weight, waist measurement, blood pressure, and serum glucose and lipid levels at time of initiation and periodically throughout treatment (ADA 2004).

In general, a lower-dose strategy in the elderly has been recommended for most atypical antipsychotics (Alexopoulos et al. 2004) though this may be less of a concern in the acute state.

Olanzapine

Olanzapine is FDA-approved for the treatment of bipolar mania and maintenance phases. The combination pill of olanzapine and fluoxetine was approved for the acute treatment of bipolar mania. There are limited data on its use in late-life BD. Two subanalyses have been conducted evaluating the efficacy and tolerability of older adults that were included in mixed-age, double-blind, placebo-controlled trials of olanzapine in acute mania. Street and colleagues (2000) found that in a subset of eight older manic patients (ages 61–67), those treated with olanzapine improved while those treated with placebo worsened. Beyer and colleagues (2001) conducted a pooled subanalysis of subjects over the age of 50 in three double-blind, placebo-controlled acute bipolar mania clinical trials with olanzapine and valproate. Of the 94 older adults (mean age 57), the 78 treated with either olanzapine or valproate demonstrated a significant improvement compared with those on placebo. Olanzapine and valproate were noted to be equally effective for the treatment of acute mania. The mean daily dose of olanzapine was 15.8 mg (range 5–20 mg), and the mean daily dose of valproate was 1354 mg (range 500–2500 mg). The side effects experienced by the older group were comparable with that seen in younger patients, the most common of which were dry mouth, somnolence, asthenia, and headache.

Quetiapine

Quetiapine has been FDA-approved for the treatment of acute mania and depression in BD. However, again, there are limited data concerning treatment response in elderly BD patients. Madhusoodanan and colleagues (2000) reported on a series of elderly patients with psychosis, some of whom had BD, who were treated successfully with quetiapine. Sajatovic and colleagues (2004) reported on a subanalysis of 59 older adults (mean age 63) from two 12-week double-blind, placebo-controlled studies of quetiapine in bipolar mania. They noted that both older and younger subjects responded compared to placebo, but that the older subjects had a particularly rapid and sustained reduction of symptoms apparent by Day 4. Most common adverse effects were dry mouth, somnolence, postural hypotension, insomnia, weight gain, and dizziness. Few side effects or extrapyramidal symptoms were noted. The dosing recommendations for quetiapine in bipolar depression is between 300 and 600 mg, while in mania it is up to 800 mg. Due to the occurrence of common side effects such as sedation, dizziness, and postural hypotension, geriatric patients may be started with lower doses and titrated as tolerated.

Other Atypical Antipsychotics

Risperidone is approved by the FDA for the treatment of acute bipolar mania, though again, there are very limited data regarding late-life BD use.

Madhusoodanan and colleagues (1995, 1999) reported on retrospective case reviews noting efficacy in elderly BD patients. Significant adverse events included postural hypotension and dose-dependent extrapyramidal symptoms. It is recommended that for elderly or debilitated patients, risperidone be initiated in doses of 0.5 mg once or twice a day (Sajatovic et al. 2005a) and titrated carefully.

Aripiprazole has been FDA-approved for the treatment of mania and the maintenance phase of BD. While this medication may be advantageous for use in the elderly due to less common propensity for dyslipidemias and orthostasis, the initial registration clinical trials did not include sufficient numbers of subjects aged 65 and over to determine whether they respond differently from younger patients (Kohen et al. 2010). Sajatovic and colleagues (2008) conducted a 12-week, open-label, augmentation trial with aripiprazole for 20 older adult patients with BD who had not optimally responded to their prescribed medications. They found that the addition of aripiprazole (average daily dose was 10 mg) significantly reduced mean depression and mania scores and improved overall functioning.

Asenapine is the most recent second-generation antipsychotic approved in the treatment of BD, though its FDA approval has been designated only for treatment in depressive episodes. As with all the other similar medications, there have been no double-blind, placebo-controlled studies in elderly BD patients. There have been two small, open-label studies. Baruch and colleagues (2013) treated 11 consecutively admitted elderly bipolar manic patients with asenapine 10 mg twice a day. They noted that all subjects had responded by week four and that 64 % had achieved remission. Sajatovic and colleagues (2014) reported on an open-label augmentation trial for 15 sub-optimally responding elderly BD patients. Seventy-three percent of the patients completed the study and demonstrated significant improvements in mood and functioning. Mean daily dose of asenapine augmentation was 11.2 mg. In both studies, asenapine was well tolerated though GI discomfort was reported in 33 % of patients in the augmentation study.

Clozapine is not FDA-approved for use in BD; however, it has been reported to be helpful in the treatment of bipolar mania, rapid cycling, and treatment-resistant disease. There are some limited case reports of its successful use in geriatric BD patients (Shulman et al. 1997; Frye et al. 1996). However, the adverse effects of particular concern in the elderly include sedation, postural hypotension, anticholinergic effects, and risk for seizures. Further, the potential for agranulocytosis has effectively limited its use to refractory conditions.

13.3.2.4 Electroconvulsive Therapy

ECT has been demonstrated to be very effective in the treatment of mania and mixed affective states (Mukherjee et al. 1994; Valentí et al. 2008). However, there are very limited data on the use of ECT in elderly BD patients, especially when compared with the literature on MDD (Wilkins et al. 2008). McDonald and Thompson (2001) reported on a case series of three elderly manic patients who also had some dementia who were resistant to pharmacotherapy, but did respond to

ECT treatment. Little and colleagues (2004) reported on a case series of five elderly patients with bipolar depression treated with bifrontal ECT. They found this method could be effective though a third experienced cognitive side effects. Tsao and colleagues (2004) reported on a case of a man with refractory mania who responded to acute and maintenance ECT. Frequent side effects noted in the elderly include confusion, memory impairment, and hypertension (Kujala et al. 2002).

13.3.3 Current Treatment Patterns

Evident in the above discussion is that there are very limited data available to guide an evidence-based approach to the treatment of BD in late life. Further, the data that are available focus almost exclusively on just one phase of the disorder: treatment of mania. Therefore, treatment of the depressed phases, hypomanic phases, and maintenance phases must be extrapolated from studies in younger populations. Also, with the limited data, it is unclear if certain clinical factors (such as late- versus early-onset) or biological and genetic markers may modify treatment response.

There are some data available that are descriptive of the current state of treatment in late life. Beyer and colleagues (2008) reviewed the treatment of 138 late-life BD patients experiencing an affective episode. Mood stabilizers remained the highest proportion of medications used (68 % of patients), though atypical antipsychotics were frequently used as well (54 % of patients). The latter appears to be a higher percentage of use than found in younger populations. The researchers also noted that despite there being no data on combination treatment in late life, they found polypharmacy was almost twice as common as monotherapy. This involved the use of some combination of lithium, mood stabilizers, antipsy- chotics, or antidepressants. Finally, despite using "good clinical practice," by the end of the treatment period (mean 342 days), 67 % met criteria for treatment response, but only 35 % of the patients progressed to remission.

13.3.4 Treatment Recommendations

1. In general, the history of treatment response and tolerability to specific medica- tions will present the best data for guiding current treatment selection and dosing. It is therefore essential that a good history of illness be obtained, including adverse events and related doses/concentrations (Young et al. 2004).
2. Elderly patients (especially those with a new onset of illness) should have a thorough physical and neurological exam. Laboratory evaluations should include basic metabolic panels, complete blood counts, thyroid studies, and liver function tests. Consideration should be given to vitamin B_{12} and folate

levels. Vital signs, including an orthostatic blood pressure and pulse, weight, and waist measurement, should be taken.

3. Elimination of unnecessary psychotropic agents along with conservative management may be an effective intervention by itself.

4. In the treatment of mania, monotherapy with a mood stabilizer is a reasonable first approach. The minimal duration for a medication effectiveness trial is three to four weeks (Young et al. 2004).

5. Lithium and valproate are the primary first choice options for the treatment of late-life mania. Classic manic symptoms may be more responsive to lithium, while atypical or rapid-cycling mania may be more responsive to valproate. Clinicians should target moderate concentration ranges initially (lithium 0.4–0.8; valproate 50–100); however, higher concentration ranges may be more effective acutely (lithium 0.8–1.0, valproate 65–100). Carbamazepine may be used as a second-line agent. Valproate or carbamazepine may be preferred treatments when neurological disease is present. Atypical antipsychotic medications (particularly olanzapine and quetiapine) have shown efficacy as monotherapy treatments and are increasingly being used. Special caution may be required if using these agents in elderly patients with dementia.

6. If monotherapy is only partially effective, consideration should be given to the addition of an atypical antipsychotic or another mood stabilizer.

7. In the treatment of an acute bipolar depression, monotherapy with a mood stabilizer is preferred. Lamotrigine, asenapine, and quetiapine may be especially useful for bipolar depression. Antidepressants may be used to augment the mood stabilizer or atypical antipsychotic, but should not be used as monotherapy or in rapid-cycling BD (see Table 13.1).

8. ECT should be considered in patients in an acute affective illness when they have been shown to be treatment-resistant or are suicidal and require critical intervention.

9. Effective acute treatment should be continued for six to 12 months. Ongoing treatment with a mood stabilizer is essential. If remission is sustained, a slow discontinuation of the augmenting agents may be considered. In cases of late-onset mania without previous episodes, the optimal duration of treatment is unknown (Young et al. 2004).

Table 13.1 Side effects of concern in late-life bipolar patients

Lithium: cognitive impairment, ataxia, urinary frequency, weight gain, edema, tremor, worsening of psoriasis/arthritis, diabetes insipidus, hypothyroidism
Valproate: nausea, somnolence, weight gain, hair thinning, gait disturbances, thrombocytopenia, hepatotoxicity, pancreatitis
Carbamazepine: sedation, ataxia, nystagmus/blurred vision, leucopenia, hyponatremia, agranulocytosis
Lamotrigine: headache, nausea, Stevens–Johnson Syndrome
Atypical antipsychotics: sedation, akathisia, weight gain, diabetes, dyslipidemia, stroke

13.4   Conclusions

Much is still unknown about late-life BD, especially late-onset BD (Charney et al. 2003). Generally speaking, late-life BD is a fairly common presentation to psychiatric practitioners and treatment facilities, despite the prevalence being fairly low in the community. This suggests that the disease may be difficult to manage and recurrences are not uncommon. Late-onset illness may be etiologically different than early-onset and may be related to the medical and neurological problems that can occur with aging or to an underlying progression of neurological illness associated with certain bipolar disorders early in life. The concept of "vascular mania" may be of both heuristic and treatment value in the future.

Treatment of late-life BD requires knowledge of "best treatment" practices and an understanding of the effect aging has on psychopharmacotherapy. Adequate clinical trials are not currently available to provide good evidence-based treatment recommendations for late-life BD, requiring extrapolation from trials in mixed-age populations and adaptation to the older patient.

Disclosures   The authors have no relevant disclosures to report, financial or otherwise.

References

Al Jurdi RK, Marangell LB, Petersen NJ et al (2008) Prescription patterns of psychotropic medications in elderly compared with younger participants who achieved a "recovered" status in the Systematic Treatment Enhancement Program for Bipolar Disorder. Am J Geriatr Psychiatry 16:922–933

Aldenkamp AP, De Krom M, Reijs R (2003) Newer antiepileptic drugs and cognitive issues. Epilepsia 44(Suppl 4):21–29

Alexopoulos GS, Streim J, Carpenter D, Docherty JP, Expert Consensus Panel for Using Antipsychotic Drugs in Older Patients (2004) Using antipsychotic agents in older patients. J Clin Psychiatry 65(Suppl 2):5–99

Almeida OP, Fenner S (2002) Bipolar disorder: similarities and differences between patients with illness onset before and after 65 years of age. Int Psychogeriatr 14(3):311–322

Altshuler LL, Curran JG, Hauser P, Mintz J, Denicoff K, Post R (1995) T2 hyperintensities in bipolar disorder: magnetic resonance imaging comparison and literature meta-analysis. Am J Psychiatry 152(8):1139–1144

American Diabetes Association; American Psychiatric Association; American Association of Clinical Endocrinologists; North American Association for the Study of Obesity (2004) Consensus development conference on antipsychotic drugs and obesity and diabetes. J Clin Psychiatry 65(2):267–272

American Psychiatric Association (2002) Practice guideline for the treatment of patients with bipolar disorder (revision). Am J Psychiatry 159(4 Suppl):1–50

American Psychiatric Association (2013) Diagnostic and statistical manual of mental disorders, 5th edn. American Psychiatric Association, Arlington, VA

Angst J (1978) The course of affective disorders. II. Typology of bipolar manic-depressive illness. Arch Psychaitr Nervenkr 226:65–73

Angst J, Preisig M (1995) Course of a clinical cohort of unipolar, bipolar and schizoaffective patients. Results of a prospective study from 1959 to 1985. Schweiz Arch Neurol Psychiatr 146 (1):5–16

Angst J, Baastrup P, Grof P, Hippius H, Poldinger W, Weis P (1973) The course of monopolar depression and bipolar psychoses. Psychiat Neurol Neurochir 76:489–500

Baron M, Mendlewicz J, Klotz J (1981) Age-of-onset and genetic transmission in affective disorders. Acta Psychiatr Scand 64(5):373–380

Baruch Y, Tadger S, Plopski I, Barak Y (2013) Asenapine for elderly bipolar manic patients. J Affect Disord 145(1):130–132

Battino D, Croci D, Rossini A, Messina S, Mamoli D, Perucca E (2003) Serum carbamazepine concentrations in elderly patients: a case-matched pharmacokinetic evaluation based on therapeutic drug monitoring data. Epilepsia 44(7):923–929

Bellivier F, Golmard J-L, Henry C, Leboyer M, Schurhoff F (2001) Admixture analysis of age at onset in bipolar I affective disorder. Arch Gen Psychiatry 58:510–512

Bellivier F, Golmard J-L, Rietschel M, Schulze TG, Malafosse A, Priesig M, McKeon P, Mynett-Johnson L, Henry C, Leboyer M (2003) Age at onset in bipolar I affective disorder: further evidence for three subgroups. Am J Psychiatry 160:999–1001

Beyer JL, Siegal A, Kennedy JS, Kaiser C, Tohen K, Berg P, Street J, Baker R, Tohen, M, Tollefson G, Breier A (2001) Olanzapine, divalproex, and placebo treatment non-head-to-head comparisons of older adult acute mania. Presented at the International Psychogeriatric Association, Nice, France

Beyer JL, Kuchibhatla M, Cassidy F, Looney C, Krishnan KRR (2003) Social support in elderly patients with bipolar disorder. Bipol Disord 5:22–27

Beyer JL, Kuchibhatla M, Payne ME et al (2004) Hippocampal volume measurement in older adults with bipolar disorder. Am J Geriatr Psychiatry 12:613–620

Beyer JL, Kuchibhatla M, Gersing K, Krishnan KRR (2005) Medical comorbidity in an outpatient bipolar clinical population. Neuropsychopharmacology 30(2):401–404

Beyer JL, Burchitt B, Gersing K, Krishnan KRR (2008) Patterns of pharmacotherapy and treatment response in elderly adults with bipolar disorder. Psychopharm Bull 41:102–114

Beyer JL, Young R, Kuchibhatla M, Krishnan KR (2009) Hyperintense MRI lesions in bipolar disorder: a meta-analysis and review. Int Rev Psychiatry 21:394–409

Biffin F, Tahtalian S, Filia K, Fitzgerald PB, de Castella AR, Filia S, Berk M, Dodd S, Callaly P, Berk L, Kelin K, Smith M, Montgomery W, Kulkarni J (2009) The impact of age at onset of bipolar I disorder on functioning and clinical presentation. Acta Neuropsychiatr 21(4):191–196

Bowden CL, Singh V (2005) Valproate in bipolar disorder: 2000 onwards. Acta Psychiatr Scand Suppl 426:13–20

Bowden CL, Lawson DM, Cunningham M et al (2002) The role of divalproex in the treatment of bipolar disorder. Psychiatr Ann 32(12):742–750

Broadhead J, Jacoby R (1990) Mania in old age: a first prospective study. Int J Geriatr Psychiatry 5:215–222

Calabrese JR, Shelton MD, Rapport DJ, Kimmel SE (2002) Bipolar disorders and the effectiveness of novel anticonvulsants. J Clin Psychiatry 63(Suppl 3):5–9

Carlson GA, Davenport YB, Jamison K (1977) A comparison of outcome in adolescent- and later-onset bipolar manic-depressive illness. Am J Psychiatry 134(8):919–922

Carlson GA, Bromet EJ, Sievers S (2000) Phenomenology and outcome of subjects with early- and adult-onset psychotic mania. Am J Psychiatry 157(2):213–219

Cassidy F, Carroll BJ (2002) Vascular risk factors in late onset mania. Psychol Med 32(2):359–362

Catterson ML, Preskorn SH, Martin RL (1997) Pharmacodynamic and pharmacokinetic considerations in geriatric psychopharmacology. Psychiatr Clin North Am 20(1):205–218

Charney DS, Reynolds CF III, Lewis L, Lebowitz BD, Sunderland T, Alexopoulos GS, Blazer DG, Katz IR, Meyers BS, Arean PA, Borson S, Brown C, Bruce ML, Callahan CM, Charlson ME, Conwell Y, Cuthbert BN, Devanand DP, Gibson MJ, Gottlieb GL, Krishnan KR, Laden SK, Lyketsos CG, Mulsant BH, Niederehe G, Olin JT, Oslin DW, Pearson J, Persky T, Pollock BG,

Raetzman S, Reynolds M, Salzman C, Schulz R, Schwenk TL, Scolnick E, Unutzer J, Weissman MM, Young RC, Depression and Bipolar Support Alliance (2003) Depression and Bipolar Support Alliance consensus statement on the unmet needs in diagnosis and treatment of mood disorders in late life. Arch Gen Psychiatry 60(7):664–72

Chen ST, Altshuler LL, Melnyk KA et al (1999) Efficacy of lithium vs. valproate in the treatment of mania in the elderly: a retrospective study. J Clin Psychiatry 60:181–185

Chu D, Gildengers AG, Houck PR, Anderson SJ, Mulsant BH, Reynolds CF, Kupfer DJ (2010) Does age at onset have clinical significance in older adults with bipolar disorder? Int J Geriatr Psychiatry 25(12):1266–1271

DeBattista C, Schatzberg AF (2006) Current psychotropic dosing and monitoring guidelines. Prim Psychiatry 13(6):61–81

Depp CA, Jeste DV (2004) Bipolar disorder in older adults: a critical review. Bipol Disord 6:343–367

Depp CA, Jin H, Mohamed S, Kaskow J, Moore DJ, Jeste DV (2004) Bipolar disorder I middle-aged and elderly adults: is age of onset important? J Nerv Ment Dis 192(11):796–799

Dhingra U, Rabins PV (1991) Mania in the elderly: a 5-7 year follow-up. J Am Geriatr Soc 39(6):581–583

Eagles JM, Whalley LJ (1985) Ageing and affective disorders: the age at first onset of affective disorders in Scotland, 1969-1978. Br J Psychiatry 147:180–187

Ettner S, Hermann R (1998) Inpatient psychiatric treatment of elderly Medicare beneficiaries. Psychiatr Serv 49:1173–1179

Evans DL, Byerly MJ, Greer RA (1995) Secondary mania: diagnosis and treatment. J Clin Psychiatry 56(Suppl 3):31–37

Fenn HH, Sommer BR, Ketter TA, Alldredge B (2006) Safety and tolerability of mood-stabilising anticonvulsants in the elderly. Expert Opin Drug Saf 5(3):401–416

Food and Drug Administration. Carbamazepine prescribing information to include recommendation of genetic test for patients with Asian ancestry. FDA News. December 12, 2007. Accessed February 17, 2008 at http://www.fda.gov/bbs/topics/NEWS/2007/NEW01755.html

Food and Drug Administration. FDA issues public health advisory for antipsychotic drugs used for treatment of behavioral disorders in elderly patients. FDA News. April 11 2005. Accessed March 3, 2008 at http://www.fda.gov/bbs/topics/ANSWERS/2005/ANS01350.html

Forester BP, Streeter CC, Berlow YA et al (2009) Brain lithium levels and effects on cognition and mood in geriatric bipolar disorder: a lithium-7 magnetic resonance study. Am J Geriatr Psychiatry 17:13–23

Foster JR (1992) Use of lithium in elderly psychiatric patients: a review of the literature. Lithium 3:77–93

Frye MA, Altshuler LL, Bitran JA (1996) Clozapine in rapid cycling bipolar disorder. J Clin Psychopharmacol 16(1):87–90

Ghadirian AM, Lalinec-Michaud M, Engelsmann F (1986) Early and late onset affective disorders: clinical and family characteristics. Ann R I C P 19:53–57

Gildengers AG, Chisholm D, Butters MA et al (2012) Two-year course of cognitive function and instrumental activities of daily living in older adults with bipolar disorder: evidence for neuroprogression? Psychol Med 43(4):1–11

Glasser M, Rabins P (1984) Mania in the elderly. Age Ageing 13:210–213

Goodwin GM, Consensus Group of the British Association for Psychopharmacology (2009) Evidence-based guidelines for treating bipolar disorder: revised second edition--recommendations from the British Association for Psychopharmacology. J Psychopharmacol 23(4):346–388

Goodwin FK, Jamison KR (1984) The natural course of manic depressive illness. In: Post RM, Ballenger JC (eds) Neurobiology of mood disorders. Williams & Wilkens, Baltimore, pp 20–37

Grunze H, Vieta E, Goodwin GM, Bowden C, Licht RW, Moller HJ, Kasper S (2009) The World Federation of Societies of Biological Psychiatry (WFSBP) guidelines for the biological treatment of bipolar disorders: update 2009 on the treatment of acute mania. World J Biol Psychiatry 10(2):85–116

Grunze H, Vieta E, Goodwin GM, Bowden C, Licht RW, Möller HJ, Kasper S; WFSBP Task Force On Treatment Guidelines For Bipolar Disorders (2010) The World Federation of Societies of Biological Psychiatry (WFSBP) Guidelines for the Biological Treatment of Bipolar Disorders: Update 2010 on the treatment of acute bipolar depression. World J Biol Psychiatry 11(2):81–109

Hamshere ML, Gordon-Smith K, Forty L, Jones L, Caesar S, Fraser C, Hyde S, Tredget J, Kirov G, Jones I, Craddock N, Smith DJ (2009) Age-at-onset in bipolar-I disorder: mixture analysis of 1369 cases identifies three distinct clinical sub-groups. J Affect Disord 116(1–2):23–29

Hays JC, Krishnan KR, George LK, Blazer DG (1998) Age of first onset of bipolar disorder: demographic, family history, and psychosocial correlates. Depress Anxiety 7(2):76–82

Himmelhoch J, Neil J, May S et al (1980) Age, dementia, dyskinesias, and lithium response. Am J Psychiatry 137:941–945

Hirschfeld RM, Vornik LA (2004) Recognition and diagnosis of bipolar disorder. J Clin Psychiatry 65(Suppl 15):5–9

Hirschfeld RM, Lewis L, Vornik LA (2003) Perceptions and impact of bipolar disorder: how far have we really come? Results of the national depressive and manic-depressive association 2000 survey of individuals with bipolar disorder. J Clin Psychiatry 64(2):161–174

Hopkinson G (1964) A genetic study of affective illness in patients over 50. Br J Psychiatry 110:244–254

Hussein Z, Posner J (1997) Population pharmacokinetics of lamotrigine monotherapy in patients with epilepsy: retrospective analysis of routine monitoring data. Br J Clin Pharmacol 43 (5):457–465

James NM (1977) Early and late onset bipolar affective disorder: a genetic study. Arch Gen Psychiatry 34:715–717

Jeste DV, Rockwell E, Harris MJ, Lohr JB, Lacro J (1999) Conventional vs. newer antipsychotics in elderly patients. Am J Geriatr Psychiatry 7(1):70–76

Kando JC, Tohen M, Castillo J et al (1996) The use of valproate in an elderly population with affective symptoms. J Clin Psychiatry 57:238–240

Kemp DE, Gao K, Chan PK, Canocy SJ, Findling RL, Clabrese JR (2010) Medical comorbidity in bipolar disorder: relationship between illnesses of the endocrine/metabolic system and treatment outcome. Bipolar Disord 12(4):404–413

Kessing LV (2006) Diagnostic subtypes of bipolar disorder in older versus younger adults. Bipolar Disord 8(1):56–64

Klap R, Unroe KT, Unutzer J (2003) Caring for mental illness in the United States: a focus on older adults. Am J Geriatr Psychiatry 11(5):517–524

Kohen I, Lester PE, Lam S (2010) Antipsychotic treatments for the elderly: efficacy and safety of aripiprazole. Neuropsychiatr Dis Treat 6:47–58

Kujala I, Rosenvinge B, Bekkelund SI (2002) Clinical outcome and adverse effects of electroconvulsive therapy in elderly psychiatric patients. J Geriatr Psychiatry Neurol 15(2):73–76

Lala SV, Sajatovic M (2012) Medical and psychiatric comorbidities among elderly individuals with bipolar disorder: a literature review. J Geriatr Psychiatry Neurol 25(1):20–25

Laursen TM, Munk-Olsen T, Nordentoft M, Mortensen PB (2007) Increased mortality among patients admitted with major psychiatric disorders: a register-based study comparing mortality in unipolar depressive disorder, bipolar affective disorder, schizoaffective disorder, and schizophrenia. J Clin Psychiatry 68(6):899–907

Leboyer M, Henry C, Paillere-Martinot ML, Bellivier F (2005) Age of onset in bipolar affective disorders: a review. Bipolar Disord 7(2):111–118

Lim PZ, Tunis SL, Edell WS, Jensik SE, Tohen M (2001) Medication prescribing patterns for patients with bipolar I disorder in hospital settings: adherence to published practice guidelines. Bipolar Disord 3(4):165–173

Little JD, Atkins MR, Munday J, Lyall G, Greene D, Chubb G, Orr M (2004) Bifrontal electro-convulsive therapy in the elderly: a 2-year retrospective. J ECT 20(3):139–141

Loranger A, Levine PM (1978) Age at onset of bipolar affective illness. Arch Gen Psychiatry 35:1345–1348

Madhusoodanan S, Brenner R, Araujo L, Abaza A (1995) Efficacy of risperidone treatment for psychoses associated with schizophrenia, schizoaffective disorder, bipolar disorder, or senile dementia in 11 geriatric patients: a case series. J Clin Psychiatry 56(11):514–518

Madhusoodanan S, Brecher M, Brenner R, Kasckow J, Kunik M, Negron AE, Pomara N (1999) Risperidone in the treatment of elderly patients with psychotic disorders. Am J Geriatr Psychiatry 7(2):132–138

Madhusoodanan S, Brenner R, Alcantra A (2000) Clinical experience with quetiapine in elderly patients with psychotic disorders. J Geriatr Psychiatry Neurol 13(1):28–32

Manchia M, Lampus S, Chillotti C, Sardu C, Ardau R, Severino G, Del Zompo M (2008) Age at onset in Sardinian bipolar I patients: evidence for three subgroups. Bipolar Disord 10 (3):443–446

McDonald WM (2000) Epidemiology, etiology, and treatment of geriatric mania. J Clin Psychiatry 61(Supp 13):3–11

McDonald WM, Thompson TR (2001) Treatment of mania in dementia with electroconvulsive therapy. Psychopharmacol Bull 35(2):72–82

McIntyre RS, Soczynska JK, Beyer JL, Woldeyohannes HO, Law CW, Miranda A, Konarski JZ, Kennedy SH (2007) Medical comorbidity in bipolar disorder: re-prioritizing unmet needs. Curr Opin Psychiatry 20(4):406–416

Mendlewicz J, Fieve RR, Rainer JD, Fleiss JL (1972) Manic-depressive illness: a comparative study of patients with and without a family history. Br J Psychiatry 120:523–530

Meyer JM (2002) A retrospective comparison of weight, lipid, and glucose changes between risperidone- and olanzapine-treated inpatients: metabolic outcomes after 1 year. J Clin Psychiatry 63(5):425–433

Mirchandani IC, Young RC (1993) Management of mania in the elderly: an update. Ann Clin Psychiatry 5(1):67–77

Mukherjee S, Sackeim HA, Schnur DB (1994) Electroconvulsive therapy of acute manic episodes: a review of 50 years' experience. Am J Psychiatry 151(2):169–176

Niedermier JA, Nasrallah HA (1998) Clinical correlates of response to valproate in geriatric inpatients. Ann Clin Psychiatry 10:165–168

Noagiul S, Narayan M, Nelson CJ (1998) Divalproex treatment of mania in elderly patients. Am J Geriatr Psychiatry 6:257–262

Okuma T, Yamashita I, Takahashi R et al (1990) Comparison of the antimanic efficacy of carbamazepine and lithium carbonate by double-blind controlled study. Pharmacopsychiatry 23:143–150

Oostervink F, Boomsma MM, Nolen WA (2009) Bipolar disorder in the elderly; different effects of age and age of onset. J Affect Disord 116:176–183

Oostervink F, Nolen WA, Kik RM (2015) Two years' outcome of acute mania in bipolar disorder: different effects of age and age of onset. Int J Geriatr Psychiatry 30:201–209

Osby U, Brandt L, Correia N, Ekbom AA, Sparen P (2001) Excess mortality in bipolar and unipolar disorder in Sweden. Arch Gen Psychiatry 58(9):844–850

Oshima A, Higuchi T (1999) Treatment guidelines for geriatric mood disorders. Psychiatry Clin Neurosci 53(Suppl 3):26S–31S

Panjehshahin MR, Bowman CJ, Yates MS (1991) Effect of valproic acid, its unsaturated metab-olites and same structurally related fatty acids on the binding of warfarin and dansylsacrosine to human albumin. Biochem Pharmacol 41:1227–1233

Paton C, Barnes TRE, Shingleton-Smith A et al (2010) Lithium in bipolar and other affective disorders: prescribing practice in the UK. J Psychopharmacol 24:1739–1746

Perlis RH (2005) The role of pharmacologic treatment guidelines for bipolar disorder. J Clin Psychiatry 66(Suppl 3):37–47

Petterson U (1977) Manic-depressive illness: a clinical, social and genetic study. Acta Psychiatr Scand 269:1–93

Ponce H, Kunik M, Molinari V et al (1999) Divalproex sodium treatment in elderly male bipolar patients. J Geriatr Drug Ther 12:55–63

Posner J, Holdrich T, Crome P (1991) Comparison of lamotrigine pharmacokinetics in young and elderly healthy volunteers. J Pharm Med 1:121–128

Post RM, Altshuler LL, Kupka R, McElroy SL, Frye MA, Rowe M, Grunze H, Suppes T, Keck PE Jr, Leverich GS, Nolen WA (2016) Age of onset of bipolar disorder: combined effect of childhood adversity and familial loading of psychiatric disorders. J Psychiatr Res 81:63–70

Prien RF, Caffey EM, Klett CJ (1972) Relationship between serum lithium level and clinical response in acute mania treated with lithium. Br J Psychiatry 120:409–414

Puryear LJ, Kunik ME, Workman R (1995) Tolerability of divalproex sodium in elderly psychiatric patients with mixed diagnoses. J Geriatr Psychiatry Neurol 8:234–237

Reed RC, Dutta S (2006) Does it really matter when a blood sample for valproic acid concentration is taken following once-daily administration of divalproex-ER? Ther Drug Monit 28 (3):413–418

Regenold WT, Prasad M (2001) Uses of intravenous valproate in geriatric psychiatry. Am J Geriatr Psychiatry 9(3):306–308

Roose SP, Bone S, Haidorfer C et al (1979) Lithium treatment in older patients. Am J Psychiatry 136:843–844

Sachs GS (2003) Unmet clinical needs in bipolar disorder. J Clin Psychopharmacol 23(Suppl 1): S2–S8

Sajatovic M (2002) Treatment of bipolar disorder in older adults. Int J Geriatr Psychiatry 17 (9):865–873

Sajatovic M, Mullen J, Calabrese JR (2004) Quetiapine for the treatment of bipolar mania in older adults. Presented at the American Association of Geriatric Psychiatry, Baltimore, MD

Sajatovic M, Madhusoodanan S, Coconcea N (2005a) Managing bipolar disorder in the elderly: defining the role of the newer agents. Drugs Aging 22(1):39–54

Sajatovic M, Gyulai L, Calabrese JR, Thompson TR, Wilson BG, White R, Evoniuk G (2005b) Maintenance treatment outcomes in older patients with bipolar I disorder. Am J Geriatr Psychiatry 13(4):305–311

Sajatovic M, Blow FC, Ignacio RV, Kales HC (2005c) New-onset bipolar disorder in later life. Am J Geriatr Psychiatry 13(4):282–289

Sajatovic M, Blow FC, Ignacio RV (2006) Psychiatric comorbidity in older adults with bipolar disorder. Int J Geriatr Psychiatry 21:582–587

Sajatovic M, Ramsay E, Nanry K, Thompson T (2007) Lamotrigine therapy in elderly patients with epilepsy, bipolar disorder or dementia. Int J Geriatr Psychiatry 22(10):945–950

Sajatovic M, Coconcea N, Ignacio RV, Blow FC, Hays RW, Cassidy KA, Meyer WJ (2008) Aripiprazole therapy in 20 older adults with bipolar disorder: a 12-week, open-label trial. J Clin Psychiatry 69(1):41–46

Sajatovic M, Gildengers A, Al Jurdi RK, Gyulai L, Cassidy KA, Greenberg RL, Bruce ML, Mulsant BH, Ten Have T, Young RC (2011) Multisite, open-label, prospective trial of lamotrigine for geriatric bipolar depression: a preliminary report. Bipolar Disord 13 (3):294–302

Sajatovic M, Forester BP, Gildengers A, Mulsant BH (2013) Aging changes and medical complexity in late-life bipolar disorder: emerging research findings that may help advance care. Neuropsychiatry (London) 3(6):621–633

Sajatovic M, Dines P, Fuentes-Casiano E, Athey M, Cassidy KA, Sams J, Clegg K, Locala J, Stagno S, Tatsuoka C (2014) Asenapine in the treatment of older adults with bipolar disorder. Int J Geriatr Psychiatry 30(7):710–719

Schaffer CB, Garvey MJ (1984) Use of lithium in acutely manic elderly patients. Clin Gerontol 3:58–60

Shulman KI, Mackenzie S, Hardy B (1987) The clinical use of lithium carbonate in old age: a review. Prog Neuropsychopharmacol Biol Psychiatry 11:159–164

Shulman KI, Tohen M, Satlin A, Mallya G, Kalunian D (1992) Mania compared with unipolar depression in old age. Am J Psychiatry 149(3):341–345

Shulman RW, Singh A, Shulman KI (1997) Treatment of elderly institutionalized bipolar patients with clozapine. Psychopharmacol Bull 33(1):113–118

Shulman KI, Rochon P, Sykora K, Anderson G, Mamdani M, Bronskill S, Tran CT (2003) Changing prescription patterns for lithium and valproic acid in old age: shifting practice without evidence. BMJ 326(7396):960–961

Sibisi CDT (1990) Sex differences in the age of onset of bipolar affective illness. Br J Psychiatry 156:842–845

Snowdon J (1991) A retrospective case-note study of bipolar disorder in old age. Br J Psychiatry 158:485–490

Spicer CC, Hare EH, Slater E (1973) Neurotic and psychotic forms of depressive illness: evidence from age-incidence in a national sample. Br J Psychiatry 123:535–541

Sproule BA, Hardy BG, Shulman KI (2000) Differential pharmacokinetics of lithium in elderly patients. Drugs Aging 16(3):165–177

Starkstein SE, Robinson RG (1989) Affective disorders and cerebral vascular disease. Br J Psychiatry 154:170–182

Starkstein SE, Robinson RG, Price TR (1987) Comparison of cortical and subcortical lesions in the production of post-stroke mood disorders. Brain 110:1045–1059

Starkstein SE, Mayberg HS, Berthier ML, Fedoroff P, Price TR, Dannals RF, Wagner HN, Leiguarda R, Robinson RG (1990) Mania after brain injury: neuroradiological and metabolic findings. Ann Neurol 27(6):652–659

Starkstein SE, Fedoroff P, Berthier ML, Robinson RG (1991) Manic-depressive and pure manic states after brain lesions. Biol Psychiatry 29(2):149–158

Steffens DC, Krishnan KR (1998) Structural neuroimaging and mood disorders: recent findings, implications for classification, and future directions. Biol Psychiatry 43(10):705–712

Stenstedt A (1952) Study in manic-depressive psychosis: clinical, social, and genetic investigations. Acta Psychiatr Neurol Scand Suppl 79:1–111

Stone K (1989) Mania in the elderly. Br J Psychiatry 155:220–224

Street J, Tollefson GD, Tohen M et al (2000) Olanzapine for the psychotic conditions in the elderly. Psychiatr Ann 30(3):191–196

Subramaniam H, Dennis MS, Byrne EJ (2007) The role of vascular risk factors in late onset bipolar disorder. Int J Geriatr Psychiatry 22(8):733–737

Tamashiro JH, Zung S, Zanetti MV, de Castro CC, Vallada H, Busatto GH, de Toledo Ferraz Alves TC (2008) Increased rates of white matter hyperintensities in late-onset bipolar disorder. Bipolar Disord 10(7):765–775

Taylor MA, Abrams R (1973) Manic states: a genetic study of early and late onset affective disorders. Arch Gen Psychiatry 28:656–658

Tohen M, Shulman KI, Satlin A (1994) First-episode mania in late life. Am J Psychiatry 151 (1):130–132

Tredget J, Kirov A, Krov G (2010) Effects of chronic lithium treatment on renal function. J Affect Disord 126:436–440

Tsao CI, Jain S, Gibson RH, Guedet PJ, Lehrmann JA (2004) Maintenance ECT for recurrent medication-refractory mania. J ECT 20(2):118–119

Tueth MJ, Murphy TK, Evans DL (1998) Special considerations: use of lithium in children, adolescents, and elderly populations. J Clin Psychiatry 59(Suppl 6):66–73

Umapathy C, Mulsant BH, Pollock BG (2000) Bipolar disorder in the elderly. Psychiatr Ann 30:473–480

Unutzer J, Simon G, Pabiniak C, Bond K, Katon W (1998) The treated prevalence of bipolar disorder in a large staff-model HMO. Psychiatr Serv 49(8):1072–1078

Valentí M, Benabarre A, García-Amador M, Molina O, Bernardo M, Vieta E (2008) Electroconvulsive therapy in the treatment of mixed states in bipolar disorder. Eur Psychiatry 23(1):53–56
van der Velde CD (1970) Effectiveness of lithium carbonate in the treatment of manic-depressive
  illness. Am J Psychiatry 123:345–351
Van Gerpen MW, Johnson JE, Winstead DK (1999) Mania in the geriatric patient population: a
  review of the literature. Am J Geriatr Psychiatry 7(3):188–202
Weissman MM, Leaf PJ, Tischler GL, Blazer DG, Karno M, Bruce ML, Florio LP (1988)
  Affective disorders in five United States communities. Psychol Med 18(1):141–153
Weissman MM, Bland RC, Canino GJ, Faravelli C, Greenwald S, Hwu HG, Joyce PR, Karam EG,
  Lee CK, Lellouch J, Lepine JP, Newman SC, Rubio-Stipec M, Wells JE, Wickramaratne PJ,
  Wittchen H, Yeh EK (1996) Cross-national epidemiology of major depression and bipolar
  disorder. JAMA 276(4):293–299
Wijeratne C, Malhi GS (2007) Vascular mania: an old concept in danger of sclerosing? A clinical
  overview. Acta Psychiatr Scand Suppl 434:35–40
Wilkins KM, Ostroff R, Tampi RR (2008) Efficacy of electroconvulsive therapy in the treatment
  of nondepressed psychiatric illness in elderly patients: a review of the literature. J Geriatr
  Psychiatry Neurol 21(1):3–11
Winokur G (1975) The Iowa 500: heterogeneity and course in manic-depressive illness (Bipolar).
  Comp Psychiatry 16(2):125–131
Wylie M, Mulsant B, Pollock B et al (1999) Age at onset in geriatric bipolar disorder. Am J Geriatr
  Psychiatry 7:77–83
Yatham LN, Kennedy SH, Parikh SV, Schaffer A, Beaulieu S, Alda M, O'Donovan C,
  Macqueen G, McIntyre RS, Sharma V, Ravindran A, Young LT, Milev R, Bond DJ, Frey
  BN, Goldstein BI, Lafer B, Birmaher B, Ha K, Nolen WA, Berk M (2013) Canadian Network
  for Mood and Anxiety Treatments (CANMAT) and International Society forBipolar Disorders
  (ISBD) collaborative update of CANMAT guidelines for the management of patients with
  bipolar disorder: update 2013. Bipolar Disord 15(1):1–44
Young RC, Kalayam B, Tsuboyama G et al (1992) Mania: response to lithium across the age
  spectrum (abstract). Soc Neurosci 18:669
Young RC, Gyulai L, Mulsant BH, Flint A, Beyer JL, Shulman KI, Reynolds CF III (2004)
  Pharmacotherapy of bipolar disorder in old age: review and recommendations. Am J Geriatr
  Psychiatry 12(4):342–357
Young RC, Schulberg HC, Gildengers AG, Sajatovic M, Mulsant BH, Gyulai L, Beyer J,
  Marangell L, Kunik M, Ten Have T, Bruce ML, Gur R, Marino P, Evans JD, Reynolds CF
  III, Alexopoulos GS (2010) Conceptual and methodological issues in designing a randomized,
  controlled treatment trial for geriatric bipolar disorder: GERI-BD. Bipolar Disord 12(1):56–67
Zis AP, Grof P, Goodwin FK (1979) The natural course of affective disorders: Implications for
  lithium prophylaxis. In: Cooper TB, Gershon S, Kline NS, Shou M (eds) Lithium: controversies and unsolved issues. Exerpta Medica, Amsterdam, pp 381–398

Chapter 14
Treatment of Childhood-Onset Bipolar Disorder

Dana Baker Kaplin and Robert L. Findling

Abstract Pediatric bipolar disorder (PBD) is a serious condition, and a substantive need exists for evidence-based treatments for children and adolescents suffering from this condition. A fundamental intervention used in this patient population is pharmacotherapy. Despite the importance of medication treatment in this patient population, only limited amounts of methodologically stringent data exist pertaining to this form of intervention. The evidence that does exist suggests that some psychotropic medications can provide salutary effects for youths suffering from bipolar I disorder (BD-I). Also relevant is the pharmacotherapy of genetically at-risk children suffering from bipolar spectrum disorder and treatment of psychiatric comorbidities. Further placebo-controlled trials are needed in order to better characterize the efficacy and safety of psychotropic medications in this population.

Keywords Pediatric bipolar disorder • Pharmacotherapy • Acute treatment • Maintenance therapy • Children • Adolescents

14.1 Introduction

Pediatric bipolar disorder (PBD) is a chronic and pernicious condition (Birmaher and Axelson 2006; Findling et al. 2013b). It is associated with serious psychosocial dysfunction, substantive affective symptomatology, and high rates of psychiatric

D.B. Kaplin, MPH
Division of Child and Adolescent Psychiatry, Department of Psychiatry and Behavioral Sciences, Johns Hopkins University School of Medicine, Baltimore, MD, USA

R.L. Findling, MD, MBA (✉)
Division of Child and Adolescent Psychiatry, Department of Psychiatry and Behavioral Sciences, Johns Hopkins University School of Medicine, Baltimore, MD, USA

The Johns Hopkins Hospital, Bloomberg Children's Center; Kennedy Krieger Institute, 1800 Orleans Street, Baltimore, MD 21287, USA
e-mail: rfindli1@jhmi.edu

© Springer International Publishing Switzerland 2016   315
C.A. Zarate Jr., H.K. Manji (eds.), *Bipolar Depression: Molecular Neurobiology, Clinical Diagnosis, and Pharmacotherapy*, Milestones in Drug Therapy,
DOI 10.1007/978-3-319-31689-5_14

comorbidity (McClellan et al. 2007; Pavuluri et al. 2005a). For these reasons, evidence-based treatments for this condition and the comorbid illnesses that accompany it are needed.

Pharmacological therapy is considered to be a fundamental intervention for young people suffering from PBD (Kowatch et al. 2005; McClellan et al. 2007). In order to provide evidence-based care, clinicians who treat children and adolescents should be familiar with pharmacotherapy studies that have been specifically conducted in pediatric patients rather than assume that research that has been performed in adults is necessarily applicable to juveniles. The efficacy and safety profiles of psychotropic agents are frequently different in young patients and adults (Wiznitzer and Findling 2003).

Similarly, the manifestations of bipolarity may differ across the life cycle. Kraepelin initially observed that the amount of time patients spent in the depressed phase of bipolar disorder (BD) increased from adolescence through adulthood (Kraepelin 1921). Recent data from pediatric studies have reported that most youth suffering from BD do not, in fact, frequently suffer from major depressive episodes, but most commonly experience manic and mixed states (Pavuluri et al. 2005a). Although ideally this chapter should focus on pediatric bipolar depression, there is such a paucity of data on this topic for this age group that the chapter will instead predominantly focus on our current knowledge of pharmacotherapy in PBD. The knowledge gleaned from this work may be helpful in eventually designing such studies for pediatric bipolar depression in particular.

14.2   Acute Monotherapy Treatment Studies of Manic or Mixed States

14.2.1   Lithium, Anticonvulsants, and Omega-3 Fatty Acids

Open-label data from multiple reports suggest that lithium may have salutary effects in youths suffering from manic or mixed states (Findling et al. 2008, 2011). Until recently, there were no scientifically rigorous studies with adequate statistical power specifically testing the role of lithium in the treatment of acute manic states in PBD. However, recent results from an NICHD-sponsored multicenter, randomized, placebo-controlled study—the Collaborative Lithium Trials (CoLT)—found that lithium significantly improved symptoms of BD-I in youth (Findling et al. 2015b). Specifically, this clinical trial examined the efficacy of lithium vs. placebo in youth ages seven to 17 diagnosed with BD-I with manic or mixed states (Findling et al. 2015b).

One relatively large prospective open-label study evaluated carbamazepine. Results from that trial suggest possible benefits in the acute treatment of symptomatic youths suffering from PBD (Findling and Ginsberg 2014). Although some studies have suggested that divalproex sodium may be helpful in the treatment of youths with manic or mixed states (DelBello et al. 2006; Wagner et al. 2002;

Table 14.1 Selected prospective, acute, placebo-controlled, mood stabilizer studies in PBD

Study	Medication	Dose	Patient diagnoses	Sample size (each arm)	Treatment duration	Results
Wagner et al. (2009)	Divalproex sodium (DVPX)	Target serum level 80–125 mcg/mL	BD-I, manic or mixed	DVPX = 76 Placebo = 74	4 weeks	DVPX not superior to placebo; ≥50 % YMRS improvement: DVPX = 24 %; placebo = 23 %
Kowatch et al. (2007)	DVPX; lithium	Target DVPX serum level 85–110 mcg/mL	BD-I, manic or mixed	DVPX = 56 Lithium = 66 Placebo = 31	8 weeks	DVPX superior to placebo; CGI-I of 1 or 2: DVPX = 53 % Lithium = 42 % Placebo = 29 %
Findling et al. (2015b)	Lithium	Target lithium level <1.4 mEq/L	BD-I, manic or mixed	Lithium = 53 Placebo = 28	8 weeks	Lithium superior to placebo
DelBello et al. (2005)	Topiramate	Target dose 400 mg/day	BD-I, manic or mixed	Topiramate = 29 Placebo = 27	4 weeks	Inconclusive owing to early study termination; however, possible benefit from active treatment was observed
Wagner et al. (2006)	Oxcarbazepine	Mean dose 1515 mg/day	BD-I, manic or mixed	Oxcarbazepine = 59 placebo = 57	7 weeks	No significant difference on change in YMRS from baseline to endpoint
Haas et al. (2009)	Risperidone	0.5–2.5 mg/day 3–6 mg/day	BD-I, manic or mixed	Risperidone low = 50 Risperidone high = 61 Placebo = 58	3 weeks	Both doses of risperidone superior to placebo; ≥50 % YMRS improvement: Risperidone low = 59 % Risperidone high = 63 % Placebo = 26 %

(continued)

Table 14.1 (continued)

Study	Medication	Dose	Patient diagnoses	Sample size (each arm)	Treatment duration	Results
Tohen et al. (2007)	Olanzapine	2.5–20 mg/day; mean daily dose: 8.9 mg	BD-I, manic or mixed	Olanzapine = 107 Placebo = 54	3 weeks	Olanzapine superior to placebo; ≥50 % YMRS improvement: Olanzapine = 48.6 % Placebo = 22.2 %
Pathak et al. (2013)	Quetiapine	400 mg/day 600 mg/day	BD-I, manic	Quetiapine 400 = 93 Quetiapine 600 = 95 Placebo = 89	3 weeks	Both doses of quetiapine superior to placebo; ≥50 % YMRS improvement: Quetiapine 400 = 64 % Quetiapine 600 = 58 % Placebo = 37 %
Findling et al. (2013b)	Aripiprazole	10 or 30 mg	BD-I, manic or mixed	Aripiprazole 10 = 98 Aripiprazole 30 = 99 Placebo = 99	4 weeks	Both doses of aripiprazole superior to placebo; >50 % YMRS improvement: Aripiprazole 10 = 44.8 % Aripiprazole 30 = 63.6 % Placebo = 26.1 %
Findling et al. (2015a)	Asenapine	2.5 mg or 5 mg or 10 mg/day	BD-I, manic or mixed	Asenapine = 403	3 weeks	All doses of asenapine were effective in reducing symptoms associated with manic or mixed episodes; no significant differences in participants with or without ADHD

| DelBello et al. (2006) | DVPX + quetiapine | DVPX serum level 80–130 mg/dL, quetiapine 450 mg/day | BD-I, manic or mixed | DVPX + quetiapine = 15 DVPX + placebo = 15 | 6 weeks | DVPX plus quetiapine more effective in manic symptom reduction than DVPX monotherapy; YMRS response rate: DVPX + quetiapine = 87 % DVPX + placebo = 53 % |
| Detke et al. (2015) | Olanzapine/ fluoxetine | 3 mg olanzapine/25 mg fluoxetine; 6 mg olanzapine/25 mg fluoxetine, day 3; 6 mg olanzapine/50 mg fluoxetine, wk 1; 12 mg olanzapine 50 mg fluoxetine, wk 2 | BD-I, depression | OFC = 170 placebo = 85 | 8 weeks | OFC superior to placebo |

BD-I = bipolar I disorder; DVPX = divalproex sodium; YMRS = Young Mania Rating Scale; CGI-I = Clinical Global Impressions-Improvement Scale

Pavuluri et al. 2005b), one placebo-controlled study reported that it was not superior to placebo (Wagner et al. 2009) (see Table 14.1). Conversely, a double-blind, placebo-controlled trial comparing divalproex sodium and lithium showed that while divalproex sodium was more efficacious than placebo, lithium was not. However, lithium was associated with a greater response, albeit not statistically significant, compared to placebo. It is possible that lithium did not demonstrate superiority to placebo owing to the doses employed in the trial and based on the relatively modest sample size (Kowatch et al. 2007).

Data from a placebo-controlled trial suggest that topiramate might have efficacy in PBD (Delbello et al. 2005). However, this topiramate study was statistically underpowered because it was stopped prior to its completion, when adult trials failed to demonstrate topiramate's efficacy.

Another double-blind and placebo-controlled study of oxcarbazepine showed no superiority for active treatment when compared to placebo (Wagner et al. 2006). There are no published double-blind or prospective studies that have examined other anticonvulsants in the acute treatment of mixed or manic states in PBD.

One published, prospective, open-label study examined omega-3 fatty acids in this patient population and found that omega-3 fatty acids may have modest salutary effects on manic symptomatology (Wozniak et al. 2007). Subsequently, results from a randomized, placebo-controlled study of flax oil, which contains omega-3 fatty acids, suggest a decrease in the severity of symptoms associated with PBD (Gracious et al. 2010).

14.2.2 Atypical Antipsychotics

The medications that have the best data to support their use in the acute treatment of PBD are the atypical antipsychotics (see Table 14.1). There is evidence from double-blind and placebo-controlled studies that risperidone (Haas et al. 2009), olanzapine (Tohen et al. 2007), quetiapine (Pathak et al. 2013), aripiprazole (Findling et al. 2013b), and, most recently, asenapine (Findling et al. 2015a) are all effective in the treatment of acute manic/mixed states for older children and adolescents (10–17 years).

There are also data from an open-label study to suggest that ziprasidone at doses of 80 or 160 mg/day may be effective in youths ages 10–18 years of age suffering from symptoms of mania (Findling et al. 2013a). Additionally, data from other open-label studies and case reviews suggest that risperidone (Biederman et al. 2005b, c), olanzapine (Biederman et al. 2005b; Frazier et al. 2001), quetiapine (Marchand et al. 2004), ziprasidone (Barnett 2004), and aripiprazole (Barzman et al. 2004; Biederman et al. 2005a) may all be associated with reductions in affective symptoms and are reasonably well tolerated in the short term. Data from a double-blind study indicated that ziprasidone may have efficacy and acceptable tolerability in this patient population (Findling et al. 2013a). A recently published open-label study in this patient population found that paliperidone was beneficial in treating bipolar spectrum disorders and other associated illness such as attention deficit hyperactivity disorder (ADHD) (Joshi et al. 2013).

For those drugs and patient populations for which placebo-controlled data are not available, the question of whether or not these agents truly have acute efficacy remains unanswered. In addition, owing to metabolic and tolerability considerations associated with the atypical antipsychotics, it is not entirely clear whether these drugs' side effect profiles are acceptable over the long term. These are key empiric questions that deserve further study.

14.3   Acute Combination Pharmacotherapy Studies

As indicated above, there are data of varying degrees of methodological stringency to support the assertion that drug monotherapy with certain agents may benefit youths with PBD. However, open-label and double-blind studies consistently report that many youths remain substantively symptomatic despite benefiting from treatment with one drug. As a result, investigators have begun to explore the safety and tolerability of combination thymoleptic drug strategies in the treatment of youths with BD.

One study examined combination pharmacotherapy using a prospective, randomized, controlled design. In that clinical trial, treatment with divalproex sodium plus quetiapine was reported to be associated with superior symptomatic relief when compared to treatment with divalproex sodium plus placebo (DelBello et al. 2002). It should be noted that the youths who received both active medications had a higher rate of sedation than those who received divalproex sodium monotherapy. Other combinations that have been studied in open-label prospective studies include lithium plus divalproex sodium (Kowatch et al. 2003; Findling et al. 2003, 2006), lithium plus either a neuroleptic or risperidone (Kafantaris et al. 2003; Pavuluri et al. 2004), and divalproex sodium plus risperidone (Pavuluri et al. 2004). These data provide preliminary evidence to support the use of combination thymoleptic treatment strategies in PBD.

14.4   Acute Treatment of Depressive Episodes

As mentioned above, the presentation of major depressive episodes appears to be substantially less common in PBD than manic or mixed episodes. This phenomenon probably partly explains why there is a paucity of data regarding the treatment of the depressed phase of PBD.

One open-label study noted that lithium monotherapy was efficacious in the treatment of 27 adolescents with bipolar depression (Patel et al. 2006). In addition, another open-label prospective study found that lamotrigine, either as monotherapy or as an adjunctive treatment, was associated with salutary effects in a group of

20 adolescents with bipolar depression (Chang et al. 2006). The fact that high placebo-response rates have been observed in youths with major depressive disorder makes the open-label design of these two studies a key methodological consideration. Further, it should be noted that these youths, although suffering from prominent and substantive symptoms of depression, also were experiencing concomitant manic symptomatology at study baseline. In both clinical trials, average pretreatment Young Mania Rating Scale (YMRS) (Young et al. 1978) scores were above 15.

Although the depressed phase of illness appears to be less common in PBD than in adults with BD, findings from several recent placebo-controlled clinical trials have been published. In two double-blind studies of quetiapine in treating the depressed phase of PBD, both failed to demonstrate separation from placebo with quetiapine, and both had a substantive placebo response (DelBello et al. 2009; Findling et al. 2014b).

Conversely, in a randomized, double-blind, placebo-controlled study of youth aged 10–17 years in an acute phase of bipolar depression, treatment with an olanzapine/fluoxetine combination (OFC; brand name Symbyax®) was found to be effective (Detke et al. 2015). Although OFC was generally well-tolerated, weight gain was found to be a significant side effect of treatment (Detke et al. 2015). At present, OFC is the only agent with FDA approval for the treatment of bipolar depression in pediatric-aged patients (see also http://www.fda.gov/down loads/Drugs/DrugSafety/ucm089140.pdf).

In the absence of more stringent methodological treatment data in youths suffering from the depressed phase of PBD, recommended therapeutic approaches have also focused on concerns regarding the potential for mania, mixed states, and cycling acceleration associated with antidepressants (Kowatch et al. 2002). It has been suggested that antidepressants should not be prescribed to youths unless the patient is being prescribed a concomitant mood stabilizer (Kowatch et al. 2005).

14.5 Maintenance Therapy: Randomized Studies

Because PBD is a chronic condition, it is likely that these patients will receive long-term pharmacotherapy. Unfortunately, there are limited data pertaining to maintenance drug treatment in this patient population.

In one discontinuation study, 40 adolescents (ages 12–18 years) who responded to treatment with lithium were randomized to either continued lithium treatment or placebo substitution for two weeks (Kafantaris et al. 2004). Overall, the authors found that 57.5 % of youths experienced a significant clinical exacerbation. Moreover, the rate of exacerbation did not differ between the lithium- and placebo-treated groups.

In another study that employed a discontinuation paradigm, investigators took 60 children and adolescents (ages five to 17 years) who achieved clinical stabilization while being treated with a combination regimen of lithium and divalproex sodium and randomized them in equal numbers to receive either drug monotherapy

(Findling et al. 2005a). The investigators found that both agents were equally effective in maintaining clinical stability. However, it is noteworthy that approximately half of the subjects ended study participation within approximately 90 days of randomization.

An additional study randomized, in approximately equal numbers, 296 patients (ages 10–17 years) with pediatric bipolar mania to four weeks of acute treatment with aripiprazole 10 mg, aripiprazole 30 mg, or placebo followed by a 26-week continuation phase (Findling et al. 2013b). Over 30 weeks of treatment (four weeks acute and 26 weeks continuation), both doses of aripiprazole (10 and 30 mg) demonstrated superiority to placebo in the treatment of these youths (Findling et al. 2013b).

There is currently one adjunctive, randomized, placebo-controlled, double-blind withdrawal study of lamotrigine in youth ages 10–17 already receiving pharmacological treatment for BD-I (Findling et al. 2014a). Although the overall study results did not show separation of active medication from placebo, the data suggest that adjunctive maintenance treatment with lamotrigine delayed the time to recurrence of a mood episode in a study subgroup of 13–17 year olds, but not in the younger patients (Findling et al. 2014a).

14.6 Genetically At-Risk Populations

A treatment strategy that has received some scientific investigation has been the evaluation of medication therapy early in the course of PBD, before youths develop BD-I or BD-II. The focus of this research is particularly interesting, because if effective early interventions can be identified for these patients, it is hoped that: (1) the dysfunction and symptomatology that occur during this period of time may be reduced (Findling et al. 2005b) and (2) that diagnostic evolution to more malignant forms of bipolarity can be prevented.

In one study, 30 depressed children ages six to 12 years who were at risk for developing BD were randomized to receive either lithium or placebo for six weeks (Geller et al. 1998b). Symptom amelioration was found both for youths who received placebo and for those who received active medication. Perhaps more importantly, the degree of clinical improvement did not differ between the two treatment groups.

In another study, 24 youths ages six to 18 years who were the offsprings of a parent with BD and who also suffered from at least mild affective symptomatology were treated with open-label divalproex sodium for up to 12 weeks (Chang et al. 2003). In that trial, the authors found that 78 % of the study subjects improved. In a randomized, double-blind study, Findling and colleagues treated 56 symptomatic at-risk youths who were between the ages of five and 17 years with either divalproex sodium or placebo (Findling et al. 2007a). The authors found that treatment with either placebo or divalproex therapy was associated with substantive clinical benefit. In addition, no between-group differences in treatment response were found. The results of this latter study highlight the need to conduct placebo-controlled studies in this patient population.

Finally, in another study, 20 adolescents ages 12–18 years who were both affectively symptomatic and who also had a first degree relative with BD-I benefited from 12 weeks of treatment with quetiapine (DelBello et al. 2007). These findings suggest that further research with quetiapine in this patient population is indicated.

14.7 Treatment of Psychiatric Comorbidities

As noted above, psychiatric comorbidity is quite common in PBD. Unfortunately, there are limited double-blind data about the treatment of comorbid conditions in these youths. At present, clinical trials that have considered interventions in this patient population have been limited to those youths suffering either from concomitant ADHD or substance abuse.

14.7.1 ADHD

ADHD is a common comorbid condition in PBD. Two published studies have examined the acute safety and efficacy of adjunctive psychostimulants in these youths.

In one randomized, placebo-controlled trial, Scheffer and colleagues treated 40 patients between the ages of six and 17 years suffering from PBD and ADHD with open-label divalproex sodium (Scheffer et al. 2005). Using a cross-over design, the effects of mixed amphetamine salts (MAS) were compared to placebo. The authors found that treatment with MAS was superior to placebo in reducing ADHD symptoms. Moreover, the authors also found that worsening of manic symptoms did not occur with MAS administration.

In another study, 16 euthymic youths who were receiving mood stabilizer treatment received various doses of methylphenidate as well as placebo in a crossover design (Findling et al. 2007b). The authors found that treatment with methylphenidate was superior to placebo in treating ADHD symptoms. In addition, the authors found that methylphenidate administration was not associated with mood destabilization.

Finally, in their aforementioned maintenance trial, Findling and colleagues noted that adjunctive treatment with psychostimulants was not associated with detrimental effects during drug monotherapy with either lithium or divalproex sodium (Findling et al. 2005a). When these studies are considered as a group, they provide preliminary evidence to support the use of adjunctive psychostimulants in the treatment of ADHD in youths with ADHD who are already receiving mood stabilizers.

14.7.2 Substance Abuse

Geller and colleagues treated a group of 25 teenagers with BD and secondary substance abuse with either placebo or lithium for up to six weeks (Geller et al. 1998a). The investigators found that treatment with lithium was associated with greater reductions in substance abuse and superior improvements in overall functioning compared to placebo.

14.8 Conclusions

Until recently, there has only been a modest amount of research pertaining to the pharmacotherapy of PBD. Several recently conducted placebo-controlled trials have failed to show efficacy for some drugs. Other placebo-controlled trials, particularly those that have studied the atypical antipsychotics, have identified drugs that are superior to placebo in the acute symptomatic amelioration of manic symptoms. However, there is still a substantive need to evaluate the long-term safety and efficacy of psychotropic agents in these vulnerable youths. Besides long-term safety, other areas of research deserving further study include the depressed phase of illness and the treatment of psychiatric comorbidity in PBD. Hopefully, even more research will become available in the near future so that clinicians can have adequate information to provide care to their patients based on methodologically stringent research.

Acknowledgments/Disclosures Dr. Findling receives or has received research support, acted as a consultant, and/or served on a speaker's bureau for Alcobra, American Academy of Child & Adolescent Psychiatry, American Physician Institute, American Psychiatric Press, AstraZeneca, Bracket, Bristol-Myers Squibb, CogCubed, Cognition Group, Coronado Biosciences, Dana Foundation, Elsevier, Forest, GlaxoSmithKline, Guilford Press, Johns Hopkins University Press, Johnson and Johnson, Jubilant Clinsys, KemPharm, Lilly, Lundbeck, Merck, NIH, Neurim, Novartis, Noven, Otsuka, Oxford University Press, Pfizer, Physicians Postgraduate Press, Purdue, Rhodes Pharmaceuticals, Roche, Sage, Shire, Sunovion, Supernus Pharmaceuticals, Transcept Pharmaceuticals, Validus, and WebMD. Ms. Kaplin has no conflict of interest to disclose, financial or otherwise.

References

Barnett MS (2004) Ziprasidone monotherapy in pediatric bipolar disorder. J Child Adolesc Psychopharmacol 14(3):471–477
Barzman DH, DelBello MP, Kowatch RA, Gernert B, Fleck DE, Pathak S, Rappaport K, Delgado SV, Campbell P, Strakowski SM (2004) The effectiveness and tolerability of aripiprazole for pediatric bipolar disorders: a retrospective chart review. J Child Adolesc Psychopharmacol 14 (4):593–600

Biederman J, McDonnell MA, Wozniak J, Spencer T, Aleardi M, Falzone R, Mick E (2005a) Aripiprazole in the treatment of pediatric bipolar disorder: a systematic chart review. CNS Spectr 10(2):141–148

Biederman J, Mick E, Hammerness P, Harpold T, Aleardi M, Dougherty M, Wozniak J (2005b) Open-label, 8-week trial of olanzapine and risperidone for the treatment of bipolar disorder in preschool-age children. Biol Psychiatry 58(7):589–594

Biederman J, Mick E, Wozniak J, Aleardi M, Spencer T, Faraone SV (2005c) An open-label trial of risperidone in children and adolescents with bipolar disorder. J Child Adolesc Psychopharmacol 15(2):311–317

Birmaher B, Axelson D (2006) Course and outcome of bipolar spectrum disorder in children and adolescents: a review of the existing literature. Dev Psychopathol 18(4):1023–1035

Chang KD, Dienes K, Blasey C, Adleman N, Ketter T, Steiner H (2003) Divalproex monotherapy in the treatment of bipolar offspring with mood and behavioral disorders and at least mild affective symptoms. J Clin Psychiatry 64(8):936–942

Chang K, Saxena K, Howe M (2006) An open-label study of lamotrigine adjunct or monotherapy for the treatment of adolescents with bipolar depression. J Am Acad Child Adolesc Psychiatry 45(3):298–304

DelBello MP, Schwiers ML, Rosenberg HL, Strakowski SM (2002) A double-blind, randomized, placebo-controlled study of quetiapine as adjunctive treatment for adolescent mania. J Am Acad Child Adolesc Psychiatry 41(10):1216–1223

Delbello MP, Findling RL, Kushner S, Wang D, Olson WH, Capece JA, Fazzio L, Rosenthal NR (2005) A pilot controlled trial of topiramate for mania in children and adolescents with bipolar disorder. J Am Acad Child Adolesc Psychiatry 44(6):539–547

DelBello MP, Kowatch RA, Adler CM, Stanford KE, Welge JA, Barzman DH, Nelson E, Strakowski SM (2006) A double-blind randomized pilot study comparing quetiapine and divalproex for adolescent mania. J Am Acad Child Adolesc Psychiatry 45(3):305–313

DelBello MP, Adler CM, Whitsel RM, Stanford KE, Strakowski SM (2007) A 12-week single-blind trial of quetiapine for the treatment of mood symptoms in adolescents at high risk for developing bipolar I disorder. J Clin Psychiatry 68(5):789–795

DelBello MP, Chang K, Welge JA, Adler CM, Rana M, Howe M, Bryan H, Vogel D, Sampang S, Delgado SV, Sorter M, Strakowski SM (2009) A double-blind, placebo-controlled pilot study of quetiapine for depressed adolescents with bipolar disorder. Bipolar Disord 11(5):483–493

Detke HC, DelBello MP, Landry J, Usher RW (2015) Olanzapine/Fluoxetine combination in children and adolescents with bipolar I depression: a randomized, double-blind, placebo-controlled trial. J Am Acad Child Adolesc Psychiatry 54(3):217–224

Findling RL, Ginsberg LD (2014) The safety and effectiveness of open-label extended-release carbamazepine in the treatment of children and adolescents with bipolar I disorder suffering from a manic or mixed episode. Neuropsychiatr Dis Treat 10:1589–1597

Findling RL, McNamara NK, Gracious BL, Youngstrom EA, Stansbrey RJ, Reed MD, Demeter CA, Branicky LA, Fisher KE, Calabrese JR (2003) Combination lithium and divalproex sodium in pediatric bipolarity. J Am Acad Child Adolesc Psychiatry 42(8):895–901

Findling RL, McNamara NK, Youngstrom EA, Stansbrey R, Gracious BL, Reed MD, Calabrese JR (2005a) Double-blind 18-month trial of lithium versus divalproex maintenance treatment in pediatric bipolar disorder. J Am Acad Child Adolesc Psychiatry 44(5):409–417

Findling RL, Youngstrom EA, McNamara NK, Stansbrey RJ, Demeter CA, Bedoya D, Kahana SY, Calabrese JR (2005b) Early symptoms of mania and the role of parental risk. Bipolar Disord 7(6):623–634

Findling RL, McNamara NK, Stansbrey R, Gracious BL, Whipkey RE, Demeter CA, Reed MD, Youngstrom EA, Calabrese JR (2006) Combination lithium and divalproex sodium in pediatric bipolar symptom re-stabilization. J Am Acad Child Adolesc Psychiatry 45(2):142–148

Findling RL, Frazier TW, Youngstrom EA, McNamara NK, Stansbrey RJ, Gracious BL, Reed MD, Demeter CA, Calabrese JR (2007a) Double-blind, placebo-controlled trial of divalproex monotherapy in the treatment of symptomatic youth at high risk for developing bipolar disorder. J Clin Psychiatry 68(5):781–788

Findling RL, Short EJ, McNamara NK, Demeter CA, Stansbrey RJ, Gracious BL, Whipkey R, Manos MJ, Calabrese JR (2007b) Methylphenidate in the treatment of children and adolescents with bipolar disorder and attention-deficit/hyperactivity disorder. J Am Acad Child Adolesc Psychiatry 46(11):1445–1453

Findling RL, Frazier JA, Kafantaris V, Kowatch R, McClellan J, Pavuluri M, Sikich L, Hlastala S, Hooper SR, Demeter CA, Bedoya D, Brownstein B, Taylor-Zapata P (2008) The Collaborative Lithium Trials (CoLT): specific aims, methods, and implementation. Child Adolesc Psychiatry Ment Health 2(1):21

Findling RL, Kafantaris V, Pavuluri M, McNamara NK, McClellan J, Frazier JA, Sikich L, Kowatch R, Lingler J, Faber J, Rowles BM, Clemons TE, Taylor-Zapata P (2011) Dosing strategies for lithium monotherapy in children and adolescents with bipolar I disorder. J Child Adolesc Psychopharmacol 21(3):195–205

Findling RL, Cavus I, Pappadopulos E, Vanderburg DG, Schwartz JH, Gundapaneni BK, DelBello MP (2013a) Efficacy, long-term safety, and tolerability of ziprasidone in children and adolescents with bipolar disorder. J Child Adolesc Psychopharmacol 23(8):545–557

Findling RL, Correll CU, Nyilas M, Forbes RA, McQuade RD, Jin N, Ivanova S, Mankoski R, Carson WH, Carlson GA (2013b) Aripiprazole for the treatment of pediatric bipolar I disorder: a 30-week, randomized, placebo-controlled study. Bipolar Disord 15(2):138–149

Findling RL, Chang K, Robb A, Foster VJ, Horrigan JP, Krishen A, Wamil AW, DelBello MP (2014a) Double-blind, placebo-controlled, 36-week maintenance study of lamotrigine in youth with bipolar i disorder. In: Annual meeting of the American Academy of Child and Adolescent Psychiatry, San Diego, CA

Findling RL, Pathak S, Earley WR, Liu S, DelBello MP (2014b) Efficacy and safety of extended-release quetiapine fumarate in youth with bipolar depression: an 8 week, double-blind, placebo-controlled trial. J Child Adolesc Psychopharmacol 24(6):325–335

Findling RL, Landbloom RL, Szegedi A, Koppenhaver J, Braat S, Zhu Q, Mackle M, Chang K, Mathews M (2015a) Asenapine for the acute treatment of pediatric manic or mixed episodes of bipolar I disorder. J Am Acad Child Adolesc Psychiatry 54(12):1032–1041

Findling RL, Robb A, McNamara NK, Pavuluri MN, Kafantaris V, Scheffer R, Frazier JA, Rynn M, DelBello MP, Kowatch RA, Rowles BM, Lingler J, Martz K, Anand R, Clemons TE, Taylor-Zapata P (2015b) Lithium in the acute treatment of bipolar I disorder: a placebo-controlled study. Pediatrics 136(5):885–894

Frazier JA, Biederman J, Tohen M, Feldman PD, Jacobs TG, Toma V, Rater MA, Tarazi RA, Kim GS, Garfield SB, Sohma M, Gonzalez-Heydrich J, Risser RC, Nowlin ZM (2001) A prospective open-label treatment trial of olanzapine monotherapy in children and adolescents with bipolar disorder. J Child Adolesc Psychopharmacol 11(3):239–250

Geller B, Cooper TB, Sun K, Zimerman B, Frazier J, Williams M, Heath J (1998a) Double-blind and placebo-controlled study of lithium for adolescent bipolar disorders with secondary substance dependency. J Am Acad Child Adolesc Psychiatry 37(2):171–178

Geller B, Cooper TB, Zimerman B, Frazier J, Williams M, Heath J, Warner K (1998b) Lithium for prepubertal depressed children with family history predictors of future bipolarity: a double-blind, placebo-controlled study. J Affect Disord 51(2):165–175

Gracious BL, Chirieac MC, Costescu S, Finucane TL, Youngstrom EA, Hibbeln JR (2010) Randomized, placebo-controlled trial of flax oil in pediatric bipolar disorder. Bipolar Disord 12(2):142–154

Haas M, Delbello MP, Pandina G, Kushner S, Van Hove I, Augustyns I, Quiroz J, Kusumakar V (2009) Risperidone for the treatment of acute mania in children and adolescents with bipolar disorder: a randomized, double-blind, placebo-controlled study. Bipolar Disord 11(7):687–700

Joshi G, Petty C, Wozniak J, Faraone SV, Spencer AE, Woodworth KY, Shelley-Abrahamson R, McKillop H, Furtak SL, Biederman J (2013) A prospective open-label trial of paliperidone monotherapy for the treatment of bipolar spectrum disorders in children and adolescents. Psychopharmacology (Berl) 227(3):449–458

Kafantaris V, Coletti DJ, Dicker R, Padula G, Kane JM (2003) Lithium treatment of acute mania in adolescents: a large open trial. J Am Acad Child Adolesc Psychiatry 42(9):1038–1045

Kafantaris V, Coletti DJ, Dicker R, Padula G, Pleak RR, Alvir JM (2004) Lithium treatment of acute mania in adolescents: a placebo-controlled discontinuation study. J Am Acad Child Adolesc Psychiatry 43(8):984–993

Kowatch RA, DelBello MP, Findling RL (2002) Depressive episodes in children and adolescents with bipolar disorders. Clin Neurosci Res 2:158–160

Kowatch RA, Sethuraman G, Hume JH, Kromelis M, Weinberg WA (2003) Combination pharmacotherapy in children and adolescents with bipolar disorder. Biol Psychiatry 53 (11):978–984

Kowatch RA, Fristad M, Birmaher B, Wagner KD, Findling RL, Hellander M, Child Psychiatric Workgroup on Bipolar D (2005) Treatment guidelines for children and adolescents with bipolar disorder. J Am Acad Child Adolesc Psychiatry 44(3):213–235

Kowatch RA, Findling RL, Scheffer RE, Stanford K Pediatric Bipolar Collaborative Mood Stabilizer Trial (2007) In: 54th annual meeting of the American academy of child and adolescent psychiatry, Boston, MA, October 23–28, 2007

Kraepelin E (1921) Manic-depressive insanity and paranoia. E&S Livingstone, Edinburgh

Marchand WR, Wirth L, Simon C (2004) Quetiapine adjunctive and monotherapy for pediatric bipolar disorder: a retrospective chart review. J Child Adolesc Psychopharmacol 14 (3):405–411

McClellan J, Kowatch R, Findling RL, Work Group on Quality I (2007) Practice parameter for the assessment and treatment of children and adolescents with bipolar disorder. J Am Acad Child Adolesc Psychiatry 46(1):107–125

Patel NC, DelBello MP, Bryan HS, Adler CM, Kowatch RA, Stanford K, Strakowski SM (2006) Open-label lithium for the treatment of adolescents with bipolar depression. J Am Acad Child Adolesc Psychiatry 45(3):289–297

Pathak S, Findling RL, Earley WR, Acevedo LD, Stankowski J, Delbello MP (2013) Efficacy and safety of quetiapine in children and adolescents with mania associated with bipolar I disorder: a 3-week, double-blind, placebo-controlled trial. J Clin Psychiatry 74(1):e100–e109

Pavuluri MN, Henry DB, Carbray JA, Sampson G, Naylor MW, Janicak PG (2004) Open-label prospective trial of risperidone in combination with lithium or divalproex sodium in pediatric mania. J Affect Disord 82(Suppl 1):S103–S111

Pavuluri MN, Birmaher B, Naylor MW (2005a) Pediatric bipolar disorder: a review of the past 10 years. J Am Acad Child Adolesc Psychiatry 44(9):846–871

Pavuluri MN, Henry DB, Carbray JA, Naylor MW, Janicak PG (2005b) Divalproex sodium for pediatric mixed mania: a 6-month prospective trial. Bipolar Disord 7(3):266–273

Scheffer RE, Kowatch RA, Carmody T, Rush AJ (2005) Randomized, placebo-controlled trial of mixed amphetamine salts for symptoms of comorbid ADHD in pediatric bipolar disorder after mood stabilization with divalproex sodium. Am J Psychiatry 162(1):58–64

Tohen M, Kryzhanovskaya L, Carlson G, Delbello M, Wozniak J, Kowatch R, Wagner K, Findling R, Lin D, Robertson-Plouch C, Xu W, Dittmann RW, Biederman J (2007) Olanzapine versus placebo in the treatment of adolescents with bipolar mania. Am J Psychiatry 164 (10):1547–1556

Wagner KD, Weller EB, Carlson GA, Sachs G, Biederman J, Frazier JA, Wozniak P, Tracy K, Weller RA, Bowden C (2002) An open-label trial of divalproex in children and adolescents with bipolar disorder. J Am Acad Child Adolesc Psychiatry 41(10):1224–1230

Wagner KD, Kowatch RA, Emslie GJ, Findling RL, Wilens TE, McCague K, D'Souza J, Wamil A, Lehman RB, Berv D, Linden D (2006) A double-blind, randomized, placebo-controlled trial of oxcarbazepine in the treatment of bipolar disorder in children and adolescents. Am J Psychiatry 163(7):1179–1186

Wagner KD, Redden L, Kowatch RA, Wilens TE, Segal S, Chang K, Wozniak P, Vigna NV, Abi-Saab W, Saltarelli M (2009) A double-blind, randomized, placebo-controlled trial of divalproex extended-release in the treatment of bipolar disorder in children and adolescents. J Am Acad Child Adolesc Psychiatry 48(5):519–532

Wiznitzer M, Findling RL (2003) Why do psychiatric drug research in children? Lancet 361 (9364):1147–1148

Wozniak J, Biederman J, Mick E, Waxmonsky J, Hantsoo L, Best C, Cluette-Brown JE, Laposata M (2007) Omega-3 fatty acid monotherapy for pediatric bipolar disorder: a prospective open-label trial. Eur Neuropsychopharmacol 17(6-7):440–447

Young RC, Biggs JT, Ziegler VE, Meyer DA (1978) A rating scale for mania: reliability, validity and sensitivity. Br J Psychiatry 133:429–435

Chapter 15
The Management of Bipolar Disorder During and After Pregnancy

Jennifer L. Payne

Abstract This chapter will discuss the issues of psychiatric management of bipolar disorder (BD) both during and after pregnancy. The risks and benefits of medication use during pregnancy will be discussed along with the risks of discontinuation of medications for pregnancy. Pharmacological treatments will be discussed in detail. The postpartum time period and its risks for depression, mania, and psychosis will be described, and recommendations for management during this critical time period will be detailed. Medication use during breastfeeding will also be discussed.

Keywords Pregnancy • Postpartum depression • Postpartum psychosis

15.1   Prevalence of Bipolar Relapse During and After Pregnancy

The risk of relapse of bipolar disorder (BD) during pregnancy appears to be approximately the same as at any other time in a woman's life. Viguera and colleagues (Viguera et al. 2000) retrospectively compared the risks of recurrence in pregnant and nonpregnant women with BD when tapered off lithium. They found no difference in the risks of recurrence between pregnant and nonpregnant women over the same time period with rates of 52 % in pregnant women and 58 % in nonpregnant women, thus demonstrating that pregnancy does not appear to increase the risk of relapse in women with BD (Viguera et al. 2000).

Although pregnancy appears to be risk-neutral for recurrence of mood disorders, it is not risk-free. This is especially true in women with preexisting mood disorders. Many women experience relapse during pregnancy—both on and off medication. In one study, approximately 50 % of women with either major depressive disorder (MDD) or BD reported significant mood symptoms during, after, or both during and after pregnancy (Payne et al. 2007). Further, the risk for relapse during pregnancy increases in the setting of discontinuation of medications. In women with BD,

J.L. Payne, MD (✉)
Johns Hopkins School of Medicine, 550 N. Broadway, Suite 305, Baltimore, MD 21205, USA
e-mail: Jpayne5@jhmi.edu

© Springer International Publishing Switzerland 2016   331
C.A. Zarate Jr., H.K. Manji (eds.), *Bipolar Depression: Molecular Neurobiology,*
Clinical Diagnosis, and Pharmacotherapy, Milestones in Drug Therapy,
DOI 10.1007/978-3-319-31689-5_15

Viguera reported that pregnant women who discontinued mood stabilizers had a recurrence risk of 81–85.5 % while women who continued mood stabilizer treatment had a much lower risk of 29–37 % (Viguera et al. 2000, 2007). Similarly, the recurrence risk in a study comparing discontinuation of any mood stabilizer to lamotrigine treatment during pregnancy was 100 % in the group of women who discontinued medications (Andersson et al. 2003). The finding that at least 80 % of women with BD will relapse when taken off of medication suggests that treatment during pregnancy for many patients is necessary in order to prevent recurrence of psychiatric illness.

The risk of relapse for BD during the postpartum time period is clearly elevated, with many women developing a postpartum depressive, hypomanic, manic, or psychotic episode. In general, there are three types of postpartum mood disorders: Postpartum Blues, Postpartum Depression (PPD), and Postpartum Psychosis. Some women with BD may also develop a postpartum hypomanic episode. Postpartum Blues is a relatively common phenomenon occurring in up to 80 % of women, generally within a few days of labor and delivery. It is usually a self-limited process, resolving over the course of several days. Symptoms include tearfulness, mood lability, and feelings of being overwhelmed, but can also include more positive feelings of happiness or elation (Payne 2003). Postpartum Blues generally requires only supportive interventions such as social support, getting adequate sleep, and time to care for oneself. PPD, in contrast, is less common, occurring in 10–20 % of the general population. PPD meets DSM criteria for a major depressive episode lasting for at least two weeks (Campbell and Cohn 1991). The risk for PPD is increased in women with a history of MDD (Frank et al. 1987), BD, or PPD after previous pregnancies (Cox et al. 1993). While the etiology of PPD is not known, it is likely to be multifactorial with psychological factors, biological factors (including hormonal changes), and social factors all playing a role. Finally, Postpartum Psychosis is a rare phenomenon, occurring in approximately 0.1 % of all births (Kendell et al. 1987). It is more common in women with BD-I, occurring in up to 30 % of those who have children. Postpartum Psychosis is considered a psychiatric emergency and resembles a manic or mixed episode with decreased sleep, psychosis, and agitation (Jones et al. 2010).

15.2 Controversies Surrounding Psychiatric Medication Use During and After Pregnancy

The treatment of psychiatric disorders during pregnancy is complicated by a dearth of studies on what medications work, how changes in body weight and metabolism affect dosing, how to manage medications both during and after pregnancy, and also on what the long-term effects of exposure may be on the developing fetus. There has been a long and appropriate tradition of minimizing the use of medications during pregnancy—however, unlike medications used for purely medical conditions such as asthma and hypertension, psychiatric medications are often considered expendable and are recommended for discontinuation. Abrupt discontinuation of psychiatric medication can result not only in withdrawal symptoms, but in relapse of the psychiatric illness; multiple studies have demonstrated that exposure to psychiatric illness in utero results in poorer outcomes for both mother and child. Relapse is especially common in women with BD.

One significant limitation of the literature is the fact that the use of psychiatric medications during pregnancy is essentially a "marker" for a population of women who have different risk factors than the general population of pregnant women, and these risk factors may influence the outcomes of studies attempting to examine the risks for a child exposed in utero to a particular psychiatric medication. For example diabetes, obesity, smoking, and substance use are more common in the psychiatric population than in the general population as a whole. Studies that have not controlled for the underlying psychiatric illness and its attendant risks may find associations between psychiatric medications and outcomes that are not due to exposure to the medication itself, but to other risk factors that are highly prevalent in the population of patients who take psychiatric medications during pregnancy. This not only complicates interpretation of the literature but also complicates recommendations for women who have psychiatric illness but no other inherent risk factors or behaviors. Studies in different populations of women with psychiatric illness and different levels of associated risk factors and behaviors need to be conducted with the goal of being able to make intelligent recommendations for individual patients.

Another area that is frequently overlooked in the risk-benefit analysis of whether or not to use psychiatric medications during pregnancy is the risk to the fetus and newborn associated with untreated maternal psychiatric illness. There is a strong literature demonstrating that, in addition to presenting risks to the mother, untreated maternal psychiatric illness during pregnancy is associated with poorer outcomes for the exposed child. For example, depression during pregnancy has been associated with low maternal weight gain, increased rates of preterm birth (Li et al. 2009), low birth weight, increased rates of cigarette, alcohol, and other substance use (Zuckerman et al. 1989), increased ambivalence about the pregnancy, and overall worse health status (Orr et al. 2007), including higher rates of preeclampsia and gestational diabetes (Field et al. 2006, 2010). In addition, prenatal exposure to maternal stress has been shown to have consequences for the development of infant temperament (Davis et al. 2005). Children exposed to perinatal (either during pregnancy or postpartum) maternal depression have higher cortisol levels than infants of mothers who were not depressed (Ashman et al. 2002; Diego et al. 2004; Essex et al. 2002; Halligan et al. 2004), and this finding continues through adolescence (Halligan et al. 2004). Importantly, treatment of depression during pregnancy appears to help normalize infant cortisol levels (Brennan et al. 2008). These findings may partially explain the mechanism for increased vulnerability to psychopathology in children exposed to depression in utero (O'Connor et al. 2005).

The literature regarding maternal illness during the postpartum period is even stronger. Adverse outcomes associated with PPD include lower IQ, slower language development, increased risk of Attention Deficit Hyperactivity Disorder, increased risk of behavioral issues, and increased risk of psychiatric illness in the exposed offspring (Grace et al. 2003). These findings often seem to get lost or even ignored in the debate of whether to use psychiatric medications during pregnancy. The inclination is to compare the risk of exposure to psychiatric medication to the risk of no exposure to medication rather than comparing the risk associated with medication exposure to the risk associated with maternal psychiatric illness which clearly also has consequences for the child. Psychiatric illness during pregnancy should be considered an exposure for the child in the same way that medication use during pregnancy is an exposure for the child.

15.3   FDA Categories: Past and Future

In December 2014, the US Federal Drug Administration (FDA) published the final version of the "Pregnancy and Lactation Labeling Rule" mandating changes to the content and format of prescription drug labeling as they pertain to use during pregnancy and lactation. The labeling changes went into effect on June 30, 2015—immediately for products submitted for FDA approval after that date and to be phased in for all other medications and products. The new labeling will contain Pregnancy and Lactation subsections, each of which will have three principal components: a risk summary, clinical considerations, and a data section. There will also be a new subsection entitled "Females and Males of Reproductive Potential." The goal is to provide relevant information that will help providers make prescribing decisions and counsel women regarding the use of the medication during pregnancy and lactation. This system attempts to include all currently available information to help the clinician weigh the risks and benefits of prescribing a particular drug during pregnancy.

Because the "Rule" will be phased in over time, we will provide a brief summary of the former FDA categories here. Categories include A, B, C, D, and X (as well as N for Not Rated), and classification is based on the amount of evidence for safety in animal and human studies (see Table 15.1). The system uses evidence from animal studies as part of the definitions of three out of the five categories. Many clinicians assume that there is an increasing level of risk from category A to X, which is inaccurate. For example, category B medications simply do not have adequate studies in humans to place them in category A as safe or in categories C, D, or X depending on the level of risk in humans. For instance, oral contraceptives are in category X because there is no reason to use them in pregnancy, not because there is evidence of associated birth defects. Further, the level of investigation used to categorize individual medications varies from medication to medication as does the level of risk actually imposed by a particular drug. As a result, medications may be placed in the same category but have vastly different levels of risk and/or different

Table 15.1 Psychotropic medication used for bipolar disorder in pregnancy

Medication	FDA category	Potential complications	Comments and recommendations
Antidepressants			
Selective serotonin reuptake inhibitors (SSRIs)	Citalopram = C Escitalopram = C Fluoxetine = C Fluvoxamine = C Paroxetine = D Sertraline = C Vilazodone = C	– Modest increased risk of spontaneous abortion – Modest increased risk of preterm birth and low birth weight – No confirmed risk of birth defects except for small absolute increased risk of cardiac defects (2/1000 births) with *paroxetine* with first trimester exposure – Poor Neonatal Adaptation Syndrome with third trimester exposure – Conflicting evidence for small increased risk of Persistent Pulmonary Hypertension with third trimester exposure	– Best studied class of antidepressants – Most studies confounded by indication (i.e., not controlled for the underlying psychiatric illness). – Behaviors and risk factors associated with the psychiatric illness might influence some of the associations – Large studies that attempt to control for the underlying psychiatric illness generally suggest no increased risks – High relapse rate in women who stop their antidepressants for pregnancy – Avoid use of paroxetine during pregnancy if possible
Serotonin and norepinephrine reuptake inhibitors (SNRIs)	Duloxetine = C Desvenlafaxine = C Venlafaxine = C	– Fewer data available – Modest increased risk of spontaneous abortion – Modest increased risk of preterm birth and low birth weight – No confirmed risk of birth defects – Poor neonatal adaptation syndrome with third trimester exposure – Conflicting evidence for small absolute increased risk of persistent pulmonary hypertension with third trimester exposure	– Most studies are confounded by not controlling for the underlying psychiatric illness

(continued)

Table 15.1 (continued)

Medication	FDA category	Potential complications	Comments and recommendations
Other antidepressants	Bupropion = C Mirtazapine = C Trazodone = C	– Fewer data available	– Most studies are confounded by not controlling for the underlying psychiatric illness
Tricyclic antidepressants	Amitriptyline = C Clomipramine = C Desipramine = N Doxepin = N Imipramine = N Nortriptyline = N	– Fewer data available	– Therapeutic drug monitoring allows monitoring of serum levels and appropriate dose adjustments during pregnancy
Monoamine oxidase inhibitors	Selegiline Transdermal = C Phenelzine = C Tranylcypromine = N	– Significantly fewer data available	– Orthostatic hypotension may be pronounced in pregnancy
Mood stabilizers			
Lamotrigine	C	– No increased risk of major congenital malformations – One early and small study found an association with cleft palate that has not been replicated	– Serum levels should be monitored and maintained during pregnancy as levels usually decrease as pregnancy progresses
Valproic acid	D	– Associated with up to 10 % rate of malformations. Neural tube defects, effects on cognition and brain volume, craniofacial anomalies, cardiac defects, cleft palate, and hypospadias have been described – Recently linked to autism	– Generally should not be used during pregnancy. – High dose folate (4 mg) supplementation is recommended.
Carbamazepine	D	– Increased risk of malformations including spina bifida, other neural tube defects, facial abnormalities, skeletal abnormalities, hypospadias, and diaphragmatic hernia – Increased risk of neonatal hemorrhage	– Generally should not be used during pregnancy – High dose folate (4 mg) supplementation is recommended

(continued)

Table 15.1 (continued)

Medication	FDA category	Potential complications	Comments and recommendations
Oxcarbamazepine	C		
Lithium	D	– 1/1000 develop Ebstein's anomaly – No cognitive or behavioral effects in exposed children	– Lithium levels should be followed closely during pregnancy – The dose should be held or reduced with the initiation of labor – Postpartum, the dose should be reduced to pre-pregnancy levels (if it was increased during pregnancy) – Fetal echocardiagram in the first trimester is recommended
Antipsychotic medications			
First generation antipsychotics	Chlorpromazine = N Fluphenazine = N Haloperidol = C Loxapine = N Perphenazine = C Trifluoperazine = C Thiothixene = N Fluphenazine = N	– No major congenital malformations have been demonstrated – Associated with low birth weight and pre-term delivery – No difference in IQ or behavior in exposed children – Exposure in third trimester associated with transient extrapyramidal and withdrawal symptoms in the infant	– Most studies are confounded by not controlling for the underlying psychiatric illness – High potency antipsychotics are preferred over low potency due to anticholinergic, hypotensive, and antihistaminergic side effects
Second generation antipsychotics	Aripiprazole = C Asenapine = C Clozapine = B Lurasadone = B Olanzapine = C Paliperidone = C Quetiapine = C Risperidone = C Ziprasidone = C	– No major congenital malformations have been demonstrated – May increase maternal weight gain – May increase risk of gestational diabetes – May increase size of the baby – Neurodevelopmental delays found at six months but resolved by 12 months	– Most studies are confounded by indication (i.e., not controlled for the underlying psychiatric illness). – There are fewer data available for clozapine and lurasadone – Glucose monitoring recommended – Routine ultrasound monitoring of fetal size in late pregnancy should be obtained – Clozapine has been associated with floppy baby syndrome, and

(continued)

Table 15.1 (continued)

Medication	FDA category	Potential complications	Comments and recommendations
			exposed infants should be monitored for agranulocytosis weekly for six months
Antianxiety medications			
Benzodiazepines	Alprazolam = D Chlordiazepoxide = D Clonazepam = D Diazepam = D Lorazepam = D Oxazepam = D	– May induce perinatal toxicity: temperature dysregulation, apnea, lower APGAR scores, hypotonia, and poor feeding – Use just before delivery associated with floppy baby syndrome – Some studies suggest oral cleft palate defects; others are negative	– Consider tapering benzodiazepines prior to delivery – Intermittent use is unlikely to induce withdrawal symptoms in the newborn
Gabapentin	C	– No increased risk of major congenital malformations – One study found increased risk of preterm birth, low birth weight, and NICU admission	
Pregabalin	C	– Evidence of congenital malformations and growth restriction in animals; no evidence of congenital malformations in humans, but there are few studies	
Buspirone	B	– No evidence of congenital malformations in animals – No data in humans	
Adjunctive psychotropic medications			
Antihistamines	Diphenhyrdramine = B Doxylamine = A Hydroxyzine = C Pheniramines = C	– Limited data – Commonly used – Conflicting data on pregnancy outcomes	– There are few data on chronic versus intermittent use
Sleep agents	Eszopiclone = C Ramelteon = C Zolpidem = C	– No evidence of major congenital malformations – May increase risk of low birth weight and preterm delivery	– There are few data on chronic versus intermittent use.

levels of evidence supporting their categorization. The FDA Pregnancy Categories therefore can provide a "quick and dirty" assessment but often do not provide enough information that is useful in planning clinical care. It is hoped that by providing all available information about a particular medication on the label itself rather than grossly classifying the information into a categorical system, clinicians and patients will make more informed decisions based on the literature as a whole.

15.4   Changes in Metabolism and Drug Clearance During Pregnancy

In pregnancy, major adaptive physiologic changes occur in a woman's gastrointestinal, cardiovascular, renal, and hepatic systems. These changes have a significant impact on the pharmacokinetic processes of drug absorption, distribution, metabolism, and excretion. Because of the impact of pregnancy on these basic physiologic and pharmacokinetic processes, adjustments in medication dosing are often required in pregnancy.

Physiological changes throughout pregnancy ultimately result in approximately a 50 % increase in plasma volume, increased body fat, and increased medication distribution volume. Renal blood flow, glomerular filtration rate, and medication elimination also increase (Seeman 2004) and changes in liver enzyme activation occur (e.g., CYP1A2 activity decreases; the activity of CYP2D6 and CYP3A increases) (Tracy et al. 2005). These liver enzyme changes, which are mostly hormone-dependent, can result in either increased or decreased medication clearance and are highly relevant to many psychiatric medications (Pavek et al. 2009; DeVane et al. 2006).

All psychotropic medications are able to move freely between the maternal and placental systems. The rate and extent of exchange between the maternal and placental systems is variable and poorly defined, and genetic factors that may play a role in variations in distribution are poorly understood. There are few data to help guide clinicians make decisions regarding medication dosing in pregnancy. The scant evidence that does exist is based primarily on observational studies and, given the reported high interindividual variability, should be cautiously interpreted (Pavek et al. 2009).

When available, therapeutic monitoring of serum levels can help guide decisions regarding medication dosing. Otherwise, clinicians must rely on what is generally known regarding pharmacokinetic changes in pregnancy and some basic principles. These principles include making sure a woman is taking the lowest beneficial dose of medication (i.e., the dose that provides sufficient maternal benefit while minimizing fetal exposure) during her pregnancy (DeVane et al. 2006). Because many of the changes in pharmacokinetics evolve over the duration of the pregnancy, this means that clinicians will need to monitor the woman's mental state more frequently and adjust the dose of psychiatric medication(s) accordingly and perhaps repeatedly (DeVane et al. 2006; Tracy et al. 2005).

15.5 Managing Medications for Bipolar Disorder During and After Pregnancy

The ideal situation is to begin planning for pregnancy prior to pregnancy. It is important to assume that every woman of childbearing age will get pregnant and to discuss medication use during pregnancy and use of birth control measures as part of their ongoing treatment. If a woman is taking a medication that should not be used during pregnancy, such as valproic acid, a discussion should be held with the woman and, if possible, her partner to discuss this fact and to plan what should be done in case of accidental pregnancy. As many as 50 % of pregnancies are still unplanned in the USA (Mosher and Bachrach 1996); thus, discussing ideal and contingency plans ahead of time would minimize the chance that psychiatric medications will be abruptly discontinued and the patient will relapse.

The patient's past psychiatric history, severity of symptoms, and history of medication response all play a role in designing the course of clinical care during pregnancy. For example, while fluoxetine and sertraline are considered appropriate antidepressant choices during pregnancy, if a woman has a history of not responding to either of these medications they cannot be part of the treatment plan. Severity of illness is important to take into account: a case in which the psychiatric symptoms were mild, responded well to medication, and had not recurred could be considered for discontinuation of the medication prior to pregnancy. In contrast, a patient whose psychiatric symptoms were severe, dangerous, and required hospitalization several times would not be a candidate for medication discontinuation. Similarly, a woman's reproductive age should play a role in deciding when and if to attempt to change medications prior to pregnancy; an older woman may not have the reproductive time to experiment with a different medication prior to pregnancy.

At the same time, the patient and her partner's wishes regarding medication use during pregnancy should be taken into account when designing a treatment plan. If one or the other is strongly against medication use during pregnancy, it is best for the treatment provider to make sure they understand the risks of no treatment to both the mother and the baby, the rates of relapse, and to outline a course of close follow-up during and after pregnancy rather than insist on the use of medication during pregnancy. It is important to maintain a partnership characterized by good communication with the patient and her partner so that if there is a relapse the patient will remain safe and is more likely to seek care and treatment.

While each case should be considered individually and there are no hard and fast rules, there are some "rules of thumb" that can be used when designing a treatment strategy (modified from Payne and Meltzer-Brody (2009)):

1. All medication changes should be done prior to pregnancy if possible. This minimizes the number of exposures to the baby and promotes mood stability for the mother.
2. Ideally the patient should be stable psychiatrically for at least three months before attempting pregnancy. This is not always practical but should provide some evidence and reassurance that the patient's mood is stable prior to entering pregnancy.
3. Use medications that we know something about; older medications are usually better. If a medication has been available for a long period of time, there will be more evidence to support its safety or teratogenicity in the literature.
4. Use medications that have been tried in larger, nonpsychiatric populations. For instance, antiseizure medications have been tested in the larger population of women with seizure disorders. While many antiseizure medications are teratogenic, some with psychiatric indications have been shown to be safer during pregnancy (e.g., gabapentin and lamotrigine).
5. Minimize the number of exposures for the baby. Try to minimize the number of medications used but consider exposure to psychiatric illness as well. One common scenario is for a woman on a newer antidepressant to become pregnant and receive the recommendation to switch antidepressants to an older medication that has more evidence for safety during pregnancy. While this might have made sense prior to pregnancy, this plan would actually increase the exposures for the baby significantly. First, the baby has already been exposed to the newer antidepressant and switching to a second medication would be another exposure. In addition, the likelihood that the patient would relapse while switching is high, thus exposure to the mood disorder would be a third exposure for the child. Changing medications for breastfeeding also increases the number of exposures.
6. Monitor blood levels of medications when possible. As noted, changes in drug metabolism and weight gain during pregnancy may alter blood levels of psychiatric medications as pregnancy progresses. Medications that can be monitored by blood levels offer an advantage in this situation as changes in dosing can be made prophylactically. Medications for which laboratory monitoring is not available must be managed based on clinical response. Many women need adjustment of doses during the late second and third trimesters and should be evaluated more often during this time period.
7. Consider breastfeeding when planning for pregnancy. Consider whether the medication should be used during breastfeeding and what the plan would be for monitoring the medication during breastfeeding.
8. If a baby was exposed to a medication during pregnancy, it may not make sense to discontinue the medication (or, alternatively, not breastfeed) for breastfeeding. The baby was exposed to a larger concentration of the drug in utero compared to the concentration found in breast milk. That being said there are certain medications that might be difficult to justify the use of drug during breastfeeding. For example, clozapine, with its risk of neutropenia, should not be used during breastfeeding. Overall, the benefits of breastfeeding are

considered to be large; thus, recommending to a woman whose infant was already exposed to a medication in utero that she not breastfeed rarely makes sense, with the obvious exception of when the mother does not feel comfortable breastfeeding while taking a medication, or if the baby appears to be having side effects (e.g., sedation) from the medication.

9. Use a team approach. This rule of thumb applies to two groups: (1) family and (2) other doctors involved in the patient's care. Educating the family regarding the risks and benefits of treatment and no treatment, as well as signs and symptoms to be aware of for relapse is essential to providing good care for both mother and child. Similarly, communicating directly with the patient's other treatment providers will minimize miscommunication and differences of opinion and maximize treatment outcomes for the patient.

10. Be supportive if your patient doesn't accept your recommendations. There are many reasons why a particular patient may choose to go against her treatment provider's advice, particularly regarding medication use during pregnancy. Many women will feel guilty if they take any medication during pregnancy and underestimate the risks of untreated mood disorder during pregnancy. Many women are also under pressure from significant others, friends, and family members, or even other doctors, to discontinue medication during pregnancy. It is important as the treatment provider to continue to provide support to the patient despite disagreements regarding treatment during pregnancy. Again, using a team approach will often help avoid disagreements, and providing as much information as possible regarding the risks of untreated mood disorder during pregnancy can also help. In addition to talking directly to family and other treatment providers, in this situation it is often most appropriate to offer close follow-up care so that if a relapse occurs it is caught early and treatment can be offered. It is also important to keep in mind that the patient and her family must feel comfortable with the treatment used during pregnancy so they do not look back and regret decisions made during this critical time.

11. If a woman has stopped her medications during pregnancy, encourage restarting them postpartum. Several studies have shown that restarting psychiatric medications, particularly lithium, postpartum reduces the risk for psychiatric relapse (Stewart et al. 1991; Cohen et al. 1995; Austin 1992; van Gent and Verhoeven 1992; Bergink et al. 2012), particularly in BD. Emphasizing the increased risk for relapse during the postpartum time period can be helpful for this discussion.

12. Monitor women during the postpartum period closely. Because the postpartum period has an elevated risk for relapse, women should be seen frequently during this time period in order to promote early intervention.

13. Promote healthy sleep habits during the postpartum time period. Studies indicate that decreased sleep for women with BD is associated with relapse during the postpartum time period (Sharma et al. 2004; Sharma 2003). Emphasizing the need for regular sleep with the patient and the family may help minimize this trigger for relapse.

14. If a woman required an increased dose of mood stabilizer during pregnancy, consider decreasing it postpartum. For example, many women require increased doses of lamotrigine during pregnancy (Clark et al. 2013). Postpartum, lamotrigine concentrations have been shown to increase rapidly, possibly resulting in toxicity (Clark et al. 2013). Blood levels should be monitored closely during the postpartum time period in order to prevent toxicity. No studies have demonstrated whether psychiatric medications that do not have blood levels available should be decreased postpartum. For example, it is unclear whether an antidepressant increased during the third trimester should be decreased postpartum. In general, given the high risk of relapse postpartum, most treatment providers continue a woman on the higher dosage that was required during pregnancy. Patients should be followed closely, however, for the emergence of side effects, which would indicate a need to decrease the dosage to the previously effective dose.

We turn now to a discussion of the use of specific psychiatric medications during and after pregnancy (see Table 15.1).

15.6 Mood Stabilizers

15.6.1 Lithium

Lithium use during the first trimester has been associated with an increased risk of a serious congenital heart defect known as Ebstein's anomaly, which occurs in approximately one out of 1000 live births. The risk for Ebstein's anomaly with first trimester exposure was originally thought to be much higher (400 times higher than baseline), but a pooled analysis of lithium-exposed pregnancies found that this defect only occurs in one out of 1000–2000 exposed children (Cohen et al. 1994). This translates to less than 1 % of exposed children developing the anomaly. Lithium has also been associated with perinatal toxicity, including case reports of hypotonia, cyanosis, neonatal goiter, and neonatal diabetes insipidus. For women with severe BD, the risk of recurrence during pregnancy may overshadow the relatively small risk of Ebstein's anomaly. For such women, maintenance lithium therapy during pregnancy may be the most appropriate course. On the other hand, for women with significant periods of euthymia and few past mood episodes, slowly tapering off lithium and reintroducing lithium after the first trimester may help reduce the risk of relapse during the postpartum period. There are limited data on the long-term outcomes of children exposed in utero, but a follow-up of children up to age five demonstrated no evidence of cognitive or behavioral issues in a small sample of children (Jacobson et al. 1992). Lithium levels should be followed closely during pregnancy and the dose should be held or reduced with the initiation of labor. Hydration during delivery should be adequate and the dosage should be reduced to pre-pregnancy levels (if it was increased during pregnancy) with close monitoring of serum levels postpartum (Pearlstein 2013).

15.6.2 Valproic Acid

Valproic acid is associated with a high rate of malformations with first trimester exposure (Pearlstein 2013). As many as 10 % of exposed children are born with neural tube defects, effects on cognition and brain volume, craniofacial anomalies, cardiac defects, cleft palate, and/or hypospadias (Pearlstein 2013). Valproic acid exposure has also been recently linked to autism (Bromley et al. 2013; Christensen et al. 2013). Providers should encourage pregnant women who elect to continue any anticonvulsant to take high-dose folate (4 mg per day) for the theoretical benefit of reducing the risk of neural tube defects and to undergo a second trimester ultrasound to screen for major congenital anomalies. Blood levels of valproic acid should also be followed carefully. In general, when prescribing valproic acid to a woman of childbearing age, a discussion of the risks of valproic acid exposure during pregnancy should be conducted.

15.6.3 Carbamazepine

Carbamazepine also carries an increased risk of malformations, primarily of spina bifida as well as other neural tube defects, facial abnormalities, skeletal abnormalities, hypospadias, and diaphragmatic hernia (Pearlstein 2013). Carbamazepine is also a competitive inhibitor of prothrombin precursors and may increase the risk of neonatal hemorrhage. As with valproic acid, high-dose folate should be taken and screening for malformations as well as therapeutic blood monitoring should be done.

15.6.4 Lamotrigine

According to the manufacturer-sponsored Lamotrigine Pregnancy Registry and other published studies (Cunnington and Tennis 2005), there appeared to be no increased risk of congenital defects above the baseline risk with lamotrigine monotherapy; however, when combined with valproic acid in pregnancy, the risk estimate was found to be elevated to above 10 %. While these initial findings seemed to offer women a relatively safe alternative to other anticonvulsants in pregnancy, the North American Antiepileptic Drug Pregnancy Registry found that infants exposed to lamotrigine monotherapy during pregnancy had a much higher risk of oral cleft defects (Holmes et al. 2008). However, a more recent and larger study failed to find an association (Dolk et al. 2008). Dolk and colleagues assessed the association between oral cleft palate and exposure to lamotrigine using a population-based case-control design using data from the EUROCAT congenital malformation registries. The study population included 3.9 million births from 19 registries between 1995 and 2005. The authors identified 5511 cases of non-syndromic oral cleft. The control group consisted of 80,052 cases of non-chromosomal, non-oral cleft malformations. In this study, there was no evidence of an increased risk of isolated oral clefts relative to other malformations. Lamictal levels may decrease over the course of pregnancy and thus should be followed and adjusted if needed (Clark et al. 2013).

15.7 Antipsychotics

A 2004 Cochrane report (Webb et al. 2004) on the use of antipsychotics for primary (non-affective) psychosis in pregnancy found no trials meeting their inclusion criteria and concluded "continued use of antipsychotic drugs in these women in pregnancy and lactation without sound evidence raises serious clinical and ethical concerns." This report led many clinicians to recommend the discontinuation of antipsychotic medications during pregnancy. However, as more evidence has accumulated in the past decade, it appears that antipsychotics are—for the most part—relatively safe to use in pregnancy. Furthermore, *not* using these medications when indicated for serious mental illness poses a much greater risk to both mother and child, including the risks of suicide and infanticide (Robinson 2012).

Although antipsychotic use in pregnancy has not been definitively associated with an increased risk of congenital anomalies or any other adverse outcomes (Einarson and Einarson 2009; Einarson and Boskovic 2009), very few rigorously designed prospective studies that control for the underlying psychiatric illness have examined their safety in pregnancy. Although studies that control for potential confounders—including smoking—indicate persistent risks for adverse pregnancy outcomes (particularly low birth weight and preterm delivery) among women with an episode of schizophrenia in pregnancy compared to controls, it cannot be determined whether these outcomes are due to antipsychotic use (Nilsson et al. 2002), genetic vulnerabilities, or associated behaviors. However, a recent study examined birth outcomes in a matched cohort of women who used antipsychotics in pregnancy ($n = 1021$) and those who did not ($n = 1021$) and an unmatched cohort of women who used antipsychotics in pregnancy ($n = 1200$) and those who did not ($n = 40,000$). The analysis revealed no increased risk of adverse outcomes (preterm birth, gestational diabetes, hypertension, and large for gestational age infants) in the matched cohorts that controlled for the underlying illness and risk factors (Khalifeh et al. 2015).

When prescribing antipsychotics in pregnancy, pharmacokinetics must also be considered. Because CYP1A2 enzymes are downregulated with advancing pregnancy, doses of olanzapine and clozapine may need to be decreased, while doses of medications that are metabolized by upregulated enzymes may need to be increased (Seeman 2013). Of note, quetiapine, risperidone, haloperidol, and olanzapine have been shown to exhibit the lowest placental transfer from mother to fetus (Newport et al. 2007).

Normal metabolic changes associated with pregnancy may increase the risk for gestational diabetes in conjunction with the use of antipsychotics. In fact, many antipsychotics, particularly second generation antipsychotics (SGAs), are associated with excessive maternal weight gain, increased infant birth weight, increased risk of gestational diabetes, and infants being born large for gestational age (Seeman 2013; Newham et al. 2008). Several cases of gestational diabetes associated with the use of antipsychotics, including clozapine and olanzapine, have been reported (Barnes 2011; Gentile 2010; Reis and Kallen 2008). This suggests that routine ultrasound monitoring of fetal size in late pregnancy might be beneficial for women taking these medications in pregnancy or for women who gain substantial weight (Newham et al. 2008; Paton 2008).

There is also a lack of evidence regarding late pregnancy exposure to antipsychotics, including little on longer-term developmental outcomes, and so the risks remain unclear. Behaviors observed in infants exposed to antipsychotics in utero include motor restlessness, dystonia, hypertonia, and tremor (Gentile 2010; Coppola et al. 2007). The few studies examining the relationship between in utero exposure to First Generation Antipsychotics (FGAs) and neurodevelopment have shown no difference in IQ or behavioral functioning at five years (Barnes 2011; Altshuler et al. 1996; Thiels 1987). Studies of SGAs have shown associated neurodevelopmental delays at six months of age (Peng et al. 2013; Johnson et al. 2012). However, in a case-control prospective study, these delays were no longer evident at 12 months (Pearlstein 2013; Peng et al. 2013). The American College of Obstetricians and Gynecologists states that "no significant teratogenic effect has been documented with chlorpromazine, haloperidol, and perphenazine" and suggests that the "use of piperazine phenothiazines (e.g., trifluoperazine, perphenazine) may have especially limited teratogenic potential" (ACOG Practice Bulletin: Clinical management guidelines for obstetrician-gynecologists number 92, April 2008 (replaces practice bulletin number 87, November 2007). Use of psychiatric medications during pregnancy and lactation 2008).

More recently, in utero exposure to FGAs has been associated with an increased risk of premature delivery (Habermann et al. 2013), and exposure specifically in the third trimester has been associated with transient extrapyramidal and withdrawal symptoms. The FDA issued a drug safety communication in 2011 for all antipsychotics noting the potential risks of abnormal muscle movements and withdrawal symptoms (Communication 2011). However, given that these extrapyramidal/withdrawal reactions are usually self-limited, the American Academy of Pediatrics Committee on Drugs guidelines recommend the preferential use of high-potency FGAs in order to minimize maternal anticholinergic, hypotensive, and antihistaminergic effects of the low-potency antipsychotics (Use of psychoactive medication during pregnancy and possible effects on the fetus and newborn. Committee on Drugs. American Academy of Pediatrics 2000). These 2000 guidelines also recommended against the use of any depot preparations of antipsychotics due to lack of flexibility in dosing and in order to limit exposure to the neonate of prolonged potential toxic effects.

SGAs have no evidence of being safer to use in pregnancy than FGAs. The best studied is olanzapine, which has global safety pregnancy surveillance data suggesting no difference in outcomes with fetal exposure to olanzapine compared to the general population (Brunner et al. 2013). However, there is concern that fetal exposure to these newer medications may increase infant birth weight and the risk of being born large for gestational age (Newham et al. 2008). Several studies suggest increased risk of hypoglycemia associated with the use of SGAs in pregnancy (Gentile 2004). However, this may be due to higher baseline rates of diabetes in women prescribed antipsychotics (Vigod et al. 2015; Khalifeh et al. 2015), reinforcing the need for a thorough evaluation and appropriate glucose monitoring for women prescribed these medications.

Clozapine has been associated with floppy baby syndrome. In addition, infants with in utero exposure to clozapine should be monitored for agranulocytosis weekly for the first six months of life (Gentile 2010).

15.8 Antidepressants

Antidepressants are the most commonly prescribed psychotropic medication during pregnancy (Hanley and Oberlander 2014). Although they are not consistently used in BD, many patients with BD require the use of antidepressants, and one study demonstrated that as many as 80 % of patients with BD have taken an antidepressant at some point during their treatment (Ghaemi et al. 2000). The literature examining antidepressant use during pregnancy and pregnancy outcomes is large and exemplifies the problems outlined previously: a number of possible negative outcomes have been identified over the past 10 years by studies that did not control for the underlying psychiatric illness; subsequent, more properly controlled studies have either shown no increased risk of adverse outcomes or yielded conflicting data. Much of the work examining the association between antidepressant medications and outcomes of exposed pregnancies has focused on the selective serotonin reuptake inhibitors (SSRI) class of antidepressants. The term "antidepressant" indicates that a mixture of antidepressant classes was examined in the study.

The baseline rate of major birth defects or malformations is approximately 3 % in the general population and less than 1 % of these are thought to be secondary to an exposure to a medication (Cunningham et al. 2010). The literature examining the rate of major birth defects and antidepressant use in pregnancy is complicated by small samples, surveillance bias, and lack of controls for the underlying psychiatric illness and associated risk factors (Byatt et al. 2013); overall they have been conflicting and inconsistent. A small increase in the absolute risk of rare defects with SSRI exposure has been reported (Alwan et al. 2007), but four meta-analyses examining the risk of major malformation with first trimester SSRI exposure found no statistically significant increased risk (Rahimi et al. 2006; Addis and Koren 2000; Einarson and Einarson 2005; O'Brien et al. 2008). Compared with the SSRIs, there are limited data on major organ malformations for other types of antidepressants. Most studies examining the risk of congenital malformations with tricyclic antidepressant (TCA) exposure found no increased risk of malformations (Davis et al. 2007; Nulman et al. 1997; Pastuszak et al. 1993; Simon et al. 2002; Ramos et al. 2008) though one large epidemiological study found a significant increase in severe malformations (OR 1.36, 1.07–1.72) (Reis and Kallen 2010). With the possible exception of heart defects (see below), bupropion has not been associated with major malformations in several studies (Chun-Fai-Chan et al. 2005; Cole et al. 2007; Alwan et al. 2010). The data available for other types of antidepressants are small but reassuring (reviewed in Byatt et al. (2013) and Yonkers et al. (2014)).

The literature has not consistently identified an association between the use of antidepressants during pregnancy and cardiovascular malformations. Paroxetine use during the first trimester has been associated with a higher risk of cardiac malformations by some studies (Kallen and Otterblad Olausson 2006, 2007) but not others (Alwan et al. 2007; Louik et al. 2007) and remains the most controversial antidepressant in terms of recommendations for pregnancy (Byatt et al. 2013; Yonkers et al. 2014). Studies of bupropion have also yielded conflicting results. The GlaxoSmithKline Pregnancy Registry found increased risk of cardiovascular malformations in both retrospective and prospective reports, and a second retrospective case-control study found a higher rate of left outflow tract heart defects (Alwan and Friedman 2009; Alwan et al. 2010). However, other studies are reassuring (Chun-Fai-Chan et al. 2005; Cole et al. 2007) and the risk, if real, is quite small (<1 % of exposed infants) (Alwan and Friedman 2009). Importantly, one study (Reis and Kallen 2013) found that the combination of a benzodiazepine and an SSRI, but not an SSRI alone, increased the incidence of congenital heart defects, and much of the literature has not controlled for other medication exposures (Byatt et al. 2013; Yonkers et al. 2014). The overall consensus in the field is that the risk of major organ malformations, if it exists, is small in the setting of antidepressant monotherapy (Yonkers et al. 2009).

15.8.1 Spontaneous Abortion

Although the studies in this area are also plagued by lack of controlling for the underlying psychiatric illness and associated risk factors, the overall results suggest that the use of antidepressants in early pregnancy is associated with a modestly elevated risk of spontaneous abortion (Yonkers et al. 2014; Ross et al. 2013; Hemels et al. 2005; Nakhai-Pour et al. 2010). Reported ORs generally range from 1.4 to 1.6 (Pearlstein 2013).

15.8.2   Preterm Birth and Birth Weight

To summarize a large literature, the rate of preterm birth is higher among mothers who take antidepressants. However, most studies did not control for the severity of psychiatric illness and other confounding variables found more commonly in the psychiatric population (Yonkers et al. 2014; Byatt et al. 2013). A recent systematic review and meta-analysis of 41 studies found that the pooled adjusted OR was 1.53 for use of antidepressants at any time and 1.96 for third trimester use (Huybrechts et al. 2014). Controlling for the diagnosis of depression did not eliminate the effect, but residual confounding could not be ruled out (Byatt et al. 2013; Yonkers et al. 2014). Controlling for health habits, depressive disorders, *and* psychiatric illness, one study found greater risk of preterm birth in SSRI users (Yonkers et al. 2012), suggesting some biological role. However, the duration of pregnancy was shortened only by three to five days, and the overall risk was considered modest (Yonkers et al. 2012). The literature examining the role of antidepressant use and low birth weight is similarly complicated by confounding by the underlying illness, and the results have been inconsistent (Yonkers et al. 2014; Byatt et al. 2013; Pearlstein 2013).

15.8.3   Persistent Pulmonary Hypertension of the Newborn

Persistent Pulmonary Hypertension of the Newborn (PPHN) is a failure of the pulmonary vasculature to decrease resistance at birth resulting in breathing difficulties for the infant, leading to hypoxia and often intubation. PPHN has a 10–20 % mortality rate and results in significant morbidity (Walsh-Sukys et al. 2000). It is very rare, affecting one to two infants out of 1000 in the general population (Hageman et al. 1984; Hernandez-Diaz et al. 2007). PPHN has been associated with a number of factors including maternal smoking, diabetes, sepsis, meconium aspiration, and C-section (Hernandez-Diaz et al. 2007).

To date there have been seven studies on the association between SSRIs and PPHN in the newborn with conflicting results. The first was published in 2006 (Chambers et al. 2006) and is the basis for the FDA Alert issued that year regarding the possible association of SSRIs and PPHN. This case-control study compared 377 women who had infants diagnosed with PPHN to 836 matched control women with infants without PPHN. Fourteen of the infants with PPHN had been exposed to an SSRI after the 20th week of gestation compared to six infants who did not have PPHN (Chambers et al. 2006). This generated an adjusted (for maternal diabetes, race, and body-mass index) odds ratio of 6:1. Since this first study, six additional studies have been conducted; three found no association between SSRI exposure and PPHN (Andrade et al. 2009; Wichman et al. 2009; Wilson et al. 2011) and three found an association (Kallen and Olausson 2008; Kieler et al. 2012; Huybrechts et al. 2015), although with lower odds ratios than 6:1. It is also important to keep the risk in perspective by considering the absolute risk. PPHN is an extremely rare condition, occurring in one to two infants out of 1000 in the general population (Hageman et al. 1984; Hernandez-Diaz et al. 2007). If one assumes that SSRI use increases the odds of developing PPHN at six times the rate in the general population, only six to 12 out of 1000 (0.6–1.2 %) infants exposed to SSRIs would develop PPHN. Thus, 99 % of women who take SSRIs during pregnancy would give birth to a healthy infant who does not develop PPHN.

15.8.4 Poor Neonatal Adaptation Syndrome

The first report of "withdrawal" symptoms in babies exposed to antidepressants occurred in 1973 (Webster 1973). It is unclear if "neonatal withdrawal syndrome" is actually a result of withdrawal from the antidepressant or is due to toxicity. Thus, the alternative "Poor Neonatal Adaptation Syndrome (PNAS)" may be a better description. There are a number of limitations to the studies in the available literature, including inconsistent definitions, no measurement tool, a lack of blinded ratings, and a lack of studies investigating treatment or prevention of the syndrome. Regardless, the FDA instituted a class labeling change in 2004 for both SSRI and SNRI (serotonin–norepinephrine reuptake inhibitors) antidepressants warning that third trimester exposure may be associated with PNAS. According to the label change, "reported clinical findings have included respiratory distress, cyanosis, apnea, seizures, temperature instability, feeding difficulty, vomiting, hypo-glycemia, hypotonia, hypertonia, hyperreflexia, tremor, jitteriness, irritability, and constant crying." The subsequent result has been that many practitioners have recommended tapering antidepressants prior to delivery even though it remains unclear if this decreases the risk for PNAS or is safe for the mother. Most cases of PNAS appear to be mild, self-limited, and are not associated with lasting reper-cussions (Moses-Kolko et al. 2005). Available data suggest that approximately one-third of exposed infants will have at least mild symptoms consistent with the syndrome and that this risk increases when multiple agents, particularly benzo-diazepines, are used (Oberlander et al. 2004). Clearly, larger, more rigorous studies of the syndrome as well as strategies to minimize the rate of the syndrome are needed. At this time, there is simply not enough evidence from a safety perspective to recommend tapering antidepressants in the third trimester, particularly in cases of moderate to severe maternal mental illness.

15.8.5 Risk of Autism

Several studies have examined a possible association between SSRI use during pregnancy and autism spectrum disorders (ASD). Croen and colleagues (2011) conducted a case-control study using data extracted from medical records.

Two-hundred ninety eight children with ASD were matched for gender, birth year, and hospital to 1507 controls. Antidepressant use in the year before delivery was found to double the risk of ASD in the offspring (OR = 2.0 [1.2–3.6]) with the strongest effect found with first-trimester exposure (OR = 3.5 [1.5, 7.9]). There was no increased risk for the children of mothers with a history of mental health treatment who did not use antidepressants during pregnancy. Another large, population-based, nested case-control study examined both maternal and paternal depression as well as antidepressant use during early pregnancy and the risk of ASD in a Swedish cohort of over 500,000 children (Rai et al. 2013). Maternal depression was associated with increased risk of ASD (OR = 1.49 [1.08–2.08]) while paternal depression was not (OR = 1.21 [0.75–1.96]). Maternal depression and antidepressant use did not increase the risk of ASD with intellectual disability but did increase the risk of ASD without intellectual disability (OR = 4.95 [1.85–13.23]), although not in the absence of maternal depression (OR = 2.1 [0.97–4.57]). This study was limited by not controlling for the underlying psychiatric illness. The most recent study (Hviid et al. 2013) was a cohort study of 626,875 Danish live births between 1996 and 2005 in which they were able to link information on maternal use of SSRIs before and during pregnancy with ASD diagnoses in the offspring. When compared to women who had never used SSRIs, use of SSRIs during pregnancy was not associated with an increased risk of ASD (OR = 1.20 [0.90–1.61]). In contrast, the OR for women who received SSRIs prior to but not during pregnancy was 1.46 [1.17, 1.81] indicating that the risk is likely due to the underlying illness—depression—not the use of antidepressants.

15.9 Antianxiety Agents

15.9.1 Benzodiazepines

Studies of benzodiazepine use during pregnancy have been contradictory and controversial. Benzodiazepine use during pregnancy has been associated with case reports of perinatal toxicity, including temperature dysregulation, apnea, depressed APGAR scores, hypotonia, and poor feeding. In addition, early studies revealed an elevated risk of oral cleft palate defects compared to the baseline risk in the general population. However, more recent studies have shown that the overall risk of cleft lip and palate with benzodiazepine use in pregnancy is likely quite low (Iqbal et al. 2002; Lin et al. 2004). Infants exposed to an SSRI in combination with a benzodiazepine may have a higher incidence of congenital heart defects even when controlling for maternal illness characteristics (Oberlander et al. 2009). In considering the risks and benefits of benzodiazepines, clinicians should also consider the risks of untreated insomnia and anxiety in pregnancy, which may lead to physiologic effects as well as diminished self-care, worsening mood, and impaired functioning. Given the consequences of untreated psychiatric symptoms and the limited and controversial risks associated with benzodiazepine use, some women with overwhelming anxiety symptoms or sleep disturbance may find that the benefits outweigh any theoretical risks. Breastfeeding infants should be monitored for sedation, and the lowest effective dose should be used.

15.9.2 Gabapentin

Several studies have indicated that there is no increased risk of major congenital malformations (Holmes and Hernandez-Diaz 2012; Molgaard-Nielsen and Hviid 2011). A recent study again found no increased risk of malformations but found higher rates of preterm birth, low birth weight, and need for neonatal intensive care admission (Fujii et al. 2013). In general, gabapentin is considered a safe alternative for the management of anxiety symptoms during pregnancy.

15.9.3 Pregabalin

Like gabapentin, pregabalin is not approved for the treatment of anxiety but clinically has some utility in decreasing anxiety symptoms. It is less well-studied than gabapentin but, to date, there is no known association with an increased risk of malformations.

15.9.4 Buspirone

Animal reproduction studies have found no evidence of teratogenesis, but there is no available evidence one way or the other in humans.

15.10 Sleep Aids

Antihistamines are widely available over-the-counter and are often used in early pregnancy as a treatment for nausea and vomiting and in late pregnancy for insomnia. These medications include diphenhydramine, doxylamine, hydroxyzine, and the pheniramines (latter not available in the USA). A recent systematic review of antihistamines and birth defects (Gilboa et al. 2014) identified two cohort ($N = 31$) and eight case-control ($N = 23$) studies that found an association between prenatal antihistamine exposure and congenital malformations; however, methodological concerns included study population selection, measurement of antihistamine exposure, and identification of malformations (Gilboa et al. 2014). In addition, potentially confounding factors such as presence of hyperemesis gravidum, a clinical condition for which antihistamines are often used and which is itself associated with an increased risk of adverse fetal outcomes, was not addressed in the analysis (Fejzo et al. 2013).

Although over 90 % of pregnant women report using over-the-counter antihistamines to treat insomnia (Black and Hill 2003), a recent systematic review of sleep-promoting medication use in pregnancy (Okun et al. 2015) identified only two studies on prenatal antihistamine exposure. One found no association between exposure and congenital malformations (Reis and Kallen 2013). The other study is the only randomized control trial of antihistamines in pregnancy, which compared antidepressant, antihistamine, and placebo for insomnia in the third trimester (Khazaie et al. 2013). Although this trial did not measure any neonatal outcomes, it did find that diphenhydramine was associated with significantly longer sleep duration and efficiency and fewer depressive symptoms compared to placebo. Sleep agents including eszopiclone, ramelteon, and zolpidem have not been associated with major organ malformations (Okun et al. 2015), but zolpidem use for greater than 90 days has been associated with increased risk of low birth weight, preterm birth, and cesarean delivery (Cohen et al. 2010). Thus, taken as a whole, the current limited evidence suggests that antihistamine and sleep agent use in pregnancy are not consistently associated with an increased risk of adverse pregnancy outcomes, though larger studies are needed.

15.11 Psychotropic Medications and Breastfeeding

The literature on the use of psychotropic medications during breastfeeding is relatively sparse with few long-term outcome studies and mainly case reports. Overall the data that are available to date are generally reassuring. All psychotropic medications enter breast milk. Any psychotropic medication may be associated with side effects in the breastfeeding infant, and therefore the baby should be monitored closely for adverse effects and either the medication or the breastfeeding stopped if there appear to be ill effects. Premature infants may be more vulnerable to adverse effects due to immature metabolic systems (Berle and Spigset 2011).

There have been limited reports on the use of lithium during breastfeeding and most are reassuring with undetectable or very low blood levels in the exposed children (Bogen et al. 2012). Breastfeeding on lithium, however, requires a very vigilant parent who will have a low threshold for taking the baby for an emergency evaluation in the setting of potential dehydration since lithium levels can rise and become toxic in this setting (Bogen et al. 2012). Monitoring the baby's blood level regularly is recommended.

The anticonvulsants, unlike during pregnancy, appear to be relatively safe during breastfeeding. In fact, a recent study demonstrated that children who were exposed to anticonvulsants in utero had improved developmental and intellectual outcomes if they were breastfed for at least six months compared to exposed children who were not breastfed (Meador et al. 2014). Blood levels in the exposed infant can be monitored but do not need to be regularly tested due to a low likelihood of toxicity, unlike with lithium (Pearlstein 2013).

Very few studies have examined the use of antipsychotic medications in breastfeeding and many agents have no data at all (Klinger et al. 2013). Olanzapine and quetiapine are generally considered safe for breastfeeding (Klinger et al. 2013), but there are few data to guide the use of other antipsychotic medications. There are sporadic case reports of adverse effects in antipsychotic-exposed breastfed infants, but no clear pattern of effects has emerged (Gentile 2008). Clozapine should not be used during breastfeeding due to the risk of neutropenia (Gentile 2008).

Overall, antidepressant medications are the best-studied class of psychotropic medications in breastfeeding, but again the data are limited. Overall, the data are reassuring, and it appears that only a small percentage (1–10 %) of the mother's dosage is found in breast milk. Long-term developmental studies show no differences in children exposed to antidepressants in breast milk compared to unexposed children. There is one case report of seizure in an infant exposed to bupropion (Pearlstein 2013). There is also a theoretical risk of fluoxetine toxicity due to its long half-life; sertraline, paroxetine, and nortriptyline generate the lowest serum levels in the exposed infant (Pearlstein 2013).

Benzodiazepine use during breastfeeding should be limited to the lowest effective dose and the baby monitored for sedation and lethargy, though this is rare. Long-term studies are not available.

15.12 Conclusions

Interpretation of the literature regarding the association between psychotropic medication use during pregnancy and outcomes for the exposed baby is complicated by the fact that the population of women who require psychotropic medications during pregnancy have other associated risk factors and behaviors that may also influence outcomes. Large, well-designed, and controlled studies have shown that most classes of psychotropic medications appear to be relatively safe for use during pregnancy. Untreated psychiatric disorders, including BD, during pregnancy have associated risks for both mother and child, and these risks need to be considered in the risk-benefit analysis of using psychotropic medication during pregnancy. Women with BD are at risk for relapse if medications are stopped for pregnancy and are at risk for both postpartum depression and postpartum psychosis after delivery. Psychotropic medications should not be precipitously stopped, and a comprehensive evaluation and individualized treatment plan is needed for women with BD who wish to become or are pregnant. Future work should focus on the proper management of psychotropic medications during pregnancy including prophylactic dosing strategies and management before and after delivery. Finally, improved prospective data procurement regarding potential confounders are also needed in order to truly address the question of whether exposure in utero to psychotropic medication affects outcomes for the child.

References

ACOG Practice Bulletin: Clinical management guidelines for obstetrician-gynecologists number 92, April 2008 (replaces practice bulletin number 87, November 2007). Use of psychiatric medications during pregnancy and lactation (2008). Obstet Gynecol 111(4):1001–1020

Addis A, Koren G (2000) Safety of fluoxetine during the first trimester of pregnancy: a meta-analytical review of epidemiological studies. Psychol Med 30(1):89–94

Altshuler LL, Cohen L, Szuba MP, Burt VK, Gitlin M, Mintz J (1996) Pharmacologic management of psychiatric illness during pregnancy: dilemmas and guidelines. Am J Psychiatry 153(5):592–606

Alwan S, Friedman JM (2009) Safety of selective serotonin reuptake inhibitors in pregnancy. CNS Drugs 23(6):493–509

Alwan S, Reefhuis J, Rasmussen SA, Olney RS, Friedman JM (2007) Use of selective serotonin-reuptake inhibitors in pregnancy and the risk of birth defects. N Engl J Med 356(26): 2684–2692

Alwan S, Reefhuis J, Botto LD, Rasmussen SA, Correa A, Friedman JM (2010) Maternal use of bupropion and risk for congenital heart defects. Am J Obstet Gynecol 203(1):52 e51–56.

American Academy of Pediatrics (2000) Use of psychoactive medication during pregnancy and possible effects on the fetus and newborn. Committee on Drugs. Pediatrics 105(4 Pt 1): 880–887

Andersson L, Sundstrom-Poromaa I, Bixo M, Wulff M, Bondestam K, aStrom M (2003) Point prevalence of psychiatric disorders during the second trimester of pregnancy: a population-based study. Am J Obstet Gynecol 189(1):148–154

Andrade SE, McPhillips H, Loren D, Raebel MA, Lane K, Livingston J, Boudreau DM, Smith DH, Davis RL, Willy ME, Platt R (2009) Antidepressant medication use and risk of persistent pulmonary hypertension of the newborn. Pharmacoepidemiol Drug Saf 18(3):246–252

Ashman SB, Dawson G, Panagiotides H, Yamada E, Wilkinson CW (2002) Stress hormone levels of children of depressed mothers. Dev Psychopathol 14(2):333–349

Austin MP (1992) Puerperal affective psychosis: is there a case for lithium prophylaxis? Br J Psychiatry 161:692–694

Barnes TR (2011) Evidence-based guidelines for the pharmacological treatment of schizophrenia: recommendations from the British Association for Psychopharmacology. J Psychopharmacol 25(5):567–620

Bergink V, Bouvy PF, Vervoort JS, Koorengevel KM, Steegers EA, Kushner SA (2012) Prevention of postpartum psychosis and mania in women at high risk. Am J Psychiatry 169(6): 609–615

Berle JO, Spigset O (2011) Antidepressant use during breastfeeding. Curr Womens Health Rev 7 (1):28–34

Black RA, Hill DA (2003) Over-the-counter medications in pregnancy. Am Fam Physician 67(12): 2517–2524

Bogen DL, Sit D, Genovese A, Wisner KL (2012) Three cases of lithium exposure and exclusive breastfeeding. Arch Womens Ment Health 15(1):69–72

Brennan PA, Pargas R, Walker EF, Green P, Newport DJ, Stowe Z (2008) Maternal depression and infant cortisol: influences of timing, comorbidity and treatment. J Child Psychol Psychiatry 49(10):1099–1107

Bromley RL, Mawer GE, Briggs M, Cheyne C, Clayton-Smith J, Garcia-Finana M, Kneen R, Lucas SB, Shallcross R, Baker GA (2013) The prevalence of neurodevelopmental disorders in children prenatally exposed to antiepileptic drugs. J Neurol Neurosurg Psychiatry 84(6): 637–643

Brunner E, Falk DM, Jones M, Dey DK, Shatapathy CC (2013) Olanzapine in pregnancy and breastfeeding: a review of data from global safety surveillance. BMC Pharmacol Toxicol 14:38

Byatt N, Deligiannidis KM, Freeman MP (2013) Antidepressant use in pregnancy: a critical review focused on risks and controversies. Acta Psychiatr Scand 127(2):94–114

Campbell SB, Cohn JF (1991) Prevalence and correlates of postpartum depression in first-time mothers. J Abnorm Psychol 100(4):594–599

Chambers CD, Hernandez-Diaz S, Van Marter LJ, Werler MM, Louik C, Jones KL, Mitchell AA (2006) Selective serotonin-reuptake inhibitors and risk of persistent pulmonary hypertension of the newborn. N Engl J Med 354(6):579–587

Christensen J, Gronborg TK, Sorensen MJ, Schendel D, Parner ET, Pedersen LH, Vestergaard M (2013) Prenatal valproate exposure and risk of autism spectrum disorders and childhood autism. JAMA 309(16):1696–1703

Chun-Fai-Chan B, Koren G, Fayez I, Kalra S, Voyer-Lavigne S, Boshier A, Shakir S, Einarson A (2005) Pregnancy outcome of women exposed to bupropion during pregnancy: a prospective comparative study. Am J Obstet Gynecol 192(3):932–936

Clark CT, Klein AM, Perel JM, Helsel J, Wisner KL (2013) Lamotrigine dosing for pregnant patients with bipolar disorder. Am J Psychiatry 170(11):1240–1247

Cohen LS, Friedman JM, Jefferson JW, Johnson EM, Weiner ML (1994) A reevaluation of risk of in utero exposure to lithium. JAMA 271(2):146–150

Cohen LS, Sichel DA, Robertson LM, Heckscher E, Rosenbaum JF (1995) Postpartum prophylaxis for women with bipolar disorder. Am J Psychiatry 152(11):1641–1645

Cohen LS, Wang B, Nonacs R, Viguera AC, Lemon EL, Freeman MP (2010) Treatment of mood disorders during pregnancy and postpartum. Psychiatr Clin North Am 33(2):273–293

Cole JA, Modell JG, Haight BR, Cosmatos IS, Stoler JM, Walker AM (2007) Bupropion in pregnancy and the prevalence of congenital malformations. Pharmacoepidemiol Drug Saf 16(5):474–484

Communication FDS (2011) Antipsychotic drug labels updated on use during pregnancy and risk of abnormal muscle movements and withdrawal symptoms in newborns. http://www.fda.gov/Drugs/DrugSafety/ucm243903.htm-sa. Accessed 28 Apr 15

Coppola D, Russo LJ, Kwarta RF Jr, Varughese R, Schmider J (2007) Evaluating the postmarketing experience of risperidone use during pregnancy: pregnancy and neonatal outcomes. Drug Saf 30(3):247–264

Cox JL, Murray D, Chapman G (1993) A controlled study of the onset, duration and prevalence of postnatal depression. Br J Psychiatry 163:27–31

Croen LA, Grether JK, Yoshida CK, Odouli R, Hendrick V (2011) Antidepressant use during pregnancy and childhood autism spectrum disorders. Arch Gen Psychiatry 68(11):1104–1112

Cunningham FG, Leveno KJ, Bloom SL, Hauth JC, Rouse DJ, Spong CY (2010) Teratology and medications that affect the fetus. In: Williams obstetrics. McGraw Hill, New York, pp 312–333

Cunnington M, Tennis P (2005) Lamotrigine and the risk of malformations in pregnancy. Neurology 64(6):955–960

Davis EP, Glynn LM, Dunkel Schetter C, Hobel C, Chicz-Demet A, Sandman CA (2005) Corticotropin-releasing hormone during pregnancy is associated with infant temperament. Dev Neurosci 27(5):299–305

Davis RL, Rubanowice D, McPhillips H, Raebel MA, Andrade SE, Smith D, Yood MU, Platt R (2007) Risks of congenital malformations and perinatal events among infants exposed to antidepressant medications during pregnancy. Pharmacoepidemiol Drug Saf 16(10):1086–1094

DeVane CL, Stowe ZN, Donovan JL, Newport DJ, Pennell PB, Ritchie JC, Owens MJ, Wang JS (2006) Therapeutic drug monitoring of psychoactive drugs during pregnancy in the genomic era: challenges and opportunities. J Psychopharmacol 20(4 Suppl):54–59

Diego MA, Field T, Hernandez-Reif M, Cullen C, Schanberg S, Kuhn C (2004) Prepartum, postpartum, and chronic depression effects on newborns. Psychiatry 67(1):63–80

Dolk H, Jentink J, Loane M, Morris J, de Jong-van den Berg LT (2008) Does lamotrigine use in pregnancy increase orofacial cleft risk relative to other malformations? Neurology 71(10):714–722

Einarson A, Boskovic R (2009) Use and safety of antipsychotic drugs during pregnancy. J Psychiatr Pract 15(3):183–192

Einarson TR, Einarson A (2005) Newer antidepressants in pregnancy and rates of major malformations: a meta-analysis of prospective comparative studies. Pharmacoepidemiol Drug Saf 14(12):823–827

Einarson A, Einarson TR (2009) Maternal use of antipsychotics in early pregnancy: little evidence of increased risk of congenital malformations. Evid Based Ment Health 12(1):29

Essex MJ, Klein MH, Cho E, Kalin NH (2002) Maternal stress beginning in infancy may sensitize children to later stress exposure: effects on cortisol and behavior. Biol Psychiatry 52(8): 776–784

Fejzo MS, Magtira A, Schoenberg FP, MacGibbon K, Mullin P, Romero R, Tabsh K (2013) Antihistamines and other prognostic factors for adverse outcome in hyperemesis gravidarum. Eur J Obstet Gynecol Reprod Biol 170(1):71–76

Field T, Diego M, Hernandez-Reif M (2006) Prenatal depression effects on the fetus and newborn: a review. Infant Behav Dev 29(3):445–455

Field T, Diego M, Hernandez-Reif M (2010) Prenatal depression effects and interventions: a review. Infant Behav Dev 33(4):409–418

Frank E, Kupfer DJ, Jacob M, Blumenthal SJ, Jarrett DB (1987) Pregnancy-related affective episodes among women with recurrent depression. Am J Psychiatry 144(3):288–293

Fujii H, Goel A, Bernard N, Pistelli A, Yates LM, Stephens S, Han JY, Matsui D, Etwell F, Einarson TR, Koren G, Einarson A (2013) Pregnancy outcomes following gabapentin use: results of a prospective comparative cohort study. Neurology 80(17):1565–1570

Gentile S (2004) Clinical utilization of atypical antipsychotics in pregnancy and lactation. Ann Pharmacother 38(7–8):1265–1271

Gentile S (2008) Infant safety with antipsychotic therapy in breast-feeding: a systematic review. J Clin Psychiatry 69(4):666–673

Gentile S (2010) Antipsychotic therapy during early and late pregnancy. A systematic review. Schizophr Bull 36(3):518–544

Ghaemi SN, Boiman EE, Goodwin FK (2000) Diagnosing bipolar disorder and the effect of antidepressants: a naturalistic study. J Clin Psychiatry 61(10):804–808, quiz 809

Gilboa SM, Ailes EC, Rai RP, Anderson JA, Honein MA (2014) Antihistamines and birth defects: a systematic review of the literature. Expert Opin Drug Saf 13(12):1667–1698

Grace SL, Evindar A, Stewart DE (2003) The effect of postpartum depression on child cognitive development and behavior: a review and critical analysis of the literature. Arch Womens Ment Health 6(4):263–274

Habermann F, Fritzsche J, Fuhlbruck F, Wacker E, Allignol A, Weber-Schoendorfer C, Meister R, Schaefer C (2013) Atypical antipsychotic drugs and pregnancy outcome: a prospective, cohort study. J Clin Psychopharmacol 33(4):453–462

Hageman JR, Adams MA, Gardner TH (1984) Persistent pulmonary hypertension of the newborn. Trends in incidence, diagnosis, and management. Am J Dis Child 138(6):592–595

Halligan SL, Herbert J, Goodyer IM, Murray L (2004) Exposure to postnatal depression predicts elevated cortisol in adolescent offspring. Biol Psychiatry 55(4):376–381

Hanley GE, Oberlander TF (2014) The effect of perinatal exposures on the infant: antidepressants and depression. Best Pract Res Clin Obstet Gynaecol 28(1):37–48

Hemels ME, Einarson A, Koren G, Lanctot KL, Einarson TR (2005) Antidepressant use during pregnancy and the rates of spontaneous abortions: a meta-analysis. Ann Pharmacother 39(5): 803–809

Hernandez-Diaz S, Van Marter LJ, Werler MM, Louik C, Mitchell AA (2007) Risk factors for persistent pulmonary hypertension of the newborn. Pediatrics 120(2):e272–e282

Holmes LB, Hernandez-Diaz S (2012) Newer anticonvulsants: lamotrigine, topiramate and gabapentin. Birth Defects Res A Clin Mol Teratol 94(8):599–606

Holmes LB, Baldwin EJ, Smith CR, Habecker E, Glassman L, Wong SL, Wyszynski DF (2008) Increased frequency of isolated cleft palate in infants exposed to lamotrigine during pregnancy. Neurology 70(22 Pt 2):2152–2158

Huybrechts KF, Sanghani RS, Avorn J, Urato AC (2014) Preterm birth and antidepressant medication use during pregnancy: a systematic review and meta-analysis. PLoS One 9(3):e92778

Huybrechts KF, Bateman BT, Palmsten K, Desai RJ, Patorno E, Gopalakrishnan C, Levin R, Mogun H, Hernandez-Diaz S (2015) Antidepressant use late in pregnancy and risk of persistent pulmonary hypertension of the newborn. JAMA 313(21):2142–2151

Hviid A, Melbye M, Pasternak B (2013) Use of selective serotonin reuptake inhibitors during pregnancy and risk of autism. N Engl J Med 369(25):2406–2415

Iqbal MM, Sobhan T, Ryals T (2002) Effects of commonly used benzodiazepines on the fetus, the neonate, and the nursing infant. Psychiatr Serv 53(1):39–49

Jacobson SJ, Jones K, Johnson K, Ceolin L, Kaur P, Sahn D, Donnenfeld AE, Rieder M, Santelli R, Smythe J et al (1992) Prospective multicentre study of pregnancy outcome after lithium exposure during first trimester. Lancet 339(8792):530–533

Johnson KC, LaPrairie JL, Brennan PA, Stowe ZN, Newport DJ (2012) Prenatal antipsychotic exposure and neuromotor performance during infancy. Arch Gen Psychiatry 69(8):787–794

Jones I, Heron J, Robertson E (2010) Puerperal psychosis. In: Kohen D (ed) Women and mental health. Oxford University Press, Oxford, pp 179–186

Kallen B, Olausson PO (2008) Maternal use of selective serotonin re-uptake inhibitors and persistent pulmonary hypertension of the newborn. Pharmacoepidemiol Drug Saf 17(8): 801–806

Kallen B, Otterblad Olausson P (2006) Antidepressant drugs during pregnancy and infant congenital heart defect. Reprod Toxicol 21(3):221–222

Kallen BA, Otterblad Olausson P (2007) Maternal use of selective serotonin re-uptake inhibitors in early pregnancy and infant congenital malformations. Birth Defects Res A Clin Mol Teratol 79(4):301–308

Kendell RE, Chalmers JC, Platz C (1987) Epidemiology of puerperal psychoses. Br J Psychiatry 150:662–673

Khalifeh H, Dolman C, Howard LM (2015) Safety of psychotropic drugs in pregnancy. BMJ 350: h2260

Khazaie H, Ghadami MR, Knight DC, Emamian F, Tahmasian M (2013) Insomnia treatment in the third trimester of pregnancy reduces postpartum depression symptoms: a randomized clinical trial. Psychiatry Res 210(3):901–905

Kieler H, Artama M, Engeland A, Ericsson O, Furu K, Gissler M, Nielsen RB, Norgaard M, Stephansson O, Valdimarsdottir U, Zoega H, Haglund B (2012) Selective serotonin reuptake inhibitors during pregnancy and risk of persistent pulmonary hypertension in the newborn: population based cohort study from the five Nordic countries. BMJ 344:d8012

Klinger G, Stahl B, Fusar-Poli P, Merlob P (2013) Antipsychotic drugs and breastfeeding. Pediatr Endocrinol Rev 10(3):308–317

Li D, Liu L, Odouli R (2009) Presence of depressive symptoms during early pregnancy and the risk of preterm delivery: a prospective cohort study. Hum Reprod 24(1):146–153

Lin AE, Peller AJ, Westgate MN, Houde K, Franz A, Holmes LB (2004) Clonazepam use in pregnancy and the risk of malformations. Birth Defects Res A Clin Mol Teratol 70(8):534–536

Louik C, Lin AE, Werler MM, Hernandez-Diaz S, Mitchell AA (2007) First-trimester use of selective serotonin-reuptake inhibitors and the risk of birth defects. N Engl J Med 356(26): 2675–2683

Meador KJ, Baker GA, Browning N, Cohen MJ, Bromley RL, Clayton-Smith J, Kalayjian LA, Kanner A, Liprace JD, Pennell PB, Privitera M, Loring DW, Neurodevelopmental Effects of Antiepileptic Drugs (NEAD) Study Group (2014) Breastfeeding in children of women taking antiepileptic drugs: cognitive outcomes at age 6 years. JAMA Pediatr 168:729–736

Molgaard-Nielsen D, Hviid A (2011) Newer-generation antiepileptic drugs and the risk of major birth defects. JAMA 305(19):1996–2002

Moses-Kolko EL, Bogen D, Perel J, Bregar A, Uhl K, Levin B, Wisner KL (2005) Neonatal signs after late in utero exposure to serotonin reuptake inhibitors: literature review and implications for clinical applications. JAMA 293(19):2372–2383

Mosher WD, Bachrach CA (1996) Understanding U.S. fertility: continuity and change in the National Survey of Family Growth, 1988–1995. Fam Plann Perspect 28(1):4–12

Nakhai-Pour HR, Broy P, Berard A (2010) Use of antidepressants during pregnancy and the risk of spontaneous abortion. CMAJ 182(10):1031–1037

Newham JJ, Thomas SH, MacRitchie K, McElhatton PR, McAllister-Williams RH (2008) Birth weight of infants after maternal exposure to typical and atypical antipsychotics: prospective comparison study. Br J Psychiatry 192(5):333–337

Newport DJ, Levey LC, Pennell PB, Ragan K, Stowe ZN (2007) Suicidal ideation in pregnancy: assessment and clinical implications. Arch Womens Ment Health 10(5):181–187

Nilsson E, Lichtenstein P, Cnattingius S, Murray RM, Hultman CM (2002) Women with schizophrenia: pregnancy outcome and infant death among their offspring. Schizophr Res 58(2–3): 221–229

Nulman I, Rovet J, Stewart DE, Wolpin J, Gardner HA, Theis JG, Kulin N, Koren G (1997) Neurodevelopment of children exposed in utero to antidepressant drugs. N Engl J Med 336(4): 258–262

O'Brien L, Einarson TR, Sarkar M, Einarson A, Koren G (2008) Does paroxetine cause cardiac malformations? J Obstet Gynaecol Can 30(8):696–701

Oberlander TF, Misri S, Fitzgerald CE, Kostaras X, Rurak D, Riggs W (2004) Pharmacologic factors associated with transient neonatal symptoms following prenatal psychotropic medication exposure. J Clin Psychiatry 65(2):230–237

Oberlander TF, Gingrich JA, Ansorge MS (2009) Sustained neurobehavioral effects of exposure to SSRI antidepressants during development: molecular to clinical evidence. Clin Pharmacol Ther 86(6):672–677

O'Connor TG, Ben-Shlomo Y, Heron J, Golding J, Adams D, Glover V (2005) Prenatal anxiety predicts individual differences in cortisol in pre-adolescent children. Biol Psychiatry 58(3): 211–217

Okun ML, Ebert R, Saini B (2015) A review of sleep-promoting medications used in pregnancy. Am J Obstet Gynecol 212(4):428–441

Orr ST, Blazer DG, James SA, Reiter JP (2007) Depressive symptoms and indicators of maternal health status during pregnancy. J Womens Health (Larchmt) 16(4):535–542

Pastuszak A, Schick-Boschetto B, Zuber C, Feldkamp M, Pinelli M, Sihn S, Donnenfeld A, McCormack M, Leen-Mitchell M, Woodland C et al (1993) Pregnancy outcome following first-trimester exposure to fluoxetine (Prozac). JAMA 269(17):2246–2248

Paton C (2008) Prescribing in pregnancy. Br J Psychiatry 192(5):321–322

Pavek P, Ceckova M, Staud F (2009) Variation of drug kinetics in pregnancy. Curr Drug Metab 10(5):520–529

Payne JL (2003) The role of estrogen in mood disorders in women. Int Rev Psychiatry 15(3): 280–290

Payne JL, Meltzer-Brody S (2009) Antidepressant use during pregnancy: current controversies and treatment strategies. Clin Obstet Gynecol 52(3):469–482

Payne JL, Roy PS, Murphy-Eberenz K, Weismann MM, Swartz KL, McInnis MG, Nwulia E, Mondimore FM, MacKinnon DF, Miller EB, Nurnberger JI, Levinson DF, DePaulo JR Jr, Potash JB (2007) Reproductive cycle-associated mood symptoms in women with major depression and bipolar disorder. J Affect Disord 99(1–3):221–229

Pearlstein T (2013) Use of psychotropic medication during pregnancy and the postpartum period. Womens Health (Lond Engl) 9(6):605–615

Peng M, Gao K, Ding Y, Ou J, Calabrese JR, Wu R, Zhao J (2013) Effects of prenatal exposure to atypical antipsychotics on postnatal development and growth of infants: a case-controlled, prospective study. Psychopharmacology (Berl) 228(4):577–584

Rahimi R, Nikfar S, Abdollahi M (2006) Pregnancy outcomes following exposure to serotonin reuptake inhibitors: a meta-analysis of clinical trials. Reprod Toxicol 22(4):571–575

Rai D, Lee BK, Dalman C, Golding J, Lewis G, Magnusson C (2013) Parental depression, maternal antidepressant use during pregnancy, and risk of autism spectrum disorders: population based case-control study. BMJ 346:f2059

Ramos E, St-Andre M, Rey E, Oraichi D, Berard A (2008) Duration of antidepressant use during pregnancy and risk of major congenital malformations. Br J Psychiatry 192(5):344–350

Reis M, Kallen B (2008) Maternal use of antipsychotics in early pregnancy and delivery outcome. J Clin Psychopharmacol 28(3):279–288

Reis M, Kallen B (2010) Delivery outcome after maternal use of antidepressant drugs in pregnancy: an update using Swedish data. Psychol Med 40(10):1723–1733

Reis M, Kallen B (2013) Combined use of selective serotonin reuptake inhibitors and sedatives/hypnotics during pregnancy: risk of relatively severe congenital malformations or cardiac defects. A register study. BMJ Open 2013:3(2)

Robinson GE (2012) Treatment of schizophrenia in pregnancy and postpartum. J Popul Ther Clin Pharmacol 19(3):e380–e386

Ross LE, Grigoriadis S, Mamisashvili L, Vonderporten EH, Roerecke M, Rehm J, Dennis CL, Koren G, Steiner M, Mousmanis P, Cheung A (2013) Selected pregnancy and delivery outcomes after exposure to antidepressant medication: a systematic review and meta-analysis. JAMA Psychiatry 70(4):436–443

Seeman MV (2004) Gender differences in the prescribing of antipsychotic drugs. Am J Psychiatry 161(8):1324–1333

Seeman MV (2013) Clinical interventions for women with schizophrenia: pregnancy. Acta Psychiatr Scand 127(1):12–22

Sharma V (2003) Role of sleep loss in the causation of puerperal psychosis. Med Hypotheses 61(4):477–481

Sharma V, Smith A, Khan M (2004) The relationship between duration of labour, time of delivery, and puerperal psychosis. J Affect Disord 83(2–3):215–220

Simon GE, Cunningham ML, Davis RL (2002) Outcomes of prenatal antidepressant exposure. Am J Psychiatry 159(12):2055–2061

Stewart DE, Klompenhouwer JL, Kendell RE, van Hulst AM (1991) Prophylactic lithium in puerperal psychosis. The experience of three centres. Br J Psychiatry 158:393–397

Thiels C (1987) Pharmacotherapy of psychiatric disorder in pregnancy and during breastfeeding: a review. Pharmacopsychiatry 20(4):133–146

Tracy TS, Venkataramanan R, Glover DD, Caritis SN (2005) Temporal changes in drug metabolism (CYP1A2, CYP2D6 and CYP3A Activity) during pregnancy. Am J Obstet Gynecol 192(2):633–639

van Gent EM, Verhoeven WM (1992) Bipolar illness, lithium prophylaxis, and pregnancy. Pharmacopsychiatry 25(4):187–191

Vigod SN, Gomes T, Wilton AS, Taylor VH, Ray JG (2015) Antipsychotic drug use in pregnancy: high dimensional, propensity matched, population based cohort study. BMJ 350:h2298

Viguera AC, Nonacs R, Cohen LS, Tondo L, Murray A, Baldessarini RJ (2000) Risk of recurrence of bipolar disorder in pregnant and nonpregnant women after discontinuing lithium maintenance. Am J Psychiatry 157(2):179–184

Viguera AC, Whitfield T, Baldessarini RJ, Newport DJ, Stowe Z, Reminick A, Zurick A, Cohen LS (2007) Risk of recurrence in women with bipolar disorder during pregnancy: prospective study of mood stabilizer discontinuation. Am J Psychiatry 164(12):1817–1824, quiz 1923

Walsh-Sukys MC, Tyson JE, Wright LL, Bauer CR, Korones SB, Stevenson DK, Verter J, Stoll BJ, Lemons JA, Papile LA, Shankaran S, Donovan EF, Oh W, Ehrenkranz RA, Fanaroff AA (2000) Persistent pulmonary hypertension of the newborn in the era before nitric oxide: practice variation and outcomes. Pediatrics 105(1 Pt 1):14–20

Webb RT, Howard L, Abel KM (2004) Antipsychotic drugs for non-affective psychosis during pregnancy and postpartum. Cochrane Database Syst Rev 2:CD004411

Webster PA (1973) Withdrawal symptoms in neonates associated with maternal antidepressant therapy. Lancet 2(7824):318–319

Wichman CL, Moore KM, Lang TR, St Sauver JL, Heise RH Jr, Watson WJ (2009) Congenital heart disease associated with selective serotonin reuptake inhibitor use during pregnancy. Mayo Clin Proc 84(1):23–27

Wilson KL, Zelig CM, Harvey JP, Cunningham BS, Dolinsky BM, Napolitano PG (2011) Persistent pulmonary hypertension of the newborn is associated with mode of delivery and not with maternal use of selective serotonin reuptake inhibitors. Am J Perinatol 28(1):19–24

Yonkers KA, Wisner KL, Stewart DE, Oberlander TF, Dell DL, Stotland N, Ramin S, Chaudron L, Lockwood C (2009) The management of depression during pregnancy: a report from the American Psychiatric Association and the American College of Obstetricians and Gynecologists. Gen Hosp Psychiatry 31(5):403–413

Yonkers KA, Norwitz ER, Smith MV, Lockwood CJ, Gotman N, Luchansky E, Lin H, Belanger K (2012) Depression and serotonin reuptake inhibitor treatment as risk factors for preterm birth. Epidemiology 23(5):677–685

Yonkers KA, Blackwell KA, Glover J, Forray A (2014) Antidepressant use in pregnant and postpartum women. Annu Rev Clin Psychol 10:369–392

Zuckerman B, Amaro H, Bauchner H, Cabral H (1989) Depressive symptoms during pregnancy: relationship to poor health behaviors. Am J Obstet Gynecol 160(5 Pt 1):1107–1111

Index

© Springer International Publishing Switzerland 2016
C.A. Zarate Jr., H.K. Manji (eds.), *Bipolar Depression: Molecular Neurobiology,
Clinical Diagnosis, and Pharmacotherapy*, Milestones in Drug Therapy,
DOI 10.1007/978-3-319-31689-5

Printed by Printforce, the Netherlands